W9-CUK-485

Optical Physics, Applied Physics and Materials Science

Laser Cleaning

Dr. Andrew C. Tam

Optical Physics, Applied Physics and Materials Science

Laser Cleaning

Editor

Boris Luk`yanchuk

Data Storage Institute,
Agency for Science, Technology and Research, Singapore

World Scientific
New Jersey • London • Singapore • Hong Kong

Published by

World Scientific Publishing Co. Pte. Ltd.
P O Box 128, Farrer Road, Singapore 912805
USA office: Suite 1B, 1060 Main Street, River Edge, NJ 07661
UK office: 57 Shelton Street, Covent Garden, London WC2H 9HE

British Library Cataloguing-in-Publication Data
A catalogue record for this book is available from the British Library.

LASER CLEANING

ISBN 981-02-4941-1

Printed by FuIsland Offset Printing (S) Pte Ltd, Singapore

List of Contributors

Arnold N	nikita.arnold@jku.at Johannes – Kepler - Universität, Angewandte Physik, Altenbergerstraße 69, A -4040, Linz, Austria
Bertsch M.	Micha.Bertsch@daimlercrysler.com
Boneberg J.	Johannes.Boneberg@vivastar.com Vivastar, Switzerland
Chong T. C.	towchong@dsi.nus.edu.sg Data Storage Institute, DSI Building 5, Engineering Drive 1, 117 608, Singapore
Dobler V.	Volker.Dobler@vivastar.com Vivastar, Switzerland
Dubbers O.	oliver.dubbers@physik.uni-ulm.de
Fernandes A. J.	alanna@ics.mq.edu.au Physics Department, Macquarie University, Sydney, NSW 2109, Australia
Grigoropoulos C. P.	cgrigoro@me.berkeley.edu Department of Mechanical Engineering, University of California, 6177 Etcheverry Hall, Berkeley, CA 94720-1740, USA
Halfpenny D. R.	davidh@ics.mq.edu.au SIBT, Macquarie University, Sydney, NSW 2109, Australia
Hong M. H.	dsihmh@dsi.nus.edu.sg Data Storage Institute, DSI Building 5, Engineering Drive 1, 117 608, Singapore
Huang S. M.	smhuang@dsi.nus.edu.sg Data Storage Institute, DSI Building 5, Engineering Drive 1, 117 608, Singapore
Kane D. M.	debkane@physics.mq.edu.au Physics Department, Macquarie University, Sydney, NSW 2109, Australia
Kim D.	dskim87@postech.ac.kr Department of Mechanical Engineering, Pohang University of Science and Technology, Pohang, Kyungbuk, 790-784, Korea
Leiderer P.	Paul.Leiderer@uni-konstanz.de Fakultät für Physik, Universität Konstanz, D-78457 Konstanz, Germany
Lu Y. F.	dsiluyf@dsi.nus.edu.sg Data Storage Institute, DSI Building 5, Engineering Drive 1, 117 608, Singapore
Luk'yanchuk B. S.	boris@dsi.nus.edu.sg Data Storage Institute, DSI Building 5, Engineering Drive 1, 117 608, Singapore

Mosbacher M.	Mario.Mosbacher@uni-konstanz.de Fakultät für Physik, Universität Konstanz, D-78457 Konstanz, Germany
Münzer H.- J.	Hans-Joachim.Muenzer@uni-konstanz.de Fakultät für Physik, Universität Konstanz, D-78457 Konstanz, Germany
Schilling A.	andreas.schilling@int.unine.ch
Shakhno E. A.	veiko@lastech.ifmo.ru St. Petersburg State Institute of Fine Mechanics & Optics 197101, St. Petersburg, Russia
Song W. D.	songwd@dsi.nus.edu.sg Data Storage Institute, DSI Building 5, Engineering Drive 1, 117 608, Singapore
Suzuki N.	suzuki@rcem.osaka-u.ac.jp Research Centre for Materials Science at Extreme Conditions, Osaka University, Machikaneyama 1 - 3, Toyonaka, Osaka 560-8531, Japan
Takai M.	takai@rcem.osaka-u.ac.jp Research Centre for Materials Science at Extreme Conditions, Osaka University, Machikaneyama 1 - 3, Toyonaka, Osaka 560-8531, Japan
Veiko V. P.	veiko@lastech.ifmo.ru St. Petersburg State Institute of Fine Mechanics & Optics 197101, St. Petersburg, Russia
Wang Z.B.	zbwang@dsi.nus.edu.sg Data Storage Institute, DSI Building 5, Engineering Drive 1, 117 608, Singapore
Yavas O.	oguz.yavas@infineon.com Infineon Technologies Dresden GmbH & Co. OHG Defect Density Engineering, DD T DDE, P.O. Box 10 09 40, D-01076 Dresden Germany
Zafiropulos V.	zafir@iesl.forth.gr Foundation for Research and Technology-Hellas (FO.R.T.H.), Institute of Electronic Structure and Laser (I.E.S.L.), 10 Heraklion, Crete, Greece
Zapka W.	werner.zapka@xaar.se XaarJet AB, Box 516, Elektronikhojden 10, SE-175 26 Jarfalla - Jakobsberg, Sweden
Zheng Y. W.	YuanweiZheng@gsmcthw.com Technology Development Department Grace Semiconductor Manufacturing Co. Ltd PC:201203 No. 818 Guo Shou Jin Road, Shanghai, P.R.C. China

Contents

Part 3. Steam Laser Cleaning

Part 4. Laser Cleaning of Artworks

Part 5. Applications of Laser Cleaning

Preface

Within the catastrophe theory there is a mathematical statement known as the "Principle of Fragility of Good Things" (Arnold, 1992). According to this principle the stability domain on the plane of parameters (on which the equilibrium state depends) is arranged "with corners outward", thus driving wedges into the instability domain. As Arnold wrote: "Thus for systems belonging to the singular part of the stability boundary, a small change of the parameters is more likely to send the system into the unstable region than into the stable region. This is a manifestation of a general principle stating that all good things (e.g. stability) are more fragile than bad things". It is easy to explain this principle from the common sense point of view. To be "good", regardless of how one defines this word, a number of requirements must hold *simultaneously*, while *even one* failure will suffice for a situation to be called bad.

It is rather difficult to be "good" when working at the frontiers of technologies, such as microelectronics, micromechanics, and data storage. These technologies should meet a range of specifications, which are often in competition with each other. It is surprising that it is possible to satisfy them. The restrictions of high technologies compared to conventional technologies are not related to the skill of the worker or the quality of the equipment. These restrictions are caused by the physical nature of the process (sometimes it is called a "theoretical limit"). High technologies are closely connected with high science and there is a fruitful physical idea at the basis of each technological victory. IBM, Philips and other top R&D Centers are not only famous for their technological achievements but also for their outstanding fundamental research. A textbook example is the research on semiconductors in the Bell Telephone Laboratory as well as the discovery of the transistor effect.

Having in mind the laser cleaning process one should start with the idea of adhesion. Researchers at the Philips research laboratory have contributed considerably to the development of this idea. It is sufficient to mention just two names – Hamaker and Casimir.

In 1937 Hamaker accomplished a very clear presentation of the attraction of particles to a surface by the van der Waals force. Although the corresponding potential varies rapidly with distance, as r^{-6} (Landau & Lifshitz, *Quantum Mechanics*), it represents a long-range interaction. The contribution of this interaction to the free energy is not additive; it depends on the body shape and configuration. If one considers the particle as a deformed sphere, then, according to Hamaker, the attraction force is given by

$$F = \frac{A\,a}{6h^2}\left(1 + \frac{r_c^2}{ah}\right), \tag{1}$$

where a is the radius of the particle; h is the separation distance, which is understood as an equilibrium distance originated by interplay of attraction and repulsion forces ($h \approx 4$ Å); and r_c is the radius of contact. The Hamaker constant $A \approx (0.5 - 5.0)\cdot 10^{-19}$ J depends on the properties of the particle, substrate and medium. This value has been measured for many materials. In the absence of additional load, for (sub) micrometer-sized particles the second term in the bracket is the most important, $r_c^2/ah >> 1$. The characteristic adhesion force for a 1 μm spherical particle is of the order of 0.1 dynes. It is a big force having in mind that modern techniques permit the measurement of a force a billion times smaller (Rugar, 2001). It is clear that this high loading leads to elastic or even plastic deformation of the material. These deformations define the size of the contact region.

In 1882, Hertz performed the first examination of the pressure distribution in the contact area. This distribution follows a parabolic law:

$$P(r) = P_{\max}\left(1 - \frac{r^2}{r_c^2}\right)^{1/2}, \tag{2}$$

(see, for example, the analysis of the Hertz solution in § 9 in Landau & Lifshitz "*Theory of Elasticity*"). According to (2), the maximum pressure, P_{max}, is 1.5 times higher than the average pressure in the contact area

$$P_{\max} = \frac{3}{2}\frac{F}{\pi r_c^2}. \tag{3}$$

Assuming a Hertzian pressure distribution, Derjaguin in 1934 found the relation for the radius of contact, r_c, versus particle size:

$$r_c = \left(\frac{Aa^2}{8E^* h^2} \right)^{1/3}, \quad \frac{1}{E^*} = \left(\frac{1-\sigma_p^{\ 2}}{E_p} + \frac{1-\sigma_s^{\ 2}}{E_s} \right), \tag{4}$$

where σ and E are the Poisson coefficients and Young's modulus, and subscripts "p" and "s" indicate the particle and the substrate, respectively.

From (1)-(4) one can find the relation:

$$P_{\max} = \frac{A}{4\pi h^3} + \frac{1}{\pi} \left(\frac{E^{*2} A}{h^2 a} \right)^{1/3}. \tag{5}$$

The value of the van der Waals attraction between planes (per unit of area) is given by the first term of (5).

One can see that $P_{\max}$ becomes higher for smaller particles. For particles with a size of the order of 10 nm (nanoclusters) this pressure can reach a value of the order of one hundred kbar.

The adhesion-induced deformations are quite complex, and some other factors, such as adhesion forces outside the area of contact, should be taken into account to describe the experimental data well. At present, two models of adhesion are commonly acceptable: the Derjaguin-Muller-Toropov (DMT) model for "hard" materials (Derjaguin, 1975, 1980; Muller, 1983) and the Johnson-Kendall-Roberts (JKR) model for "soft" materials (Johnson, 1971, 1976). The transition between the two models also has been discussed (Maugis, 1992, 1995; Greenwood, 1998).

One should pay attention to the phenomena of instability, which are important for the understanding of laser cleaning. Namely, the jump-like disconnection of the particle from the surface arises when the critical negative force (tensile) is applied. On the contrary, particles can approach the surface without a load but when the particle touches the surface the jump-like adhesion force arises. This hysteresis phenomenon can be verified experimentally when the load of the particle is performed with the help of an atomic force microscope (AFM). Such experiments are used to measure the Hamaker constants (see, for example, Shaefer, 1995; Mizes, 1995).

The known paradox in the history of science is that sometimes discoveries are attributed to people who did not make these discoveries. For example, the exponential decay of the intensity in absorbing media, $I = I_0 \exp[-\alpha z]$, is frequently called Beer's law. Actually this law was experimentally established by P. Bouguer in 1729 and was deduced theoretically by J. H. Lambert in 1760. In 1852 A. Beer established another law on the relation of the absorption coefficient to the density N of the absorbing particles, $\alpha = \sigma N$, where σ is the absorption cross-section. It is similar to the association of the name of Christopher Columbus with the discovery of America. Actually, America had been discovered before Columbus. Moreover, Columbus has not even rediscovered America. His caravels had only reached the islands Cuba and Haiti, while we know the continent of America due to Amerigo Vespucci.

Hamaker did not perform any calculations of the Hamaker constant for solids. In his paper he introduced the value $A = \pi q^2 \lambda$, where q is the number density of interacting atoms and λ is a London constant, which describes the dipole-dipole interaction between isolated atoms. This formula can be applied to the description of attraction within a diluted gas but not to solids, where the situation is much more complex.

Casimir carried out the first "real calculation" of the Hamaker constant for solids in 1948. He made the calculations of the attraction force between conducting plates; now this force is called the "Casimir force". This force is universal and it does not depend on the metal properties. The value of the Casimir force (per unit of area) is given by

$$F_{\infty} = \frac{\pi^2}{240} \frac{\hbar c}{h^4}. \tag{6}$$

The Casimir force arises due to quantum vacuum fluctuations of the electromagnetic field. Fluctuations outside the plates are unrestricted, whereas between the plates some restrictions occur due to the boundary conditions. The Casimir force is remarkable, because it is a more palpable consequence of the zero-point field than the Lamb shift (Bordag, 1999; Maclay, 2001). The Casimir force is very big and plays a major role in modern Micro Electro Mechanical Systems (MEMS) (Chan, 2001).

In the framework of macroscopic theory, the van der Waals interaction in a material medium is considered to be caused by a long-wavelength electromagnetic field; Lifshitz suggested this concept in 1955 (see also § 90 in Landau & Lifshitz, *Electrodynamics of Continuous Media*). It includes not only thermal fluctuations but also the zero-point oscillations of the field.

The Lifshitz theory is based on the theory of electromagnetic fluctuations developed by Leontovich (1952) and Rytov (1953). Later, the

Lifshitz formula was derived from the microscopic point of view, using the methods of the quantum theory of the field (Dzyaloshinski, 1959,1960 a, b; Abrikosov, 1965; see also §§ 80-82 in Lifshitz & Pitaevskii, *Statistical Physics, Part 2*). Josef Brodsky said that with the growth of a poet's skill the number of his readers shrinks. I understand the reason for this very well, because I remember the efforts I made during my studies to reproduce the results on the van der Waals attraction from the book of Abrikosov, Gorkov and Dzyaloshinski. In the static limit, the permittivities ε_i can be replaced by their electrostatic dielectric constants ε_{i0}. Then the formula for the attraction force can be written in simplified form:

$$F = \frac{\hbar c}{32\pi^2 h^4} \int_0^\infty dx \int_1^\infty dp \frac{x^3}{p^2} \left\{ \left[\frac{(s_{10}+p)(s_{20}+p)}{(s_{10}-p)(s_{20}-p)} e^x - 1 \right]^{-1} + \left[\frac{(s_{10}+p\varepsilon_{10})(s_{20}+p\varepsilon_{20})}{(s_{10}-p\varepsilon_{10})(s_{20}-p\varepsilon_{20})} e^x - 1 \right]^{-1} \right\}, \tag{7}$$

$$s_{10} = \sqrt{\varepsilon_{01} - 1 + p^2}, \quad s_{20} = \sqrt{\varepsilon_{02} - 1 + p^2}.$$

It is readily seen that in the limit of two good metals, $\varepsilon_{10} = \varepsilon_{20} \to \infty$, formula (7) yields the Casimir force (6). This force varies strongly with distance h. For a typical separation distance in MEMS of h = 10 nm, the value F_∞ is about 1 bar while for a typical separation thickness in adhesion of $h = 4$ Å, the value F_∞ is about 0.5 megabar.

The Lifshitz theory suggests that the van der Waals interaction between solids arises due to the exchange of electromagnetic fluctuations with characteristic energy $\langle \hbar\omega \rangle$ (this value is called the Lifshitz - van der Waals constant). The Hamaker constant is related to the Lifshitz - van der Waals constant by $A = \frac{3}{4\pi}\langle \hbar\omega \rangle$. The numerical factor $\frac{3}{4\pi}$ has no physical relevance.

A simple formula for metals follows under the approximation of the Drude theory:

$$\langle \hbar\omega \rangle \approx \frac{\pi^{3/2}}{8} \hbar\omega_p, \tag{8}$$

where ω_p is the plasma frequency, and γ is the frequency of collisions, which is considered to be small, $\gamma \ll \omega_p$. For aluminum, the only real Drude metal, formula (8) yields a value $\langle\hbar\omega\rangle$ = 8.11 eV, which is close to the value of $\langle\hbar\omega\rangle$ = 8.59 eV that follows from the integration of an experimental spectrum.

Klimchitskaya (2000), discussed the generalization of the Lifshitz formalism for the case when the spatial dispersion can be important, in addition to the frequency dependence. Another problem is the feedback caused by the variation of a dielectric constant with pressure. Here I want to draw attention to one possible effect. With higher loading pressure one has a higher ε and a higher attraction force. This positive feedback may lead to bistability in optical parameters (either in the Lifshitz-Hamaker constant, or in the attraction force) versus particle size or loading force. Formally, two curves: $F = F(\varepsilon)$ from (7) and $\varepsilon = \varepsilon(F)$ produce a diagram, similar to Semenov's diagram in the theory of combustion (Karlov, 2000). The bistability looks like an optical phase transition, which is attractive for applications to high-density information recording.

Great scientists like Casimir and those from the Landau school created the basic understanding in adhesion. They say that when Wolfgang Pauli learnt that Casimir had become director of the Philips Research Laboratories, he exclaimed: "Well, now I'll stop reading Casimir's papers, as it is clear - what papers directors write!" Probably this anecdote came from Casimir himself; he enjoyed telling anecdotes about his relationship with Pauli. In fact, Casimir wrote outstanding papers both before and during his directorship' period. He discussed the retardation effects in attraction (see e.g. Landau & Lifshitz, *Quantum Electrodynamics*) and many other perfect physics.

Adhesion by itself is not the only problem, which is necessary for the understanding of laser cleaning. We should mention Casimir's name once again because of the problem related to the diffraction of light by the particle (Bouwkamp & Casimir, 1954). In this paper it was shown that spatially distributed electrical currents could be represented as a series of multipole fields. This allows, for example, the calculation of antennas with arbitrary current distributions. The same method was used to find the exact solution for the problem of "particle on the surface" (Bobbert, 1986). Recently, a new interest in this problem has been raised by the study of near-field effects in laser cleaning (Luk`yanchuk, 2000).

No doubt, great and beautiful physical ideas encourage the advancement of new technologies. However, in the relations of science and technology a feedback, which produces the so-called "science-technology spiral", plays a very important role; Casimir discussed this problem in 1983. Russian poet Anna Akhmatova wrote: "I wish you were aware from what stray matter springs poetry to prosper without shame..." The soil of science is not nobler than that of poetry. Many problems, which lead to great discoveries started off as purely technological studies. I want to present just two examples.

The first is related to the problem of minimization of scraps of a carton sheet during its cutting solved by L. V. Kantorovich. Kantorovich was a mathematical genius (a Full Professor of mathematics at Leningrad State University at the age of 18). He did a lot of outstanding mathematical discoveries: the Kantorovich space, the Hahn Banach - Kantorovich theorem, etc. (Kantorovich, 1996). But the highest honor given to him is closely associated with the solution of the problem of scrap minimization mentioned earlier. To solve this problem Kantorovich has developed a new mathematical technique of linear programming, which later earned him the Nobel Prize in economics.

Another example is the paper by Landau & Levich, 1942, devoted to dragging of a liquid by a moving plate. This paper was also motivated by a purely technological problem (deposition of a uniform layer of photosensitive emulsion upon the cinefilm base). It is worth mentioning that the entire modern theory of dynamic wetting and interfacional hydrodynamics has come out of this work (see Landau & Lifshitz, Fluid Mechanics; Levich, 1962).

While preparing this book, I had been very interested in the history of laser cleaning. To the best of my knowledge the first papers devoted to laser cleaning of the surface from small particles were published in 1987-88 (Beklemyshev, 1987, 1990, 1991; Assenfel'ft, 1988 a, b; Petrov, 1989; Balychenko, 1991). One of the co-authors of these papers was late A. M. Prokhorov, a Nobel Prize winner. These experimental results received good theoretical support (Kolomenskii, 1991a, b). Finally, the Russian group obtained a patent for laser cleaning (Boykov, 1991). Unfortunately, Dr. Yu. N. Petrov, who was the "driving force" behind Russian laser cleaning, passed away at the beginning of the 90's. Further development of the concept of laser cleaning was accomplished by Kolomenskii (1993a, b, 1995, 1998); Mikhalevich (1994); Maznev (1995), who presented several new and interesting effects such as the excitation of nonlinear surface acoustic waves and the phonon focusing effect.

Prof. S. D. Allen is another well-known name in laser cleaning. Her group has been studying laser cleaning since 1987, while the first papers were published in 1991 (Imen, 1991a; Lee, 1991). Her group received an USA Patent on laser cleaning in 1991 (Imen, 1991b). Among the pioneering research results of S. D. Allen are a number of papers (Lee, 1992, 1993a, b).

Last but not least the group of Dr. W. Zapka should be mentioned. Probably Zapka's contribution has been the most important for industrial applications of laser cleaning. In any case, he received the first European patent on laser cleaning in 1989 (priority from 1987). Although it is not possible to restore the history with the accuracy of one day or one week, my examination shows that in 1987 three groups (Yu. Petrov in Russia, S. D. Allen in USA, and W. Zapka in Germany) discovered the laser cleaning effect simultaneously and independently. The achievement of Petrov's group was that they published the first papers on laser cleaning, while Zapka's group obtained the first patent and realized the first industrial applications. Nevertheless if somebody asked me to recommend people for an award for the discovery of laser cleaning, I would mention all the three names.

This book is dedicated to Dr. Andrew C. Tam from IBM Almaden Research Center. He was not the first to discover the laser-cleaning effect. His first publications appeared only in 1991 (Zapka, 1991 a, b). Nevertheless if one examines the citation index of papers on laser cleaning he will find that papers (Zapka, 1991 b) and (Tam, 1992) are the most cited. Other papers of A. C. Tam written in collaboration with the groups of C. P. Grigoropoulos and P. Leiderer (Park, 1992, 1996; Leung, 1992, Yavas, 1993, 1994 a, b) also have a very high citation impact. At present about half of the citations on laser cleaning are related to the papers with A. C. Tam's name. One can find a parallel in the history of science. It is well known that Galileo was not the first to discover the telescope. Nevertheless only after Galileo's demonstration of Jupiter's moons the telescope became popular throughout the whole world. Andrew Tam definitely had Galileo's talent as a popularizer.

Within 5 to 6 years after Tam's investigations scientists from many countries published more than one hundred papers devoted to laser cleaning. It is not possible to recall the history of these research results objectively. Here different storytellers state the same events in a completely different ways like in the famous novella of Ryunosuke Akutagawa (and in the not less famous film “Rashomon” of Akira Kurosawa).

At this point it is necessary to mention the work of A. C. Engelsberg (1994), which clearly presented the problem with industrial applications of laser-assisted cleaning technology. Papers of J. D. Kelly (1993) gave the foundation for the 1D model of dry laser cleaning. For 10 years this model has been the basic model for the interpretation of dry laser-cleaning experiments.

One can say also a few words about the papers of Y. F. Lu. He started with his research in the group of Prof. M. Takai (Osaka University, Japan). Later Y. F. Lu moved to Singapore where he set up a highly active research group. His group published more than 40 papers, devoted to different aspects of laser cleaning. Probably the most famous paper is related to laser cleaning of magnetic head sliders (Lu, 1996), but other papers (Lu, 1994 a, b, c; 1997a, b; 1998a, b, c; 1999, 2000a, b, c; 2001) are also well known. Papers of Y. F. Lu are the most cited after A. C. Tam’s papers.

I remember how Prof. I. Dzyaloshinski told the young PhD students in the Landau Institute: “I am very sorry that you came to physics too late when all the simple problems had been solved by your teachers. Thus you have to solve only very difficult problems!” I would say that the period of easy results in laser cleaning was over around 1995. In Lambda Highlights No. 47 (May 1995) the editor published a perfect review on “Surface cleaning by UV laser radiation becomes economical”. At the end of this review he mentioned the works of Prof. C. Fotakis’ group, devoted to laser cleaning of artworks. This paper in Lambda Highlights outlined the end of the first stage in the research on laser cleaning.

A new vortex in laser cleaning was initiated after 1998 due to very intensive work of the network “Laser cleaning” of the European Community (see, e.g. papers: Zafiropulos (1998); Dobler (1999); Mosbacher (1999, 2000); Oltra (2000); Leiderer (2000); Fourrier (2001)). This European project coordinates the efforts of 8 Universities and Institutes from 5 European countries.

New and more detailed research works were done by Halfpenny (1999, 2000); Kane (2000); Wu (1999a, b, 2000); G. Veerecke (1999 a, b, 2000) and other groups (for example, She (1999); Kim (2001); Yavas (2000); Feng (1999); Schmidt (1999); Lee (2000, 2001); Cetinkaya (2000); Zheng (2001); Luk`yanchuk (2001, 2002). In addition, D. Bäuerle (2000), presented a comprehensive review on laser cleaning. In Singapore we decided to organize an International Workshop to discuss new trends in laser cleaning. This Workshop was held in May 2001 within the frame of the 2nd International Symposium on Laser Precision Microfabrication (LPM 2001). Miyamoto (2002) published the proceedings of this Symposium.

Since 1984 I have met Dr. A. C. Tam at several international conferences and was astonished by his fast scientific reaction as well as by his nontrivial physical and technological ideas. I remember how Nikita Arnold, who was at that time a student in my group at the General Physics Institute, tried to reproduce some of Tam's results. Now Nikita presents the complete theory of 1D effects in dry laser cleaning (Chapter 2). When I started to do my own calculations in laser cleaning, Andrew Tam was among the first people with whom I discussed the results. Dr. A. C. Tam actively participated in the organization of the Workshop on Laser Cleaning and we planned his opening talk "*The emergence of laser cleaning for Hi-Tech industrial applications*". Our grief was very deep when we received the sad news about Andrew Tam's death. I am grateful to the specialists of many countries who found it possible to come to the Workshop to give tribute to Andrew Tam. In memory of A. C. Tam's input in science and technology, as well as his great contribution to laser cleaning, we decided to publish this book. It presents the review papers of the participants of the 1st International Workshop on Laser Cleaning. This book reflects recent achievements in laser cleaning and can be used as a modern introduction for master's and PhD students and engineers.

Werner Zapka (Chapter 1) introduces the history of the discovery of laser cleaning and Andrew Tam's role in these studies. The three consequent parts of the book are devoted to dry and steam laser cleaning and laser cleaning of artworks. The last part of the book illustrates some applications of laser cleaning.

Laser cleaning contains a broad range of interesting and numerous physics. It is sufficient to say that within the papers devoted to laser cleaning one can meet references to essentially all the volumes of the Landau & Lifshitz Course on Theoretical Physics. For example, "*Mechanics*" can be mentioned when one discusses oscillations of the particles at dry laser cleaning. "*Statistical Physics, Part 1*" and "*Physical Kinetics*" are more cited when one discusses the nucleation and bubble formation at steam laser cleaning.

Molière's monsieur Jourdain had no idea for 40 years that he spoke prose. I did not know either that my life for a long time produced the "Brownian pushes", encouraging me to take a step on the "road to laser cleaning". Many people supported me on my way. I am very thankful to my colleagues in the laser group of DSI (especially to W. D. Song, M. H. Hong and Y. W. Zheng) for the related discussions.

Prof. T. C. Chong supported the idea of the 1st International Workshop on Laser Cleaning and helped me with its organization. Profs. I. Miyamoto, Y. F. Lu and Dr. Koji Sugioka gave me their support as Co-chairs of LPM2001.

When preparing this book we have used "internal" reviewing of the papers. In fact, all the authors have helped me with their critical comments. Profs. S. I. Anisimov, D. Bäuerle, P. Dyer, A. Rubenchik, M. Tribelsky and Dr. E. Gatskevich made many useful suggestions as "external" referees. Profs. S. Allen, C. Grigoropoulos, S. Mikhalevich, M. Takai, H. Coufal and Drs. A. Kolomenskii and V. Pustovoy all sent me their comments on the history and the first stage in laser cleaning.

I am also very thankful to Ms. Lim Sook Cheng, Publishing Director of World Scientific Singapore and to the Editor, Mr. Yeow-Hwa Quek, for their great help and patience.

B. S. Luk`yanchuk
Data Storage Institute, Singapore

References

Abrikosov A. A., Gorkov L. P., Dzyaloshinski I. E., *Methods of Quantum Field Theory in Statistical Physics*, (Prentice-Hall, Englewood Cliffs, New Jersey, 1965)

Arnold V. I., *Catastrophe Theory*, 3rd Edition, (Springer-Verlag Berlin, Heidelberg, 1992)

Assendel'ft E. Y., Beklemyshev V. I., Makhonin I. I., Petrov Yu. N., Prokhorov A. M., Pustovoy V. I., *Optoacoustic effect on the desorption of microscopic particles from a solid surface into a liquid.* Sov. Tech. Phys. Lett., vol. **14**(6), pp. 444-445 (1988a)

Assendel'ft E. Y., Beklemyshev V. I., Makhonin I. I., Petrov Yu. N., Prokhorov A. M., Pustovoy V. I., *Photodesorption of microscopic particles from a semiconductor surfaces into a liquid*, Sov. Tech. Phys. Lett., vol. **14**(8), pp. 650-654 (1988b)

Balychenko A. A., Beklemyshev V. I., Dar'yushkin A. E., Makhonin I. I., Feizulova R. K. G., *Processes for removal of aluminum from silicon plate surfaces*, Soviet Microelectronics, vol. **20**(4), pp. 252-255 (1991)

Bäuerle D., *Laser Processing and Chemistry*, 3rd Edition (Springer-Verlag, Berlin, Heidelberg 2000)

Beklemyshev V. I., Makarov V. V., Makhonin I. I., Petrov Yu. N., Prokhorov A. M., Pustovoy V. I., *Photodesorption of metal ions in a semiconductor-water system*, JETP Letters, vol. **46**(7), pp. 347-350 (1987)

Beklemyshev V. I., Dar'yushkin A. E., Makhonin I. I., Petrov Yu. N., Prokhorov A. M., Pustovoi V. I., *Electronic Industry*, No. **4**, pp. 34-37 (1990)

Beklemyshev V. I., Kolomenskii A. A., Maznev A. A., Makhonin I. I., Mikhalevich V. G., Petrov Yu. N., Pustovoy V. I., *On the abilities on laser induced acoustic cleaning of silicon surface, covered by thin liquid layer, from the submicrometer particles.* Preprint No. 33, General Physic Institute of Russian Academy of Sciences, Moscow, Russia, 1991

Bobbert P. A., Vlieger J., *Light scattering by a sphere on a substrate*, Physica, Vol. **137A**, pp. 209-242 (1986)

Bordag M. (Ed.), *The Casimir Effect 50 Years Later*, (World Scientific, Singapore, 1999)

Bouwkamp C. J., Casimir H. B. G., *On multipole Expansions in the theory of Electromagnetic Radiation*, Physica, vol. **20**, pp. 539-554 (1954)

Boykov D. A., Kolomenskii A. A., Maznev A. A., Maishev Yu. V., Panfilov Yu. V., Ravich A. M., *USSR Patent* #4872125/25 (August, 1991)

Casimir H. B. G., *On the attraction between two perfectly conducting plates*, Proc. K. Ned. Akad. Wet. **51**, pp. 793-795 (1948)

Casimir H. B. G., *Haphazard Reality: Half a Century of Science* (Harper & Row, New York, 1983)

Casimir force: http://www.no-big-bang.com/process/casimireffect.html

Cetinkaya C., Wu C., Li C., *Laser-based transient surface acceleration of thermoelastic layers*, Journal of Sound and Vibration, vol. **231**, pp. 195-217 (2000)

Chan H. B., Aksyuk V. A., Kleiman R. N., Bishop D. J., Capasso F., *Quantum mechanical actuation of microelectromechanical systems by the Casimir force*, Science **291**, pp. 1941-1944 (2001)

Derjaguin B. V., *Untersuchungen über die Reibung und Adhäsion*, Kolloid Z., vol. **69**, pp. 155-164 (1934)

Derjaguin B. V., Muller V. M., Toporov Yu. P., *Effect of Contact Deformations on the Adhesion of Particles*, Journal of Colloid and Interface Science, **53**, pp. 314-325 (1975)

Derjaguin B. V., Muller V. M., Toporov Yu. P., *Journal of Colloid and Interface Science*, **73**, 293 (1980)

Dobler V., Oltra R., Boquillon J. P., Mosbacher M., Boneberg J., Leiderer P., *Surface acceleration during dry laser cleaning of silicon*, Appl. Phys. **A 69**, pp. 335-337 (1999)

Dzyaloshinski I. E., Pitaevskii L. P., *Van der Waals forces in an inhomogeneous dielectric,* JETP, **9**, pp. 1282 (1959)

Dzyaloshinski I. E., Lifshitz E. M., Pitaevskii L. P., *Van der Waals forces in liquid films*, JETP, **10**, pp. 161-177 (1960a)

Dzyaloshinski I. E., Lifshitz E. M., Pitaevskii L. P., *The general theory of van der Waals forces*, JETP, **10**, pp. 165-190 (1960b)

Engelsberg A. G., *Transition from laboratory to manufacturing for a dry, laser-assisted cleaning technology*, Proc. SPIE, vol. **3274**, pp. 8-15 (1994)

Feng Y., Liu Z., Vilar R. et al., *Laser surface cleaning of organic contaminants*, Appl. Surf. Sci. 150, pp. 131-136 (1999)

Fourrier T., Schrems G., Mühlberger T., Heitz J., Arnold N., Bäuerle D., Mosbacher M., Boneberg J., Leiderer P., *Laser Cleaning of Polymer Surfaces,* Appl. Phys. A 72, p 1-6 (2001)

Greenwood J. A., Johnson K. L., *An alternative to the Maugis model of adhesion between elastic spheres,* J. Phys. D. Appl. Phys., **31**, pp. 3279-3290 (1998)

Halfpenny D. R., Kane D. M., *A quantitative analysis of single pulse ultraviolet dry laser cleaning*, J. Appl. Phys., **86**, pp. 6641-6646 (1999)

Halfpenny D. R., Kane D. M., Lamb R. N., Gong B., *Surface modification of silica with ultraviolet laser radiation,* Appl. Phys. **A. 71**, pp. 147-151, (2000)

Hamaker H. C., *The London-Van der Waals attraction between spherical particles*, Physica IV, No. 10, pp. 1058-1072 (1937)

Johnson K. L., Kendall K., Roberts A. D., *Surface energy and the contact of solids*, Proc. Roy. Soc., **A 324**, pp. 301-313 (1971)

Johnson K. L., *Adhesion at the contact of solids*, In: "*Theoretical and Applied Mechanics*", Ed. by W. T. Koiter, pp. 133-143 (North-Holland 1976)

Imen K., Lee S. J., Allen S. D., *Laser Assisted Micron Scale Particle Removal,* Appl. Phys. Lett., **58**, pp. 203-205 (1991a)

Imen K., Lee S. J., Allen S. D., *U. S. Patent,* Serial No. 4, 987, 286 (January 1991b)

Kane D. M., Halfpenny D. R., *Reduced Threshold UV-Laser Ablation of Glass Substrates with Surface Particle Coverage and Associated Systematic Surface Laser Damage*, J. Appl. Phys., **87**, pp. 4548-4552 (2000)

Kantorovich L.V., *Selected works*, Parts I, II, Ed. by S. S. Kuteladze (Gordon & Breach, New York, London 1996)

Karlov N. V., Kirichenko N. A., Luk`yanchuk B. S., *Laser Thermochemistry* (Cambridge International Science Publishing, Cambridge, 2000)

Kelly J. D., Hovis F. E., *A thermal detachment mechanism for particle removal from surfaces by pulsed laser irradiation*, Microelectronic Engineering **20**, pp. 159-170 (1993)

Kim D., Park H. K., Grigoropoulos C. P., *Interferometric probing of rapid vaporization at a solid-liquid interface induced by pulsed-laser irradiation*, Int. J. Heat Mass Tran. 44, pp. 3843-3853 (2001)

Klimchitskaya G. L., Mohideen U., Mostepanenko V. M., *Casimir and van der Waals forces between two plates or a sphere (lens) above a plate made of real metals*, Phys. Rev. A, vol. **61**, pp. 062107-062112 (2000)

Kolomenskii A. A., Maznev A. A., *Observation of phonon focusing with pulsed laser excitation of surface acoustic waves in silicon*, JETP Letters, vol. **53**, pp. 423-426 (1991a)

Kolomenskii A. A., Maznev A. A., *Shaking the mechanical micro-particles off the silicon surface induced by surface acoustic-wave excited by laser-pulses,* Sov. Tech. Phys. Lett., vol. **17**, No. 13, pp. 62-66 (1991b)

Kolomenskii A. A., Maznev A. A., *Laser induced acoustic cleaning of surfaces from mechanical microparticles*, Bull. Acad. Sci. USSR, Phys. Ser., vol. **57**, No. 2, pp. 180-189 (1993a)

Kolomenskii A. A., Maznev A. A., *Phonon-focusing effect with laser-generated ultrasonic surface-waves*, Phys. Rev. B, vol. **48**, No. 19, pp. 14502-14508 (1993b)

Kolomenskii A. A., Maznev A. A., *Propagation of laser-generated surface acoustic-waves visualized by shake-off of fine particles,* J. Appl. Phys., vol. **77**, pp. 6052-6054 (1995)

Kolomenskii A. A., Schuessler H. A., Mikhalevich V. G., Maznev A. A., *Interaction of laser-generated surface acoustic pulses with fine particles: Surface cleaning and adhesion studies*, J. Appl. Phys., vol. **84**, No. 5, pp. 2404-2410 (1998)

Landau L. D., Levich V. G., *Dragging of a liquid by a moving plate,* Acta Phys. - Chem. USSR **17**, p. 42 (1942); see in Collected Papers of L. D. Landau, Ed. by D. Ter Haar (Gordon & Breach, New York, London 1965), paper 48, pp. 355-364

Landau L. D., Lifshitz E. M, *Course of Theoretical Physics* (Pergamon Press, 1994, 1998, 1999):
Vol. 1, *Mechanics*, 3rd Edition
Vol. 2, *The Classical Theory of Field*, 4th Edition
Vol. 3, *Quantum Mechanics*, 3rd Edition
Vol. 4, *Quantum Electrodynamics*, 2nd Edition
Vol. 5, *Statistical Physics*, Part 1, 3rd Edition
Vol. 6, *Fluid Mechanics*, 2nd Edition
Vol. 7, *Theory of Elasticity,* 3rd Edition
Vol. 8, *Electrodynamics of Continuous Media*, 2nd Edition
Vol. 9, *Statistical Physics*, Part 2
Vol. 10, *Physical Kinetics*

Lee S. J., Imen K., Allen S. D., *Threshold Measurements in Laser Assisted Particle Removal*, Proc. SPIE, vol. **1598**, 2 (1991)

Lee S. J., Imen K., Allen S. D., *CO_2 -laser assisted particle removal threshold measurements*, Appl. Phys. Lett., **61**, pp. 2314-2316 (1992)

Lee S. J., Imen K., Allen S. D., *Laser Assisted Particle Removal from Silicon Surfaces*, Microelectronic Engineering **20**, 145 (1993a).

Lee S. J., Imen K., Allen S. D., *Shock Wave Analysis of Laser Assisted Particle Removal*, J. Appl. Phys. **74**, 12 (1993b)

Lee J. M., Watkins K. G., Steen W. M., *Angular laser cleaning for effective removal of particles from a solid surface*, Appl. Phys. A 71, pp. 671-674 (2000)

Lee J. M., Watkins K. G., *Removal of small particles on silicon wafer by laser-induced airborne plasma shock waves*, J. Appl. Phys. **89**, pp. 6496-6500 (2001)

Leiderer P., Boneberg J., Dobler V., Mosbacher M., Münzer H.-J., Fourrier T., Schrems G., Bäuerle D., Siegel J., Chaoui N., Solis J., Afonso C.N., *Laser-Induced Particle Removal from Silicon Wafers,* Proc. SPIE **4065**, pp. 249-259 (2000)

Leontovich M. A., Rytov S. M., *On the theory of electrical fluctuations*, DAN SSSR, vol. **102**, pp. 535-539 (1952)

Leung P. T., Do N., Klees L., Leung W. P., Tong F., F. Lam F., Zapka W., Tam A. C., *Transmission studies of explosive vaporization of a transparent liquid-film on an opaque solid-surface induced by excimer-laser-pulsed irradiation*, J. Appl. Phys. **72**, pp. 2256-2263 (1992)

Levich V. G., *Physicochemical Hydrodynamics* (Englewood Cliffs, N. J., Prentice-Hall, 1962)

Lifshitz E. M., *Theory of molecular attraction forces between condensed bodies*, JETP, **29**, pp. 94-98 (1955)

Lifshitz E. M., Pitaevskii L. P., *Statistical Physics, Part 2* (Pergamon Press, 1980)- volume 9 in Landau & Lifshitz Course

Lu Y. F., Aoyagi Y., *Laser-Induced Dry Cleaning in Air - A New Surface Cleaning Technology in Lieu of Carbon Fluorochloride (CFC) solvents*, Jpn. J. Appl. Phys., Part 2, vol. **33** (3B), pp. L430-L433 (1994a)

Lu Y. F., Takai M., Komuro S., Shiokawa T., Aoyagi Y., *Surface Cleaning of Metals by Pulsed Laser Irradiations in Air*, Appl. Phys. **A 59**, pp. 281-288 (1994b)

Lu Y. F., Aoyagi Y., Takai M., Namba S., *Laser Surface Cleaning in Air—Mechanisms and Applications*, Jpn. J. Appl. Phys. Part 1, vol. **33** pp. 7138-7148 (1994c)

Lu Y. F., Song W. D., Hong M. H., Chong T. C., Teo B. S., Low T. S., *Laser removal of particles from head sliders*, J. Appl. Phys. **80**, pp. 499-504 (1996)

Lu Y. F., Song W. D., Ang B. W., Chan D. S. H., Low T. S., *A Theoretical Model for Laser Removal of Particles from Solid Surface*, Appl. Phys. **A 65**, pp. 9-13 (1997a)

Lu Y. F., Song W. D., Ye K. D., Hong M. H., Chan D. S. H., Low T. S., *Removal of Submicron Particles from Nickel-Phosphorous Surface by Pulsed Laser Irradiation*, Appl. Surf. Sci. **120**, pp. 317-322 (1997b)

Lu Y. F., Lee Y. P., Zhou M. S., *Laser cleaning of Etch-Induced Polymers from Via Holes*, J. Appl. Phys., 83, pp. 1677-1684 (1998a)

Lu Y. F., Song W. D., Tee C. K., Chan D. S. H., Low T. S., *Wavelength Effects in Laser Cleaning Processes*, Jpn J. Appl. Phys. **37**, pp. 580-584 (1998b)

Lu Y. F., Zhang Y., Song W. D., Chan D. S. H., *A Theoretical Model for Laser Cleaning of Microparticles in a thin liquid layer*, Jpn. J. Appl. Phys., vol. **37**, pp. L1330-L1332 (1998c)

Lu Y. F., Zheng Y. W., Song W. D., *An energy approach to the modelling of particle removal by pulsed laser irradiation*, Appl. Phys. A **68**, pp. 569-572 (1999)

Lu Y. F., Zheng Y. W., Song W. D., *Characterization of ejected particles during laser cleaning*, J. Appl. Phys. **87**, pp. 549-552 (2000a)

Lu Y. F., Zheng Y. W., Song W. D., *Laser induced removal of spherical particles from silicon wafers*, J. Appl. Phys. **87**, pp. 1534-1539 (2000b)

Lu Y. F., Zhang L., Song W. D., Zheng Y. W., Luk'yanchuk B. S., *Laser writing of sub-wavelength structure on silicon (100) surfaces with particle enhanced optical irradiation,* JETP Letters, vol. **72**, pp. 457-459 (2000c)

Lu Y. F., Zheng Y. W., Song W. D., *Angular effect in laser removal of spherical silica particles from silicon wafers*, J. Appl. Phys. **90**, pp. 59-63 (2001)

Luk'yanchuk B. S., Zheng Y. W., Lu Y. F., *Laser Cleaning of the surface: Optical resonance and near-field effects*, Proc. SPIE, vol. **4065**, pp. 576-587 (2000)

Luk'yanchuk B. S., Zheng Y. W., Lu Y. F., *A new mechanism of laser dry cleaning*, Proc. SPIE, vol. 4423, pp. 115-126 (2001)

Luk'yanchuk B. S., Zheng Y. W., Lu Y. F., *Basic physical problems related to dry laser cleaning*, RIKEN Review, No. **43**, pp. 28-34 (2002)

Maclay G. J., Fearn H., Milonni P. W., *Of Some Theoretical Significance: Implications of Casimir Effects*, European Journal on Physics **22**, pp. 463-469 (2001)

Maugis D., *The JRK-DMT Transition Using a Dugdale Model*, Journal of Colloid and Interface Science, **150**, pp. 243-269 (1992)

Maugis D., Gauthier-Manuel B., *JRK-DMT Transition in the Presence of Liquid Meniscus,* In "Fundamentals of Adhesion and Interfaces", Ed. by D. S. Rimai, L. P. DeMejo, K. L. Mittal (Utrecht, VSP 1995), pp. 49-60

Maznev A. A., Kolomenskii A. A., Hess P., *Time-resolved cuspidal structure in the wave-front of surface acoustic pulses on (111) gallium-arsenide,* Phys. Rev. Lett., vol. **75**, pp. 3332-3335 (1995)

Mikhalevich V. G., Kolomenskii A. A., Benck E. C., et al., *The interaction of fine particles with laser-generated nonlinear surface acoustic-waves*, J. Phys. IV 4: (C7), pp. 709-711 (1994)

Miyamoto I., Lu Y. F., Sugioka K., Dubowski, J. J. (Eds.), *Second International Symposium on Laser Precision Microfabrication*, Proc. SPIE, vol. **4426**, 498 pages (2002)

Mizes H., Loh K. G., Ott M. L., Miller R. J. D., *Polymer to Particle Adhesion Probed with Atomic Force Microscopy*, In "Particles on Surfaces: Detection, Adhesion and Removal", Ed. by K. L. Mittal (Marcel Dekker, New York, 1995), pp. 47-60

Mosbacher M., Chaoui N., Siegel J., Dobler V., Solis J., Boneberg J., Afonso C. N., Leiderer P., *A comparison of ns and ps steam laser cleaning of Si surfaces*, Appl. Phys. **A 69**, pp. 331-334 (1999)

Mosbacher M., Dobler V., Boneberg J., Leiderer P., *Universal threshold for the steam laser cleaning of submicron spherical particles from Silicon*, Appl. Phys. **A 70**, pp. 669-672 (2000)

Muller V. M., Yushchenko V. S., Derjaguin B. V., *General Theoretical Consideration of the Influence of Surface Forces on Contact Deformations and the Reciprocal Adhesion of Elastic Spherical Particles*, Journal of Colloid and Interface Science, **92**, pp. 92-101 (1983)

Oltra R., Arenholz E., Leiderer P., Kautek W., Fotakis C., Autric M., Afonso C., Wazen P., *Modelling and Diagnostic of Pulsed Laser-Solid Interactions. Application to Laser Cleaning*, Proc. SPIE, vol. **3885**, pp. 499-508 (2000)

Park H. K., Xu X., Grigoropoulos C. P., Do N., Klees L., Leung P. T., Tam A. C., *Temporal profile of optical-transmission probe for pulsed-laser heating of amorphous-silicon films*, Appl. Phys. Lett **61**, pp. 749-751 (1992)

Park H. K., Grigoropoulos C. P., Poon C. C., Tam A. C., *Optical probing of the temperature transients during pulsed-laser induced boiling of liquids*, Appl. Phys. Lett. **68**, pp. 596-598 (1996)

Petrov Yu. N., *Laser cleaning of semiconductor surface*, Proc. SPIE, vol. **1352**, pp. 266-273 (1989)

Rugar D., Stipe B. C., Mamin H. J., Yannoni C. S., Stowe T. D., Yasumura K. Y., Kenny T. W., *Adventures in attonewton force detection*, Appl. Phys. **A 72**, pp. S3-S10 (2001)

Rytov S. M., *Theory of Electromagnetic Fluctuations and Thermal Radiation*. Publ. of USSR Academy of Sciences, Moscow 1953 (English Translation, AFCRL TR 59-162)

Schmidt M. J. J., Lin L., Spencer J. T., *Characteristics of high power diode laser removal of multilayer chlorinated rubber coatings from concrete surfaces*, Opt. Laser Technol. **31**, pp. 171-180 (1999)

Shaefer D. M., Carpenter M., Gady B., Reifenberger R., DeMejo L. P., Rimai D. S., *Surface roughness and its influence on particle adhesion using atomic force techniques*, In "Fundamentals of Adhesion and Interfaces", Ed. by D. S. Rimai, L. P. DeMejo, K. L. Mittal (Utrecht, VSP 1995), pp. 35-48

She M., Kim D., Grigoropoulos C. P., *Liquid-assisted pulsed laser cleaning using near-infrared and ultraviolet radiation*, J. Appl. Phys. **86**, pp. 6519-6524 (1999)

Tam A. C., Leung W. P., Zapka W., Ziemlich W., *Laser-cleaning technology for removal of surface particulates*, J. Appl. Phys. **71**, pp. 3515-3523 (1992)

Vereecke G., Röhr E., Heyns M. M., *Laser-assisted removal of particles on silicon wafers*, J. Appl. Phys. **85**, pp. 3837-3843 (1999a)

Vereecke G., Röhr E., Heyns M. M., *Evaluation of a dry laser cleaning process for the removal of surface particles*, Solid State Phenom. **65**, pp. 187-190 (1999b)

Vereecke G., Röhr E., Heyns M. M., *Influence of beam incidence angle on dry laser cleaning of surface particles*, Appl. Surf. Sci., **157**, pp. 67-73 (2000)

Wu X., Sacher E., Meunier M., *The effects of hydrogen bonds on the adhesion of inorganic oxide particles on hydrophilic silicon surfaces*, J. Appl. Phys. **86**, pp. 1744-1748 (1999a)

Wu X., Sacher E., Meunier M., *Excimer laser induced removal of particles from hydrophilic silicon surfaces*, J. Adhesion. **70**, pp. 167-178 (1999b)

Wu X., Sacher E., Meunier M., *The modeling of excimer laser particle removal from hydrophilic silicon surfaces*, J. Appl. Phys. **87**, pp. 3618-3627 (2000)

Yavas O., Leiderer P., Park H. K., Grigoropoulos C. P., Poon C. C., Leung W. P., Do N., Tam A. C., *Optical reflectance and scattering studies of nucleation and growth of bubbles at a liquid-solid interface induced by pulsed laser-heating*, Phys. Rev. Lett. **70**, pp. 1830-1833 (1993)

Yavas O., Leiderer P., Park H. K., Grigoropoulos C. P., Poon C. C., Tam A. C., *Enhanced acoustic cavitation following laser-induced bubble formation - long-term-memory effect*, Phys. Rev. Lett. **72**, pp. 2021-2024 (1994a)

Yavas O., Leiderer P., Park H. K., Grigoropoulos C. P., Poon C. C., Leung W. P., Do N., Tam A. C., *Optical and acoustic study of nucleation and growth of bubbles at a liquid-solid interface induced by nanosecond-pulsed-laser heating*, Appl. Phys. **A 58**, pp. 407-415 (1994b)

Yavas O., Suzuki N., Takai M., et al., *Improvement of electron emission of silicon field emitter arrays by pulsed laser cleaning*, J. Vac. Sci. Technol. **B 18**, pp. 1081-1084 (2000)

Zapka W., Asch K., Meissner K., *European Patent* EP 0297506 A2, (January 1989)

Zapka W., Tam A. C., Ziemlich W., *Laser cleaning of wafer surfaces and lithography masks*, Microelectronic Engineering **13**, pp. 547-550 (1991a)

Zapka W., Ziemlich W., Tam A. C., *Efficient pulsed laser removal of 0.2 μm sized particles from a solid surface*, Appl. Phys. Lett. **58**, pp. 2217-2219 (1991b)

Zafiropulos V., Fotakis C., *Lasers in the conservation of painted artworks*, Chapter 6 in “Laser in Conservation: an Introduction”, Ed. by M. Cooper (Butterworth Heineman, Oxford 1998)

Zheng Y. W., Luk’yanchuk B. S., Lu Y. F., Song W. D., Mai Z. H., *Dry laser cleaning of Particles from Solid Substrates: Experiments and Theory*, J. Appl. Phys. **90**, pp. 2135-2142 (2001)

Part 1. History

Chapter 1

THE ROAD TO 'STEAM LASER CLEANING'

W. Zapka

To Andrew C. Tam, my colleague and friend

Laser cleaning was studied in a sequence of experiments that spanned a time frame of 15 years after it had originally been identified as a means to remove sub-micrometer particles from sensitive silicon membrane stencil masks. This paper describes how the subsequent investigations added to the understanding of the basic mechanisms of both, dry laser cleaning and steam laser cleaning, respectively.

Keywords: steam laser cleaning, organic contamination, microelectronics
PACS: 41.20.-q; 42.62.Cf; 43.35.-c; 52.70.-m; 81.15.Fg; 81.65.Cf; 85.40.-e

1. Meeting Andrew C. Tam

Andrew C. Tam was already an internationally respected expert when the author joined him as his first post-doc in 1981 at IBM's San Jose Research Laboratories. As Assistant Professor at Columbia University Andrew had discovered 'laser snow' of alkali-hydride crystals, which he produced in a laser-induced chemical reaction, and he had become famous for his pioneering work in optoacoustic spectroscopy, which he conducted at Bell Laboratories before he joined IBM in 1979.

In hindsight my post-doc fellowship with Andrew was a most influential period. On the one side it was Andrew's scientific creativity paired with his experimental skills that lead us to develop new applications of optoacoustics. On the other side it was Andrew's drive and enthusiastic support, which enabled him to form a highly motivated and creative team. It was this period of intense work and fun, which laid the ground of a friendship, that lasted almost 20 years, and gave us the opportunity to cooperate on various tasks, 'Laser Cleaning' being one of them. Fig. 1 is a photo from 1982, showing Andrew amongst his colleagues. In the following the author will give a 'historical' review of the work on 'laser cleaning', most of which was done in close cooperation with Andrew C. Tam.

Fig. 1. Andrew C. Tam in the midst of his colleagues in 1982; rear from left: Gary Bjorklund, W. E. Moerner, Andrew C. Tam, Frank Schellenberg, Peter Pokrowsky; in front the author, Werner Zapka, and daughter Karolin who co-authored a paper on laser cleaning (Zapka, 2000).

2. A cleaning method needed

The author returned to Germany in 1982 and joined IBM's German Manufacturing Technology Center GMTC. In those days IBM conducted massive research and development in lithography, working in parallel on optical and x-ray lithography, as well as on two types of electron-beam lithography, namely direct-write e-beam and masked e-beam lithography. The GMTC group focused on the latter technique. The basic idea was to produce a stencil mask, which contains the pattern to be printed as physical holes, and to 1:1 shadow project this pattern with an e-beam onto the resist wafer. A specific scanning mode with a large (1 mm) diameter e-beam and automatic image correction scheme enabled exposure with a high throughput, not attainable with standard direct-write e-beam lithography machines. This lithographic technique, which was termed EPB, Electron Beam Proximity Printing, therefore combined the e-beam's high resolution and overlay capabilities with high printing speed, and consequently it was ideally suited for high throughput manufacturing of memory chips.

The key feature of EBP lithography was obviously the stencil mask (Greschner, 1977; Nehmiz, 1985). This mask was produced from a silicon wafer, the central area of which was wet-etched to form a membrane of 3μm thickness. By means of RIE-etching the mask pattern was then formed as through holes, or stencil holes, within the membrane area. Stencil mask pattern of minimum linewidth of 0.3 μm were manufactured in this fashion. In order to maintain mechanical stability of the mask even in the case of complex memory chip pattern, the full chip pattern was split into two complementary mask patterns and both of these were produced side by side within one stencil mask. During lithographic exposure both these complementary mask pattern were sequentially printed to yield the complete chip memory pattern in the photoresist. The stencil masks were shadow projected onto the substrate wafers, positioned at 1 mm distance from the stencil mask. Due to the shadow projection mode, the stencil mask pattern had to be a 1:1 replica of the pattern to be printed. The GMTC team was successful in developing a functional lithography tool. In a test run using IBM's 1MB memory chip as a test vehicle the EBP tool capability was successfully demonstrated, and the required overlay precision was achieved both between the complementary masks and layer-to-layer, respectively (Zapka, 1991a). While the tool-related problems like overlay, image distortion, resolution and print speed could all be solved readily, a key problem, however, appeared that could not be solved with standard techniques, namely the contamination of the stencil mask. It was shown that any contamination within the stencil holes of the mask, be it particles from electrically conducting or non-conducting material, produced high contrast images in the photoresist when shadow projected with the 10 keV electron beam. As a rule of thumb any defect of size larger than ¼ of the minimum linewidth to be printed must be avoided. In view of the 1μm linewidth structures of DRAM's in the mid-1980s this meant that particles down to

250 nm diameter had to be avoided or to be removed from the 1:1 stencil masks. Such small particles are known to adhere extremely well to surfaces, due to van-der-Waals, electrostatic or capillary forces. The adhesion forces of 1 μm sized particles, for example, are of some 10^6 times their gravity (Mittal, 1988).

Contamination of a stencil mask can occur during handling or processing in the course of the manufacturing sequence. Contamination can as well occur even inside the lithography chamber. While gas pressure inside the chamber is of the order 10^{-6} mbar or lower, bursts of gas during wafer load/unload cannot totally be prevented, so that particles could reach the stencil mask. It is normal procedure to encapsulate optical lithography masks with pellicles and keep eventual particulate contamination out of focus during lithographic exposure. Since e-beams and ion-beams would be absorbed and deviated by such pellicles, the masks for e-beam and ion-beam lithography cannot be protected from contamination by particulates.

Attempts by the GMTC group to remove particles from stencil masks with standard wet cleaning techniques failed. Ultrasonic agitation in wet cleaning baths lead to immediate rupture of the membrane. It was further assumed that the probability of recontamination of the membrane mask was high when using wet cleaning baths. The task was consequently to develop a technique that allows removal of particulate contamination of sub-micrometer size from the surface as well as from the trenches of delicate silicon membrane stencil masks while avoiding damage like mask rupture or mask pattern distortion.

3. 'Laser Cleaning'

To the author's surprise the silicon membrane stencil masks proved to be stable under short pulse laser irradiation. In an early test it was possible to remove a 35 nm gold coating from such a mask by means of focused nitrogen-laser irradiation without damage to the stencil pattern. These early observations were confirmed by tests with a special 'pan-handle' mask, where a non-coated silicon stencil membrane of size 4 mm x 9 mm was supported only by a single sided cantilever beam of 20μm width, 3 μm thickness and 200 μm length. This mask survived multiple pulses of KrF excimer laser irradiation at fluences in excess of 100 mJ/cm^2. 'Laser Cleaning', i.e. removal of particles from solid surfaces by means of short-pulse laser irradiation was attempted next. Particles of different material, silicon, SiO_2, Al_2O_3, and polystyrene, and size 0.3μm to 5 μm were deposited onto bare silicon wafers and onto silicon membrane stencil masks from dilute suspensions in alcohol. Fig. 2 shows a silicon membrane stencil mask after contamination. KrF-laser radiation of 20 ns pulse length was directed onto an aperture, and imaged onto the silicon stencil mask with a single element reduction lens. The experiment was carried out at ambient atmosphere. The samples were positioned horizontally and irradiated at normal incidence. Success was obvious immediately. At laser fluences between 300 and 500 mJ/cm^2 the particles within the irradiated areas were efficiently removed with no visible damage to the mask. Fig. 3 shows areas of the same stencil mask as in Fig. 2 after 4 KrF laser pulses at 350 mJ/cm^2 at ambient air. Particles were removed from both the surface and the trenches.

Fig. 2. A silicon membrane stencil mask contaminated with Al_2O_3-spheres of size 0.3 to 1.5 μm. The pattern linewidth in the silicon stencil mask was about 1.8 μm, and the membrane thickness 3 μm.

First hints on the mechanism of the particle removal process could be obtained from these initial experiments. The boundaries between the cleaned and non-cleaned areas were clearly visible, and the cleaning action was obviously confined precisely to the irradiated area. The fact that even SiO_2 particles, which do not absorb KrF radiation, could be removed (though at a slightly higher laser fluence) indicated that the direct absorption of the laser radiation by the particle did aid the particle removal, but was not essential to the process. Consequently the driving force behind the particle ejection process had to be the absorption of the KrF laser radiation in the silicon substrate. Due to the high absorption coefficient of silicon at 248 nm the top surface layer of some 50 nm depth absorbed the laser energy, heated up, and expanded on a time scale of the order of the laser pulse length, i.e. in a few ten nanoseconds. The silicon surface would thus perform a transient motion at the frequency of several megahertz, and eject the particles.

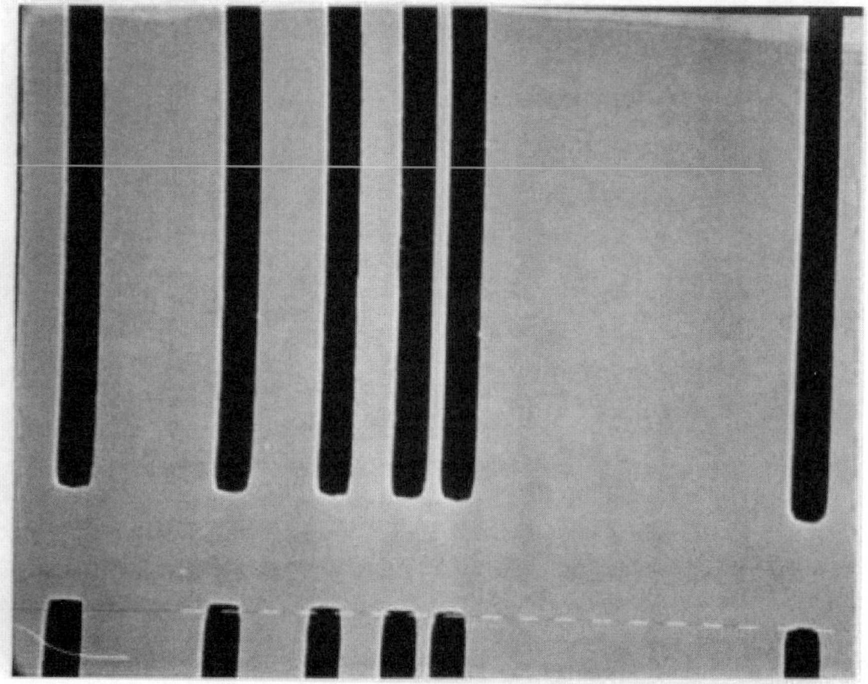

Fig. 3. Four KrF laser pulses of 350 mJ/cm^2 removed Al_2O_3 particles of size 0.3 to 1.5 μm in the irradiated area of the silicon membrane stencil mask shown in Fig. 2.

The presence of a threshold laser fluence indicated that a certain surface expansion was necessary to overcome the particle adhesion force. It was further observed that the particle removal efficiency was similar for bulk silicon wafers and for silicon membranes, which supported the model that the absorption process in the top silicon surface layer played the dominant role. Upon KrF-laser irradiation the silicon membranes vibrated with frequencies around 10 kHz. It appeared, however, that these low frequency vibrations did not result in particle ejection, because multiple clean areas with sharp mesh-type boundaries were obtained on the silicon membrane surface, when a mesh aperture was imaged with the KrF-laser onto the sample, while no cleaning action was visible in the shaded areas of the mesh image. A striking observation was that the region surrounding the cleaned spots did not appear more contaminated after the cleaning process. This was the first indication that the particles were ejected with high speed and enabled to travel 'macroscopic' distances from the substrate surface before re-depositing.

It was concluded that 'laser cleaning' might be a versatile method to clean sensitive silicon samples in a non-contact and dry fashion, which would be possible to install in a clean room or even within a lithography tool. Though these first investigations were focused on 'dry' conditions, it was realized that an appropriate 'booster' layer would further increase the acceleration and momentum transfer. The basic findings and considerations were compiled in the German patent application in 1987 (Zapka, 1987). The initial experiments were conducted with the support and in the laboratories of the Max-Planck-Institute for Biophysical Chemistry and of Lambda Physik, both in Göttingen, Germany.

4. 'Steam Laser Cleaning'

Andrew Tam had shortly before started to investigate the etching of air-bearing surfaces of magnetic sliders, and set up an excimer laser lab for this purpose at IBM's Almaden Research Center. Andrew welcomed the author to jointly study the particle ejection mechanism and to develop the technique further. For the author this meant to come 'home' again, and to enjoy Andrew's creativity and the stimulating atmosphere in his department. After repeating the initial experiments with KrF and XeCl excimer lasers (Zapka, 1990, 1992) the work was focused onto the question of the booster layer, which would enhance the cleaning efficiency further.

Water turned out to be a very versatile 'booster' layer. When silicon substrates were cleaned by UV-excimer laser radiation in the presence of a thin water film on the substrate surface, we observed high particle removal efficiencies at reduced laser fluences as compared to 'dry' laser cleaning. Thin films of water or water-alcohol mixtures were condensed onto the substrate surface with a simple home-made apparatus (see the experimental set-up in Fig. 4), which applied short bursts of warm saturated vapour onto the sample area to be irradiated just prior to the laser pulse. Before the condensed film evaporated again, typically after less than a second, the excimer laser was fired. At this instance the liquid film had a thickness of the order one micrometer. We observed the effect of laser irradiation on silicon wafers with the naked eye first. Directly after the laser pulse the irradiated area was dry, the liquid film and essentially all of the added particles were removed, while the particle density was unchanged in the non-irradiated area. Furthermore there was the appearance of a thin jet of steam originating from the center of the irradiated area and proceeding vertically from the wafer surface to a distance of some 10 to 15 mm. The jet then slowed down, obviously due to viscous drag in air, and formed a plume, which then drifted away and dissolved in the ambient air.

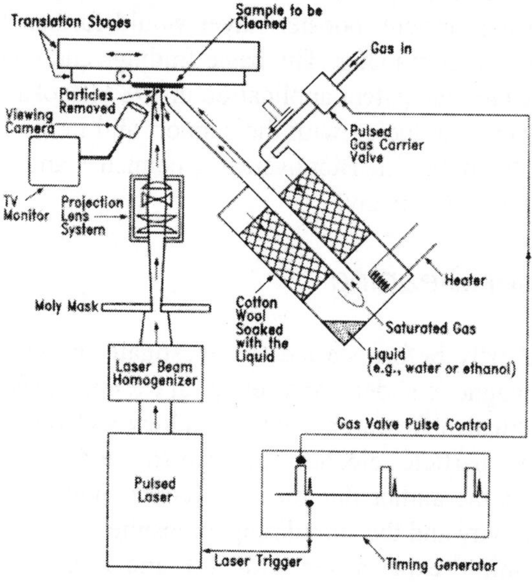

Fig. 4. Early experimental set-up for 'Steam Laser Cleaning'.

In our first publication about Steam Laser Cleaning (Zapka, 1991b) we described the experimental findings as 'liquid film enhanced laser cleaning', a term too long to memorize. It was Andrew who coined the more catchy term 'steam laser cleaning' in a later publication (Tam, 1992a).

Andrew took photos of the jet formation, while the laser and puffer were pulsing at about 1 Hz. Andrew's camera had a 200 μs flash, and he attempted to synchronize to the laser pulses by his finger click. He indeed succeeded and we obtained several photos, in which the full jet trajectories were visible, from the substrate surface to the position where they had slowed down to form a plume. The photo in Fig. 5 shows a jet originating from a 2mm x 5mm irradiated area. The steam jet was visible over a length of about 10 mm. It was not clear to us whether the photo showed the actual motion of the jet, or just showed the stationary steam formation after the passage of the steam jet. Given that the first assumption was correct, this 'quick and dirty' pocket camera experiment would give us a rough estimate of the jet speed of some 50 m/s. This estimate proved to be a fair approximation in later experiments (see part 6).

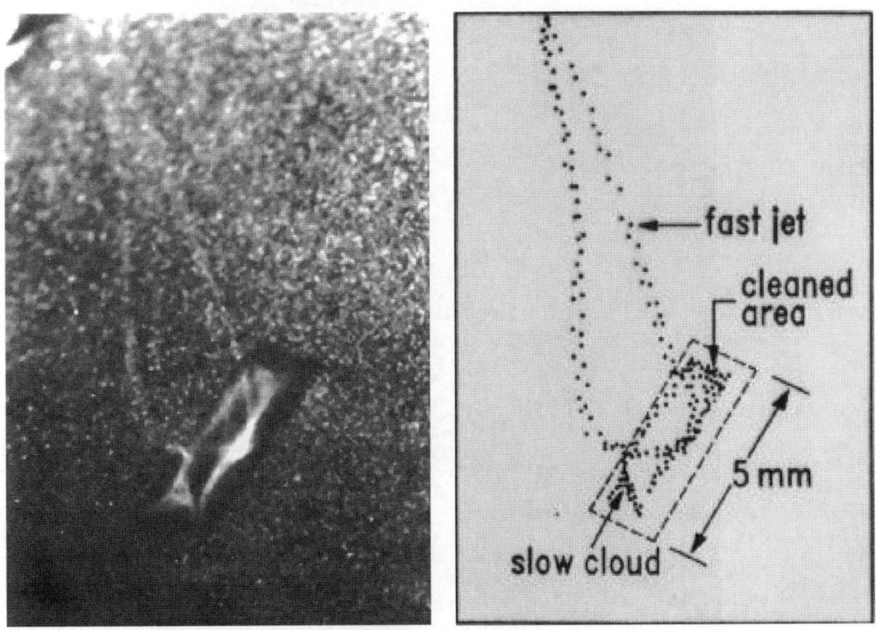

Fig. 5. The photo (a) and the schematic (b) show the formation of a fast jet in 'Steam Laser Cleaning'.

Experiments were carried out with silicon substrates, which were contaminated with 0.2 μm gold particles. It was possible to remove these particles by KrF laser pulses of 200 mJ/cm^2 fluence in the presence of a thin water film on the surface without causing melting of the gold particles (Zapka, 1991b). This clearly demonstrated the capability of the cleaning method. As a further application of this new cleaning technique we attempted to remove residual 'slurry' particles from the 'air bearing surface' of magnetic heads in magnetic data storage devices. Such slurry particles, typically Al_2O_3 particles of size 0.1 μm, would cause damage to the head or disk at the low flight heights, which are used in modern systems. Indeed such particles could be removed from the ceramic slider surface by means of Steam Laser Cleaning. Fig. 6 shows a photo of the cleaning process taken by Andrew in 1993. A similar application, namely the removal of ZrO_2 particles from air bearing surfaces was demonstrated by Y. F. Lu, 1995.

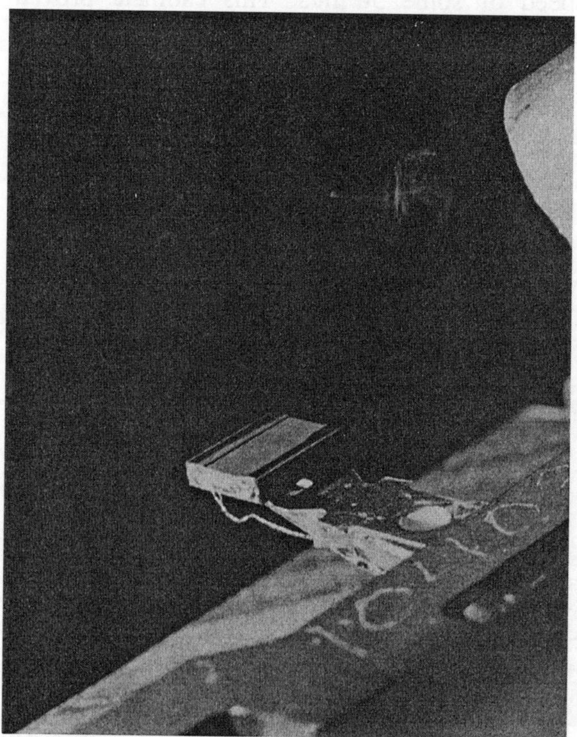

Fig. 6. Steam Laser Cleaning of the air-bearing surface of a magnetic head slider.

A liquid film was condensed onto the surface of the slider by a burst of humid gas from the nozzle on the upper right of the photo. After irradiation of the magnetic head the ceramic air-bearing surface was cleaned. A faint and narrow jet, barely visible in this photo, indicated the trace of a steam jet moving away in normal direction from the irradiated surface, before it was decelerated in ambient air and formed a stationary steam cloud.

Though the mechanism of the jet formation was not understood at that time, it was obvious that the absorption of the laser radiation in the silicon surface layer led to the forceful explosion of the liquid film. Momentum transfer from the exploding liquid to the embedded particles was assumed to be the reason for the high cleaning efficiency.

5. Steam laser cleaning with short-pulse IR-lasers

Susan Allen and coworkers were conducting steam cleaning experiments with pulsed IR-lasers in the early 1990s (Imen, 1991). A short pulse CO_2 laser was used to irradiate the substrate to be cleaned in the presence of a condensed water-film. The 10.6 μm CO_2 radiation is absorbed directly in the top 10 μm of the water film, and the formation of a steam jet is readily visible to the naked eye. Energy transfer from the substrate to the water-film is not necessary, so that this cleaning technique in principle works with all type of substrate materials.

Together with Andrew the author performed cleaning experiments with silicon substrates using a 10.6 μm CO_2 laser with 50 ns pulse length. It became apparent that the cleaning efficiencies were considerably lower than those achieved with Steam laser cleaning using excimer lasers. Similar observations were made by us when employing 50 ns pulse length Er: YAG lasers at both wavelengths, 2.70 μm and 2.94 μm, where the absorption depths in water are 20 μm and 1 μm. While it was possible to produce a steam jet when irradiating substrates coated with a condensed liquid film of water, the obtained cleaning efficiencies were low.

The superior cleaning efficiency of excimer laser based Steam laser cleaning was assigned to the fact that the explosion occurred at the interface between the liquid film and the silicon surface, i.e. beneath the particles, so that the exploding liquid interface sheet could propel these. In the case of CO_2 irradiation the explosion occurs throughout the water-film, and in the case of 2.94 μm Er: YAG irradiation the explosion occurred at the top surface of the water film.

Laser Cleaning experiments with short pulsed IR-lasers were carried out in the late 1980s by another group as well. E. Y. Assendel'ft and coworkers used a CO_2-laser of 100 ns pulse length and 300 mJ pulse energy to irradiate a silicon wafer, which was immersed in water at different depths, ranging from 0 to 5 mm. The laser radiation was focused to yield laser fluences of up to 10 J/cm^2 and created strong acoustic pressure waves of up to 10^8 Pa within the water. These pressure waves traveled along the silicon surface and could eject sub-micrometer particles from the surface of the silicon wafer in an area extending several millimeters around the irradiated area. It was observed that the particle removal efficiency increased, when the water level above the silicon wafer was reduced. The conclusion was that the pressure wave could then interact directly with the adsorbed particles and facilitate their detachment. The basic idea was to use this technique to remove particles by a 'remote' irradiation process, i.e. it was not required to directly irradiate the area to be cleaned. To avoid damage to sensitive parts of the sample the high power irradiation would have to be aimed at un-used areas of the sample surface adjacent to the area to be cleaned (Assendel'ft, 1988). Similar 'remote' laser cleaning was investigated by A. A. Kolomenskii and coworkers, who used high power Nd: YAG irradiation of up to 30 J/cm^2 to produce strong surface acoustic waves ('SAW') on dry silicon wafer surfaces in vacuum. These SAW pulses yielded high accelerations of more than 10^9 m/s^2 onto particles on the surface, and could remove 50 nm particles within distances of several millimeters from the irradiated surface area. Again, these high irradiation powers caused substrate damage, so the actual sample area could not be directly irradiated (Kolomenskii, 1998).

6. Steam laser cleaning: direct imaging of the jet motion

Andrew's snapshot with the 200 μs flash had shown the trajectories and enabled an estimation of the speed of the steam jets that were created during Steam Laser Cleaning. A time-resolved study of this motion would shed more light on the lift-off velocities and the acceleration of the liquid film during explosion. An appropriate experimental arrangement was available at IBM's Watson Research Center in Yorktown Heights, where Charles Otis and Bodil Braren performed time resolved analysis of the photo-ablation process of polyimide under excimer laser irradiation (Kelly, 1992). The author was welcomed to use their laboratory and in 1992 an excite-and-probe experiment was carried out, where one excimer laser produced the steam jet, and another excimer laser was used as a flashlight to record CCD-images of the event. Since these results have only recently been presented orally (Zapka, 2001), and the earlier report was internal to IBM, this experiment is described here in some detail. The experimental set-up is schematically depicted in Fig. 7. A trigger unit allowed to choose arbitrary delay times between the excite laser and the probe laser, and it synchronized the puffer with the excite laser so that essentially equal water-alcohol films were deposited before each excite laser pulse. A KrF excimer laser was used as the excite laser, and a XeCl excimer laser as the probe laser, with pulse lengths of 20 ns and 25 ns, respectively. The probe laser was directed at a small angle to the surface of the silicon wafer sample to illuminate the steam jet, which was imaged onto a CCD-camera. Two different projection lenses were used, one for capturing the formation of the steam jet at short delay times with high resolution and the other for imaging the jet at larger delay times as it propagated away from the surface. Unfortunately the dye-laser, which was pumped by the probe excimer laser in Charles Otis' experiments, was not functional during the time of our experiment. The time resolution of the measurements was therefore defined by the 25 ns pulse length of the probe XeCl excimer laser.

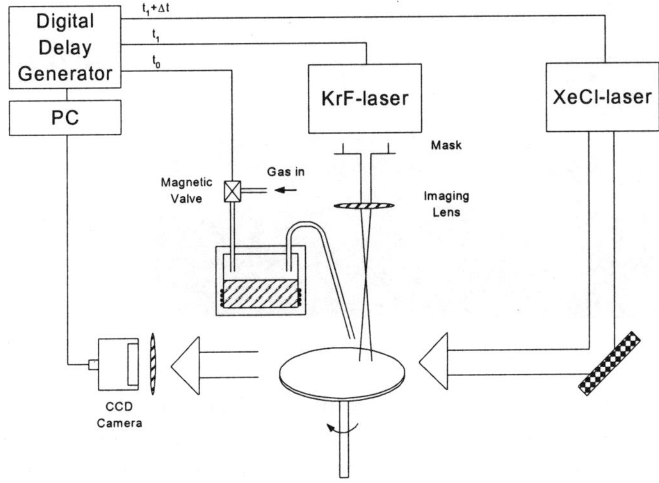

Fig. 7. The excite- and probe experiment to monitor the liquid film explosion in Steam Laser Cleaning.

Figures 8 and 9 show CCD-images taken at delay times of 1.5 μs and 8 μs between the excite- and the probe excimer laser pulses. The excite excimer laser irradiated the sample from vertically above with a spot size of 450 μm. The CCD-image shows that a very well defined thin *DISK* of liquid material was formed. The diameter of the disk corresponded to the spot size of the excite laser pulse. The excite KrF excimer laser fluence was 180 mJ/cm^2 and the water-alcohol film had a thickness of about 0.5 μm at the instance of the excite laser irradiation. Due to the small angle of the probe laser beam with the surface of the silicon wafer, the CCD-image in Fig. 8 shows both, the real disk and its mirrored image, respectively.

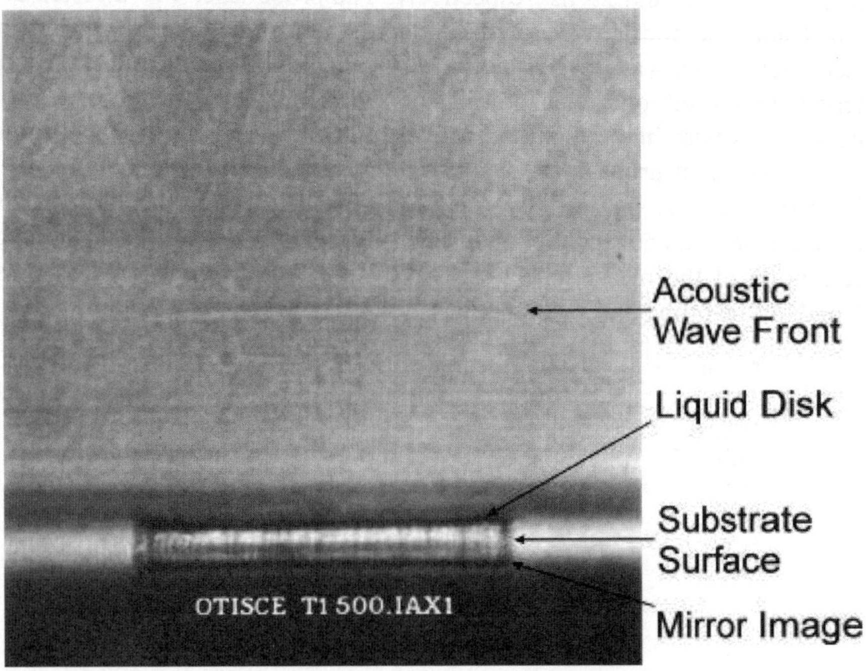

Fig. 8. A CCD-image taken at a delay time of 1.5 μs between the excite- and the probe excimer laser pulses. The excite excimer laser irradiated the silicon sample from the top. Due to the small angle of the probe laser beam with the surface of the silicon wafer, the CCD-image shows both, the real liquid disk and its mirrored image, respectively.

Fig. 9. A CCD-image taken at a delay time of 8.0 μs.

The formation of the 'liquid disk' was interpreted in the following way: the excite excimer laser radiation was absorbed in the top surface of the bare silicon wafer from where it heated up the sheet interface layer of the water-alcohol film. This sheet layer exploded and propelled the remaining 'bulk' part of the liquid film as an intact liquid disk away from the surface. Such integral disks were recorded even at delay times of 30 μs after the excite laser pulse. During their travel through the ambient air each liquid disk gradually disintegrated until it was spread into a cloud of micro-droplets at delay times of some 100 μs. The observation of a 'steam jet' as seen with naked eye, had therefore been misleading. Strictly speaking we faced the propagation of a liquid disk, which only in the course of its propagation disintegrated into 'steam'. The formation of the disk / steam jet was so repetitive that a 'slow motion movie' could be recorded by sampling CCD-images with sequentially increasing delay times.

From the CCD-images the distance traveled by the liquid disk could be plotted as a function of delay time for an excite-laser fluence of 180 mJ/cm^2, see Fig. 10. It was attempted to fit the experimental data with a theoretical equation of motion, which was derived under the following assumptions: A 'liquid' disk of thickness H, radius R, and density ρ_L is ejected with a lift-off velocity v_0 at time t_0 from the silicon wafer surface into a viscous medium of density ρ_A, the density of ambient air. The common assumption of the drag force being proportional to the square of the velocity was made, with

$$F_{drag} = \frac{1}{2} \cdot \rho_A \left(\frac{ds}{dt} \right)^2 \pi \cdot R^2 \cdot C_D . \tag{1}$$

The drag coefficient C_D for a thin round disk is close to 1 and ignored in the following. Assuming that the drag force is the only decelerating force, i.e. $F_{drag} = - m\, (d^2 s/dt^2)$, we obtained the equation of motion

$$s(t) = 2H \left(\frac{\rho_L}{\rho_A} \right) \ln \left[\frac{v_o \cdot \rho_A}{2H \cdot \rho_L} (t - t_0) + 1 \right] \tag{2}$$

A fit of the type $s(t) = a \ln [b\,(t\text{-}t_0)\text{+}1]$ to the experimental data is shown as the solid line in Fig. 10, with the time of ejection of the liquid disk as t_0 = 400 ns (see Fig. 11). The time derivative of this fit yields the velocity of the liquid disk. This is shown as the dashed line and can be extrapolated to the lift-off velocity v_0 = 37 m/s. The fit parameters a and b then allow calculation of the thickness H of the liquid disk as 870 nm, which is in fair agreement with estimates based on interference observations. The good agreement between the experimental and the modelled data supports the observation that the laser-induced explosion at the interface in fact created a 'liquid disk' and propelled it away from the surface.

Figure 8 yields further information. Above the liquid disk a Schlieren image of an acoustic wave can be seen. When plotting the distance of the acoustic wave front from the sample surface as a function of delay time between the excite- and probe lasers, as seen in fig. 11, it appears that this acoustic wave traveled with a constant velocity of 370 m/s. This increased acoustic velocity is most likely due to the local humidity and temperature right above the silicon substrate. Extrapolation back to distance zero showed that the acoustic wave originated from the surface at a delay of 400 ns after the excite excimer laser pulse.

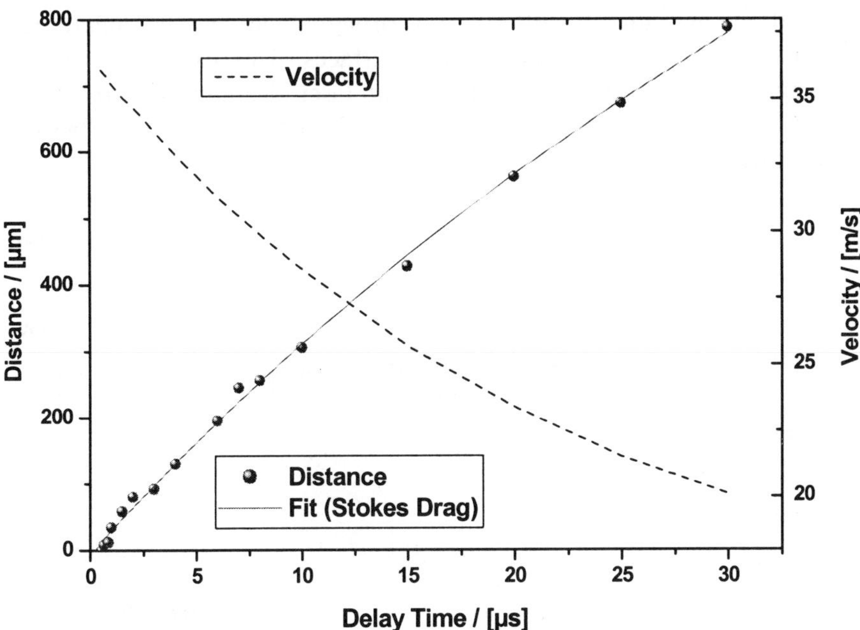

Fig. 10. The distance traveled by the liquid disk as a function of the delay time between the excite and probe laser beams. A fit assuming Stokes drag in ambient air yields the thickness of the liquid disk.

The propagation of the liquid disk as a function of time from Fig. 10 is plotted again in Fig. 11. The extrapolation of both curves indicated that the time of lift-off of the liquid film disk coincided with the formation of the acoustic wave. It was not yet clear at the time of the experiment what precisely was the reason for the time interval between the excite laser pulse and the ejection of the liquid film. Anyway this experiment allowed estimation of the acceleration of the liquid disk. Assuming that the explosion that resulted in the lift-off of the liquid disk took place within 100 ns (this is equal to the length of the 'second dip' in the cw-probe transmission signal, see further part 8), and that the liquid disk had then reached its lift-off velocity of 37 m/s this translated into an acceleration of $3.7 \cdot 10^8$ m/s^2. In a later study O. Yavas estimated the length of the pressure pulse from the explosion to be 40 ns (see part 8), which would imply even higher accelerations of 10^9 m/s^2. It was concluded that the particles embedded within the liquid film would experience the same high acceleration, and that therefore even strongly adhering sub-micrometer particles would be ejected from solid surfaces. This explained the high cleaning efficiencies observed in the Steam laser cleaning experiments and proved that the exploding liquid film indeed acted as the desired 'booster' layer for particle removal.

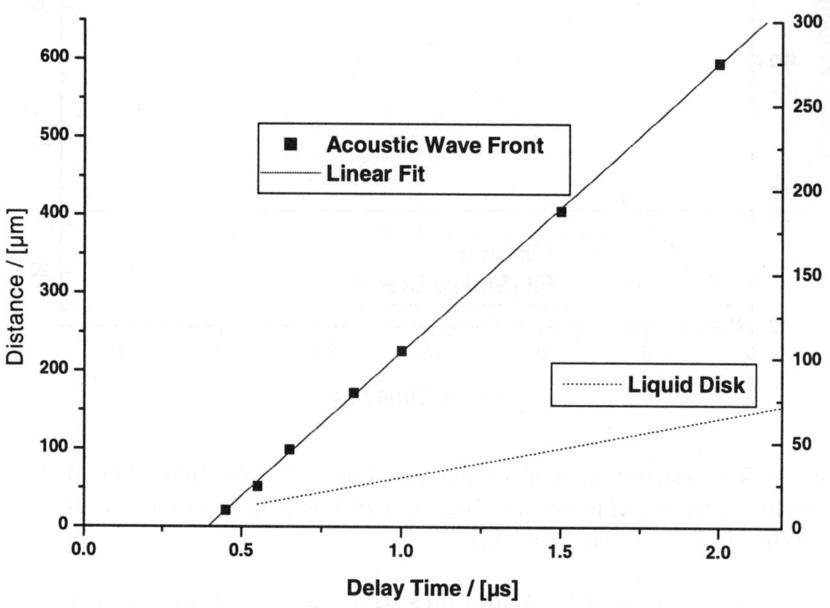

Fig. 11. At a delay time of 400 ns after the excite laser pulse an acoustic pulse originates from the surface. The Schlieren images as in figure 8 allow monitoring the propagation of the acoustic wave front as function of time. Propagation distances are indicated for the acoustic wave front on the left hand ordinate, and for the liquid disk on the right hand ordinate.

7. Opto-acoustic detection of the liquid film explosion

Andrew Tam's deep understanding of opto-acoustic effects led him to use this very technique as a further means to detect the liquid film explosion. In one of our early experiments on Steam laser cleaning he remarked that 'the fast jet also causes a shock pulse in the air, audible as a snapping sound, and also detectable by a fast piezoelectric transducer'. Together with his coworkers he employed probe-beam deflection sensing (PDS) to study the explosion of liquids on solid samples under pulsed laser irradiation. Here he kept the specific sample in a cuvette filled with the given liquid. A cw-probe laser beam passing horizontally above the sample could detect the acoustic waves originating from the sample. They were monitored by the transient probe-beam deflections that occurred when the acoustic wave propagates across the probe laser beam of small diameter. Andrew found that large amplitude acoustic waves were detected when the sample was irradiated with KrF excimer laser pulses beyond the certain threshold fluence, while only small amplitude pulses appeared below that threshold. It was concluded that the low fluence laser pulses caused thermoelastic expansion, which generated shear waves in the solid samples, while high amplitude longitudinal waves were created above threshold. Only the longitudinal waves could couple well into the liquid, where they were detected with the cw-probe beam. The occurrence of the large amplitude acoustic signal was therefore assumed to be due to the explosion of the liquid sheet layer at the interface to the solid surface. A threshold fluence of 20 mJ/cm^2 was, for example, found for the explosion of water on amorphous silicon. Andrew Tam and coworkers used this sensitive method to study the explosion of different liquids on various solid substrate materials (Tam, 1992b; Nhan Do, 1993).

8. Steam laser cleaning: study of the mechanism of the liquid film explosion

In our phenomenological studies of the Steam Laser Cleaning we had so far concentrated on effects observable *after* the explosion of the liquid film, namely the cleaning efficiency and the motion of the steam jet. Andrew's curiosity was raised to search for the basic mechanism behind the liquid film explosion itself.

In 1991 Andrew and his coworkers developed a non-contact method to measure the surface temperature of thin amorphous silicon layers with nanosecond temporal resolution. They found that the optical absorption of 752 nm radiation within a-Si varied considerably with temperature and they developed a calibration curve (Nhan Do, 1992; Park, 1992). The experiment to study the dynamics of the liquid film explosion was carried out with quartz substrates, which were coated with 200 nm thick a-Si layers. When irradiated with a pulsed excimer laser the temperature within the a-Si layer could be monitored in real-time by a cw diode laser beam of 752 nm wavelength.

Of most importance for the understanding of the liquid film explosion was the observation of a distinct 'second dip' in the transient transmission signal of the cw diode laser beam. This appeared only in the presence of a liquid film on the substrate surface and beyond a certain threshold of the excimer laser fluence, but not with dry substrates. The threshold fluences varied with the different liquids (water, solvents, mixtures thereof). With the calibration of the transient transmission signal the a-Si surface temperatures at the threshold of the 'second dip' were estimated. These threshold temperatures (e.g. 290 °C and 210 °C for water and isopropanol, respectively) were close to those at which spontaneous nucleation takes place in the corresponding pure superheated liquids. The 'second dip' in the transmission signal could therefore be interpreted as the scattering of the cw diode laser beam during spontaneous nucleation within the liquid film (Leung, 1992). The transient transmission curves showed a time lag between the excimer laser pulse and the 'second dip', which ranged from 150 ns for water to 300 ns for isopropanol. Indeed we had observed in the study of the steam jet motion (see part 6) that an acoustic wave was emitted from the substrate surface 400 ns after the excite excimer laser pulse was applied. Together with the observation of the 'second dip' at a similar delay time, this supported the assumption that the spontaneous nucleation resulted in the immediate explosion of the liquid film, and the lift-off of the liquid disk.

After the cause of the 'second dip' in the transient transmission curve of the cw-probe laser beam was identified as scattering during spontaneous nucleation a more rigorous investigation of the nucleation phenomenon was undertaken. For this purpose both the transient reflection and the scattering signals from an incident cw-probe laser beam were recorded following the irradiation of the sample with a KrF excimer laser beam of 16 ns pulse length. The sample, a 0.2 μm thick chromium film on a sapphire substrate, was kept immersed in liquids like water, alcohol or mixtures thereof. The observed reflectance transients could be fitted well with calculated data based on the 'effective medium theory' by Maxwell Garnett and assuming that the small bubbles in the initial phase of the nucleation process would cause Rayleigh scattering, while Mie scattering would dominate when the bubbles grew to diameters larger than the cw-probe wavelength. The results supported the assumption that the reflectance transient indeed represented the bubble growth dynamics. This early work of Andrew Tam produced estimates of the bubble growth velocities of 3.6 m/s and 1.9 m/s for water and isopropanol, respectively (Yavas, 1993). Combined measurements of the scattering signal and the temperature at the liquid-solid interface indicated that nucleation of bubbles starts at relatively low temperatures. It was concluded that a heterogeneous nucleation process was taking place, with microscopic features on the solid surface acting as nucleation sites (Park, 1996). This finding was of importance for the practical usage of Steam laser cleaning since in fact any particle on the surface acts as a nucleation site and facilitates the explosion of the capillary liquid between the particle and the solid surface.

While Andrew had pioneered the study of the liquid film explosion by the use of opto-acoustics, piezoelectric transducers and specular reflection, the precision of the measurement techniques was further improved by his post-doc students, who continued the research on Steam laser cleaning at their home Universities. In the course of time a number of Universities and Institutes worldwide contributed to the further understanding and development of Laser Cleaning.

Oguz Yavas and Paul Leiderer introduced a surface plasmon probe as a tool to measure bubble nucleation and growth dynamics on the nanosecond scale. Being sensitive to bubbles as small as 20 nm the surface plasmon probe allowed the formation of the bubbles to be monitored, as well as their growth velocities. Furthermore the pressure pulse, generated by the fast bubble growth, could be estimated to be of the amplitude of 1 – 5 MPa, at a pulse length of some 40 ns (Yavas, 1997). M. Mosbacher and coworkers carried out extensive cleaning experiments on silicon wafers contaminated with particles of different material, size and shape, and showed that the threshold laser fluence for efficient Steam Laser Cleaning was constant for particle sizes from tens to several hundreds of nanometers (Mosbacher, 2001). Such 'universal threshold' indeed facilitates the usage of Steam laser cleaning in industrial applications. The same group at the University of Konstanz found that picosecond-excimer laser pulses tend to damage the substrate surface (Leiderer, 2000), so that the standard excimer lasers with pulse lengths of some 20 ns appear to be the ideal light sources for Steam laser cleaning of silicon or other UV-absorbing substrates.

Theoretical models of the steam laser cleaning process were developed by several groups, who calculated the removal forces based on the growth of bubbles in the superheated liquid film and compared the results with experimental data for removal of particles with sizes down to 0.1 μm (Lu, 1998; Wu, 2000). These models predict the removal forces in Steam Laser Cleaning process to be considerably higher than in dry laser cleaning. G. Vereecke et. al. modeled the dry laser cleaning and performed 'dry' laser cleaning experiments both in ambient air, and in air saturated with moisture. For the latter they found higher cleaning efficiencies, which were attributed to the explosive evaporation of the capillary water film, which had condensed between the particle and the surface to be cleaned (Vereecke, 1999). Aspects relating to both the fundamental process and to applications of Laser Cleaning were addressed jointly by a group of eight European Universities and Institutes (Oltra, 2000). Work on Steam laser cleaning appears to be carrying on, and feasibility has been shown in a number of applications.

In summary the various findings add up to the development of the Steam laser cleaning process as follows:

- Condensation of a liquid film of micrometer thickness onto the sample surface to be cleaned just prior to laser irradiation.
- Short pulse laser irradiation (20 ns excimer laser pulses) are OK for silicon or other highly absorbing sample materials.
- The laser radiation is strongly absorbed within a surface layer of the sample.
- Heat is transferred to the liquid film.
- Superheating of the liquid sheet layer takes place at the interface between the liquid and solid surface; there is limited heat transfer to the bulk of the liquid film.
- A heterogeneous nucleation process with bubble nucleation occurs at surface sites of the solid sample; particles can act as such nucleation sites.
- Creation of a dense population of bubbles.
- Fast growth of the bubbles in the liquid sheet interface layer.
- The bubble growth generates an explosive blast wave.
- The pressure pulse from the explosion of the interface layer has a duration of some 40 ns (Yavas, 1997).

- At a time delay of 200-400 ns after the laser pulse: the bulk of the liquid film is ejected as a liquid disk and accelerated to a lift-off velocity of the order of 40 m/s during lift-off the liquid disk experiences an acceleration of some 10^9 m/s^2.
- The particles embedded within the liquid film will experience similar accelerations, overcome their high adhesion forces, and are removed from the surface.
- The liquid film and the ejected particles are propelled to macroscopic distances of more than 10 mm away from the surface before the atmosphere decelerates them.
- This provides the opportunity to collect the particles and prevent re-deposition.

Steam laser cleaning works efficiently on silicon, metal or certain ceramics, which absorb the pulsed laser radiation well. However, on non- or poorly absorbing substrates like SiO_2, Steam laser cleaning with UV or visible lasers does not perform well since the liquid layer at the interface to the solid substrate cannot be heated efficiently. For these substrates one has to resort to steam laser cleaning with short pulsed IR-laser sources, or to laser cleaning techniques with strong acoustic waves, which are produced remotely, either at other parts of the substrate or within the atmosphere.

9. Stencil masks again

Although the focus of Andrew Tam work had shifted to laser texturing of thin-film disks for magnetic recording he continued to contribute to the research on the basics of Steam laser cleaning until the mid 1990ties. At that time several groups throughout the world had picked up the topic and studied the basics and potential applications of this technique as described in part 8.

The author had also turned to another field, to inkjet printing, in 1993. IBM had decided to stop the development efforts on EBP lithography, and the cleaning of stencil masks consequently lost priority. Today, however, steam laser cleaning of silicon stencil masks has regained interest since such masks are used for Ion-Beam Projection Lithography IPL (Kaesmeier, 2000). This lithography technique is presently developed for printing of 100 nm linewidth and below by projecting a stencil mask pattern with an ion-beam. The projection demagnifies the mask pattern at a ratio of 4:1, so that the minimum linewidth in the mask is 400 nm, i.e. comparable to the earlier EBP stencil masks. Steam laser cleaning has been applied to these masks, and particles could be effectively removed (Zapka, 2000).

Acknowledgement

We gratefully acknowledge the cooperation of Chuck Otis and Bodil Braren in the experiments of the direct observation of the jet motion in Steam laser cleaning. This work was carried out at IBM's Watson Research Laboratories, Yorktown Heights, NY, USA, where Chuck and Bodil worked in 1992. Part 6 is based on an IBM-internal draft report from 1992. Bodil is presently at IBM, White Plains, and Chuck at Hewlett Packard's inkjet printer division. Funding of part of the laser cleaning project from International Sematech is gratefully acknowledged. My hearty thanks to Prof. Boris Luk'yanchuk for his constant urge to write this historical overview, and for the many fruitful discussions and his support in finalizing it.

Thanks to Andrew C. Tam

This article is intended as a tribute to Andrew C. Tam, who has contributed much to the understanding and development of Laser Cleaning. It is also a happy reminiscence of the author's many years together with Andrew, as his postdoc, colleague and friend. Andrew C. Tam died on February 23, 2001.

References

Assendel'ft E. Y., Beklemyshev V. I., Makhonin I. I., Petrov Y. N., Prokhorov A. M., Pustovoi V. I., *Russian Ultrasonics* **18**, pp. 298-302 (1988)

Greschner J., Bohlen H., Engelke H., Nehmiz P., *IBM J. Res. Develop.* **21**, pp. 514-521 (1977)

Imen K., Lee J., Allen S. D., *Appl. Phys. Lett.* **58**, pp. 203-205 (1991)

Kaesmeier R., Loeschner H., Proc. SPIE, Vol. **3997**, pp. 19-32 (2000)

Kelly R., Miotello A., Braren B., Otis C. E., *Appl. Phys. Lett.* **60**, pp. 2980-2982 (1992)

Kolomenskii A. A., Schuessler H. A., Mikhalevich V. G., Maznev A. A., *J. Appl. Phys.* **84**, (1998), pp. 2404 – 2410

Leiderer P., Boneberg J., Dobler V., Mosbacher M., Münzer H.-J., Chaqoui N., Siegel J., Solis J., Afonso C. N., Fourrier T., Schrems G., Bäuerle D., *Laser-induced particles removal from silicon wafers*, Proc. SPIE, vol. 4065, pp. 249-259 (2000)

Leung P. T., Nhan Do, Klees L., Leung W. P., Tong F., Lam L., Zapka W., Tam A. C., *J. Appl. Phys.* **72**, pp. 2256-2263 (1992)

Lu Y. F., Song W. D., Hong M. H., Chong T. C., Low T. S., *Proceedings of the 1995 MRS Fall Symposium,* MRS Materials Research Society, Pittsburgh, PA, USA, pp. 317-322 (1995)

Lu Y. F., Song W. D., Low T. S., *Materials Chemistry and Physics*, **54**, pp. 181-185 (1998)

Mittal K. L. (Ed), *Particles on Surfaces, Detection, Adhesion and Removal,* Vol. 1, Plenum Press, New York (1988)

Mosbacher M., Dobler V., Boneberg J., Leiderer P., *Universal threshold for the steam laser cleaning of submicron spherical particles from silicon*, Appl. Phys. A 72, pp. 41 - 44 (2001)

Nehmiz P., Zapka W., Behringer U., Kallmeyer M., Bohlen H., *Journal Vacuum Science and Technology B* **3**, pp. 136-139 (1985)

Nhan Do, Klees L., Leung P. T., Tong F., Leung W. P., Tam A. C., *Appl. Phys. Lett.* **60**, pp. 2186-2188 (1992)

Nhan Do, Klees L., Tam A. C., Leung P. T., Leung W. P., *J. Appl. Phys.* **74**, pp. 1534-1538 (1993)

Oltra R., Arenholz E., Leiderer P., Kautek W., Fotakis C., Autric M., Afonso C., Wazen P., *Proc. SPIE*, vol. 3885, pp. 499-508 (2000)

Park H. K., Xu X., Grigoropoulos C. P., Nhan Do, Klees L., Leung P. T., Tam A. C., *Appl. Phys. Lett.* **61**, pp. 749-751 (1992)

Park H. K., Grigoropoulos C. P., Poon C. C., Tam A. C., *Appl. Phys. Lett.* **68**, pp. 596-598 (1996)

Tam A. C., Nhan Do, Klees L., Leung P. T., Leung W. P., *Optics Letters*, Vol. **17**, pp. 1809-1811 (1992a)

Tam A. C., Leung W. P., Zapka W., Ziemlich W., *J. Appl. Phys.* **71**, pp. 3515-3523 (1992b)

Vereecke G., Röhr E., Heyns M. M., *J. Appl. Phys.*, **85**, pp. 3837 – 3843 (1999)

Wu X., Sacher E., Meunier M. M., *J. Appl. Phys.*, **87**, pp. 3618 – 3627 (2000)

Yavas O., Leiderer P., Park H. K., Grigoropoulos C. P., Poon C. C., Leung W. P., Nhan Do, Tam A. C., *Phys. Rev. Lett.*, **70**, pp. 1830-1833 (1993)

Yavas O., Schilling A., Bischof J., Boneberg J., Leiderer P., *Appl. Phys.* **A 64**, pp. 331-339 (1997)

Zapka W., Asch K., Keyser J., Meissner K., *German patent DE* 3721940 C2, Priority July 2, 1987

Zapka W., Tam A. C., *CLEO/IQEC '90 Conference Paper CWB4,* p. 226 (1990)

Zapka W., Sonchik S., Behringer U., Haug W., Meissner K., Silverman S., Ziemlich W., Bohlen H., Smith W., *Microelectronic Engineering* 13, pp. 357-360 (1991a)

Zapka W., Ziemlich W., Tam A. C., *J. Appl. Phys.* **58**, pp. 2217-2219 (1991b)

Zapka W., Tam A. C., Ayers G., Ziemlich W., *Microelectronic Engineering* **17**, pp. 473-478 (1992)

Zapka W., Gollasch C., Lilischkis R., Ehrmann A., Zapka K. F., In: *Proceedings of the 7th Intern. Symposium on 'Particles on Surfaces'*, Ed. by K. L. Mittal, Newark, NJ, USA, June 19-21 (2000)

Zapka W., Tam A. C., *2nd International Symposium on Laser Precision Microfabrication* LMP2001, Singapore, May 16-18 (2001); *Intern. Symposium on Surface Contamination and Cleaning*, Newark, NJ, USA, May 23-25 (2001)

Part 2. Dry Laser Cleaning

Chapter 2

DRY LASER CLEANING OF PARTICLES BY NANOSECOND PULSES: THEORY

N. Arnold

A model for nanosecond dry laser cleaning that treats the substrate and particle expansion on a unified basis is proposed. Formulas for the time-dependent thermal expansion of the substrate, valid for temperature-dependent parameters, are derived. Van der Waals adhesion, substrate and particle elasticity, and particle inertia are taken into account for an arbitrary temporal profile of the laser pulse. The characteristic time for the particle on the surface system is deduced. This time is related to the size of the particles as well as the adhesion and elastic constants. Cleaning proceeds in different regimes if the duration of the laser pulse is much shorter or longer than this time. Expressions for cleaning thresholds are provided and compared with experiments on the 248 nm KrF excimer-laser cleaning of Si surfaces from spherical SiO_2 particles with radii between 235 and 2585 nm in vacuum. Discrepancies between the experimental data and theoretical results seem to indicate that nanosecond dry laser cleaning cannot be explained purely on the basis of one-dimensional thermal expansion mechanism.

Keywords: Dry laser cleaning, modeling, adhesion, threshold, oscillations, SiO_2 particles, Si.

PACS: 42.62.Cf, 81.65.Cf, 68.35.Np, 81.65.-b, 85.40.-e, 81.07.Wx

1. Introduction

Particle removal by means of dry laser cleaning (DLC) (Zapka, 1991; Tam, 1991) has grown in importance during the last decade (Park, 1994; Leiderer, 1998; Heroux, 1996; Bäuerle, 2000). It is used, or consider for usage, in the fabrication of printed circuit boards (PCB), in the production of dynamic random access memory (DRAM) (SIA, 1994; Kern, 1990), in lithography (Teutsch, 1990) and epitaxial growth (Kern, 1993), for the removal of contaminations during via hole production (Lu, 1998a), for the cleaning of microoptical and micromechanical components. In damage-free DLC expansion of the substrate or particle leads to particle removal. In some cases, other mechanisms related to field enhancement and local ablation may play a role. Another technique is steam laser cleaning (SLC) (Assendel'ft, 1988; Lee, 1991, 1993), which is due to laser-induced explosive vaporization of an auxiliary liquid layer. "Ablative" cleaning is based on the removal of particles/contaminants by ablation. SLC, though more efficient (She, 1999), cannot be applied to hygroscopic materials, and is incompatible with many applications where high purity is required, as in small-scale optics and nanocluster technology. It is also a multiple-step process because of necessary liquid delivery (Halfpenny, 1999).

In this article we concentrate on dry laser cleaning. With DLC it is observed that it is more difficult to remove smaller particles. This has been explained by higher specific adhesion forces. However, accurate theoretical predictions for the dependence of the cleaning fluence on particle size have not been derived. It is not entirely clear which parameters should be optimized to improve the cleaning efficiency and decrease the cleaning threshold, especially with smaller particles. Usually cleaning forces acting on the particles are compared with measured adhesion forces. At the same time, nanosecond laser cleaning takes place over very short time scales, which -- as opposed to conventional adhesion measurements -- requires the consideration of dynamic effects.

Several models of DLC exist. Accelerations and forces due to thermal expansion of the substrate (Tam, 1992; Dobler, 1999) and particles (Lu, 1997a), elastic deformation of the particles which are compressed by the expanding substrate (Lu, 1999), cleaning via generation of surface acoustic waves (Kolomenskii, 1998), the influence of hydrogen bonds (Wu, 1999, 2000), have been considered. The behavior of the particles *after* the detachment and the redeposition has been studied (Lu, 2000a; Vereecke, 1999). These models employ many inadequate assumptions. Among these are: Thermal expansion of the substrate and particle are often treated separately and incorrectly. The temporal profile of the laser pulse is not taken into account. This assumes infinite acceleration or deceleration of the substrate in the beginning or end of the laser pulse. Deformation of the substrate and particle, their interdependence and influence of the particle on the substrate expansion are not described properly. Adhesion forces are treated separately from the elastic forces, which can lead to erroneous results. Though the importance of force and energy criteria were mentioned

(Kolomenskii, 1998), their regions of applicability are not clearly stated. The removal of absorbing particles and elasticity of the substrate (Lu, 1996, 1997b) was analyzed on the basis of force balance only (Tam, 1992; Lu, 1997a, 1998b) without taking into account particle movement. The temperatures of particles are estimated in a very crude way there. Dissipative processes are considered in ref. (Lu, 2000a) but only in the post-detachment stage. Numerical calculations do not provide formulas relating cleaning fluence to laser and material parameters.

Our goal is to develop a unified description, which will easily incorporate the influence of different cleaning mechanisms and experimental parameters, and to estimate factors that contribute to DLC, without long numerical calculations. Sound related effects that can be important in picosecond DLC (Mosbacher, 2001; Leiderer, 2000) and SLC (Mosbacher, 1999, 2000), as well as field enhancement effects (Mosbacher, 2001; Lu, 2000b; Luk'yanchuk, 2000, 2001, 2002, Zheng, 2001) are not considered.

2. Adhesion potential and equation of motion

Various forces are responsible for the adherence of particles to a substrate (Mittal, 1988, 1995; Visser, 1976). This section introduces an approximation that takes into account the Van der Waals (VdW) and elastic forces.

2.1. Model expression for VdW-elastic potential

We describe particle-substrate VdW interaction by the energy per unit area φ (work of adhesion). It can be obtained by the integration of a Lennard-Jones-like potential acting between two plane surfaces (Muller, 1983). It is related to the Hamaker constant H and Lifshitz-VdW constant ς by

$$\varphi = \frac{H}{2\pi\varepsilon^2} = \frac{3\varsigma}{8\pi^2\varepsilon^2}\,. \tag{1}$$

Here ε is the equilibrium distance for the force f or energy u (per area) between two planes:

$$f(z)=-\frac{8\varphi}{3\varepsilon}\left(\left(\frac{\varepsilon}{z}\right)^9-\left(\frac{\varepsilon}{z}\right)^3\right),\quad u(z)=\frac{\varphi}{\varepsilon}\left(\left(\frac{\varepsilon}{z}\right)^8-4\left(\frac{\varepsilon}{z}\right)^2\right). \tag{2}$$

All macroscopic results depend only on φ which can be inferred from measurements of the pullout force. We now describe a simple approximation, which we will subsequently employ, and discuss its applicability. If the centers of two spheres (or a sphere and a plane) are moved together by a distance h (see Fig. 1), the energy of the system is the sum of the adhesion energy and the elastic energy. The former can be approximated by

$$U_a=-\pi a^2\varphi\approx-2\pi rh\varphi\,. \tag{3}$$

Here r is the particle radius (reduced radius for the case of two particles) and a the contact radius. It is estimated *from geometrical considerations* as $a^2\approx 2rh$, and we assume that everywhere in the contact region the interaction energy per unit area is - φ. We estimate the elastic energy from the Hertz contact problem of elasticity theory (Landau, 1986a)

$$U_e=\frac{2\overline{Y}}{5}r^{1/2}h^{5/2},\qquad \frac{1}{\overline{Y}}=\frac{3}{4}\left(\frac{1-\sigma_p^2}{Y_p}+\frac{1-\sigma_s^2}{Y_s}\right). \tag{4}$$

Here, the effective Young's modulus $\overline{Y}$ characterizes the elastic properties of the particle and the substrate. Its value is dominated by the properties of the softer material, where most of the energy is stored. Adding both energies we get the potential and the force acting on a particle in the positive h direction, towards the substrate.

$$U=-2\pi rh\varphi+\frac{2\overline{Y}}{5}r^{1/2}h^{5/2},\quad F=-\frac{\partial U}{\partial h}=2\pi r\varphi-\overline{Y}r^{1/2}h^{3/2}. \tag{5}$$

This approximation yields equilibrium values of h_0, a_0 and U_0:

$$h_0=\left(2\pi\varphi/\overline{Y}\right)^{2/3}r^{1/3},\qquad a_0=\left(2h_0r\right)^{1/2}=\left(2^{5/2}\pi\varphi/\overline{Y}\right)^{1/3}r^{2/3}, \tag{6}$$

$$U_0=-\frac{3}{5}\left(2\pi\varphi\right)^{5/3}\overline{Y}^{-2/3}r^{4/3}. \tag{7}$$

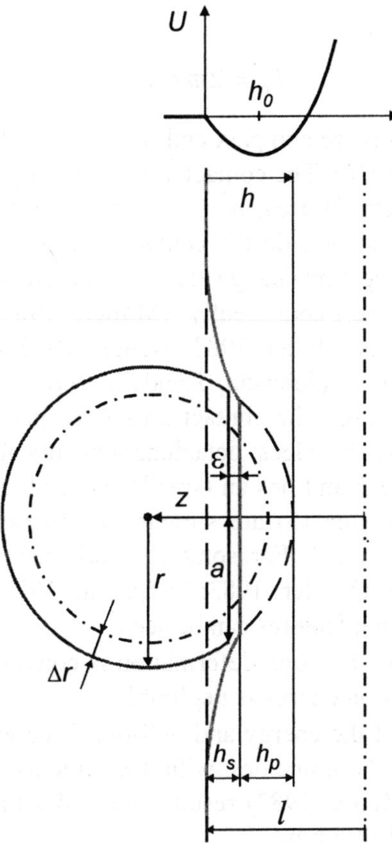

Fig. 1. Schematic of the particle-substrate deformation at a given moment of time. Solid lines - current boundaries of substrate and particles. Dashed lines - their imaginary non-deformed boundaries. Dash-dotted lines - initial position of the substrate and the surface of non-heated (but displaced) particle. l - surface displacement of the substrate, r - radius of the heated particle, Δr - particle expansion, z - position of the particle center referred to initial substrate surface without particle (note, that in general substrate is deformed even before the expansion). Overall deformation is characterized by the so-called approach h; h_s and h_p are its fractions that belong to the substrate and the particle. a - contact radius. ε - equilibrium distance between (plane) adhering surfaces. Arrows indicate positive directions for the corresponding quantities. Adhesion potential $U = U\ (h)$ is also schematically shown. For the depicted moment h is bigger than the equilibrium deformation h_0 (compression stage).

The *maximum* (pull-out, detachment) force is achieved at $h = 0$ and is given by

$$F_0 = 2\pi r\varphi . \tag{8}$$

The real situation is more complex and has been studied for a long time (Hertz, 1881, Bradley, 1932). The contact area in the Hertz problem is twice smaller than the geometrical area, $a^2 = hr$. On the other hand, attractive forces near the edges and outside the contact area increase a, modify the profile of the Hertzian gap, and change elastic deformation. Thus, one has to calculate the problem self-consistently (Muller, 1983; Maugis, 1992). Relatively compact results (Muller, 1983; Maugis, 1992; Greenwood, 1998) as well as the earlier JKR (Johnson, Kendall, Roberts) model (Johnson, 1971) predict the following: The contact area is bigger than the Hertzian; tensile stresses exist near the edges; detachment occurs abruptly (with finite contact area) at negative h and has an (small) energy barrier. Nevertheless, these models and expression (5) are similar. A comparison of calculated potentials is shown in Fig. 2. For somewhat different DMT (Derjaguin, Muller, Toropov) model (Muller, 1983; Derjaguin, 1975) the agreement is comparable. The following features can be seen:

- The elastic energy with strong deformation is conveyed correctly, as in this region the adhesion energy is negligible.
- The adhesion part of the energy and pull-out force are linear in radius. Expression (8) for F_0 coincides with the famous Bradley (Bradley, 1932) and DMT (Muller, 1983) results and is 4/3 times bigger than in the JKR (Johnson, 1971) limit.
- The functional dependences for important parameters are correct.
- The total adhesion energy and the difference between the equilibrium h_0 and h at detachment are quite similar. The potential (5) approximates the exact potential shifted by the value of h at the detachment point.
- The contact radius *a(h)* is conveyed less accurately. For this reason, we avoid using it and express everything via the approach h.

The expressions (5)-(8) should be considered as approximations valid within 20-25%. One can also use the measured pull-out force F_0 and the separation approach h_0 as main parameters of potential. In this form, results apply also to non-spherical particles. The work of adhesion φ is actually always calculated from the measured value of F_0. All measurable quantities can be expressed in terms of h_0 and φ (or F_0, or U_0, or ω_0 introduced below).

2.2. Parabolic approximation

The potential (5) is a smooth function as it is determined mainly by macroscopic elasticity. We can linearize the problem, approximating *U(h)* by the parabolic well of the same depth with the minimum at h_0.

$$U \approx U_0\left(1-\left(\frac{h-h_0}{h_0}\right)^2\right). \tag{9}$$

Oscillations within this potential have frequency and period:

$$\omega_0 = \sqrt{\frac{2|U_0|}{mh_0^2}} = \frac{3}{5^{1/2}(2\pi)^{1/3}}\left(\frac{\overline{Y}^2\varphi}{\rho_p^3 r^7}\right)^{1/6} \sim \left(\frac{v_0^4\varphi}{\rho r^7}\right)^{1/6} \sim 0.5\times 10^9 s^{-1}, \tag{10}$$

$$\tau_0 = 2\pi/\omega_0 \approx 10 \text{ ns}, \tag{11}$$

where m is mass of the particle. The estimations assume $r \sim 1$ μm and the typical values listed in Table 2. Note that the sound velocity $v_0 \sim (\overline{Y}/\rho)^{1/2}$. The approximation based on the second derivative of (JKR or DMT) potential near the equilibrium point leads to a similar frequency, as can be inferred from Fig. 2.

This frequency is *not related* to sound vibrations as the particle is treated quasi-statically. This is allowed as long as ω_0 is much smaller than the frequency of the first mode of particle sound vibrations. For the estimations we use the expression (Landau, 1986b).

$$\omega_0 << \omega_{sound} \approx \frac{\pi}{2}\frac{v_{0p}}{r} \Rightarrow \frac{6}{\pi 5^{1/2}(2\pi)^{1/3}}\left(\frac{\varphi\overline{Y}^2}{r\rho_p^3 v_{0p}^6}\right)^{1/6} << 1, \tag{12}$$

which yields $r >> \dfrac{1}{2^6}\dfrac{\varphi}{\overline{Y}} \sim 1.6\times 10^{-12}$ cm.

This is always fulfilled. As $\omega_{sound} \propto r^{-1}$ and $\omega_0 \propto r^{-7/6}$ this condition is almost independent on r.

Many conclusions follow from the linear approximation. The most convenient parameters for the analysis are the pull-out distance from equilibrium, h_0 and the internal frequency ω_0. To increase generality, we also express this frequency in terms of h_0 and measured pull-out force or total adhesion energy. In this form, subsequent results can be applied to other potentials.

$$\omega_0 = \sqrt{\frac{F_0}{mh_0}} = \sqrt{\frac{2|U_0|}{mh_0^2}}, \quad U_0 = -\frac{F_0 h_0}{2}. \tag{13}$$

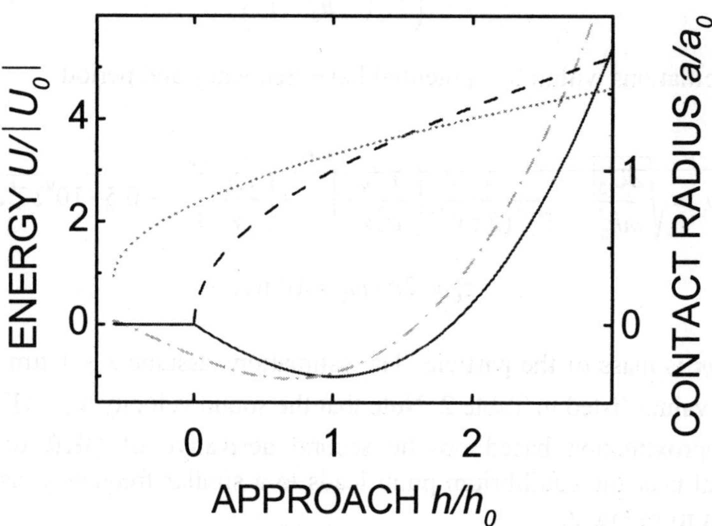

Fig. 2. Exact and approximate elastic-VdW adhesion potential and contact radius in dimensionless variables. Solid line - approximate potential (5). Dash-dotted line - exact potential from the eq. (38) in (Muller, 1983). Dashed line - approximate contact radius (see Eq. (3)). Dotted line - exact contact radius given below the eq. (38) in (Muller, 1983).

2.3. Equation for the evolution of deformation *h*

When both particle and substrate move, detachment and adhesion are determined by the approach h. Let z be the coordinate of the particle center and l the surface displacement in the *laboratory* frame, both counted from the *initial* position of the substrate surface (see Fig. 1). Then h is given by

$$h = l + r - z\,. \tag{14}$$

Here all quantities may depend on time: l and r due to the thermal expansion of the substrate and the particle, and z due to particle movement. The force balance should be written for $m\ddot{z}$ (dot stands for time derivative). Rewriting Newton's equation with the force (5) for h instead of z, with the help of (14), we obtain the equation for the evolution of h:

$$\ddot{h}+\gamma\dot{h}=\frac{1}{m}\left(2\pi r\varphi-\overline{Y}r^{1/2}h^{3/2}\right)+(\ddot{l}+\ddot{r}). \qquad (15)$$

Though the term $m\ddot{l}$ can be interpreted as the force of inertia in the frame moving with the substrate, and the term $m\ddot{r}$ cannot, both the substrate and particle expansion can be treated similarly. All other terms can be neglected in the first approximation.

2.4. Damping coefficient

With the small particles and substantial velocities involved, dissipative processes may become significant. The importance of damping in Eq. (15) is difficult to estimate.

2.4.1. Knudsen viscosity

The motion of the particle is slowed down by the ambient medium. At spatial scales smaller than the mean free path of ambient gas molecules, the damping force can be estimated as:

$$F_d\approx\frac{m_a^2 p}{3(k_BT)^2}\pi r^2 v_T^3\dot{z}\;\Rightarrow\;\gamma\sim\frac{4\sqrt{2}}{\pi\sqrt{\pi}}\frac{N\sqrt{m_a k_B T}}{\rho_p r}\sim\frac{\rho_g v_0}{\rho_p r}\overset{gas}{\sim}10^6\,\mathrm{s}^{-1}. \qquad (16)$$

Here, k_B is the Boltzmann constant, m_a is the mass of gas molecules, v_T their thermal velocity, p the pressure, T the temperature, and N the number density of the gas. The force is defined following the method given in (Lifshitz, 1981).

2.4.2. Stokes viscosity

With bigger particles and / or a (thick) liquid layer at the surface, γ can be estimated from the Stokes formula:

$$\gamma = \frac{6\pi\eta r}{m} = \frac{9}{2}\frac{\rho_l \nu}{\rho_p r^2} \overset{gas}{\sim} \frac{v_0 r_a}{r^2} \sim 10^5\,\mathrm{s}^{-1}, \tag{17}$$

where η is the dynamic viscosity of the ambient, ν the kinematic viscosity and r_a typical molecular size. Both viscous mechanisms, together with thermophoresis, seem to be more important *after* the detachment, in particular for the redeposition problem.

2.4.3. Absorption of sound

The ultrasound generated by the thermal expansion may be strongly damped. As an estimate one can use the rate of energy dissipation in the sound waves (Landau, 1986c)

$$\gamma \approx \frac{\omega_c^2}{2v_0^2}\left(\nu + D\frac{\beta v_0^2}{c}\frac{\beta\Delta T}{3}\frac{1+\sigma}{1-\sigma}\right). \tag{18}$$

Here D is the thermal diffusivity of the material, c the specific heat, β the thermal expansion coefficient, and ω_c some characteristic frequency. The temperature rise $\Delta T \sim T$. With a small particle size, the situation can become even more complicated due to reflection of the sound and temperature wave from the boundaries. In any case,

$$\gamma \sim \max\left(\nu,\, D\frac{\beta v_0^2}{c}\beta T\right)\frac{1}{d^2} \sim 10^8\,s^{-1}, \tag{19}$$

where d is some characteristic length, $d \sim r, v_0\tau, v_0/\omega_c, \sqrt{D\tau}$, depending on the parameters and geometry. The two terms in brackets are related to viscous damping and dissipation by heat conduction due to adiabatic temperature variations within the sound wave. Finally, estimation in Eq. (19) is performed with the typical values from Table 2. It agrees with logarithmic decrements of 10^{-1}-10^{-3} given in (Gray, 1972). In any case expression (19) seems to be smaller than the energy losses due to *emission* of sound.

2.4.4. Emission of sound

The particle, which oscillates on the surface with a frequency ω_0, *emits* sound into the substrate. This seems to be a primary mechanism of energy loss. Dipole approximation from ref. (Landau, 1987) for sound emission from a sphere oscillating in a liquid yields the following power:

$$P = \frac{2\pi\rho_s}{3v_0^3}u^2 \frac{r^6\omega_0^4}{4+(\omega_0 r/v_0)^4}. \tag{20}$$

Here, u is the amplitude of the velocity. The rate of loss for the energy E is given by

$$\gamma \approx P/E \quad \text{with} \quad E = \frac{mu^2}{2} \tag{21}$$

leading to

$$\gamma \approx \frac{\rho_s}{\rho_p}\frac{v_0}{r}\frac{1}{1+4(v_0/\omega_0 r)^4}. \tag{22}$$

This loss can be smaller as the particle is not completely immersed in the substrate material. As a frequency we can use ω_0 from Eq. (10). In any case all expressions for the damping coefficient are only estimates.

2.4.5. Plastic deformations

With sub-μm adhering particles, stresses near the edge of the contact area may exceed the tensile strength of the material leading to plastic deformations (Mittal, 1988; Johnson, 1971). In this case, the contact area becomes bigger than that used in the Eq. (3) and below. As a result, the expression for the adhesive force needs to be significantly modified. Another effects can be due to slow formation of covalent bonds between the particle and the surface (Wu, 2000). Phenomenologically, the can be described as an increase in work of adhesion φ. With real particles these effects usually become more pronounced with smaller particles, where stresses are larger. There are experimental indications that such effects may take place not only with soft polymer particles, but even with relatively hard sub-μm colloidal silica particles as well.

3. Thermal expansion

To study *dry* laser cleaning the thermal expansion of the substrate and the particle must be calculated. Their independent contributions are combined in Eq. (15). We start with the thermal expansion of the substrate.

3.1. Hierarchy of scales

In the present problem there exists a certain hierarchy of scales. The spot sizes typically employed in laser cleaning are rather big. As a result, the *axial* (z) extension of the thermal field is much smaller than its *lateral* (x-y) dimension (laser spot size w_0), even with "weakly absorbing" substrates.

$$l_\alpha + l_T < w_0 . \tag{23}$$

Here l_T is the heat diffusion length (Bäuerle, 2000) and l_α the absorption length. If the sound does not leave the heated region in the *axial* direction during the laser pulse, i.e., if

$$v_0 \tau < (l_\alpha + l_T) < w_0 \tag{24}$$

the *dynamic* equations of elasticity must be considered. As long as the sound wave is *within the laser spot size*

$$v_0 \tau < w_0 \tag{25}$$

the problem is one-dimensional. That is, thermal expansion is *dynamic*, but *unilateral* and only *axial* displacement $u_z \neq 0$ exists. If the sound wave leaves the heated area in z-direction, but is still within the lateral extension of the source,

$$(l_\alpha + l_T) < v_0 \tau < w_0 \tag{26}$$

the expansion is *quasi-static* and *unilateral*. This case is most applicable for dry cleaning with ns laser pulses. With $v_0 \sim 10^{6}$ cm/s, $\alpha \sim 10^{-4}$ cm^{-1} and $w_0 \sim 1$ cm, this yields 10^{-10} s $< \tau < 10^{-6}$ s. Here, quasi-static compressive stresses in x-y planes influence expansion in z direction via Poisson ratio. When sound leaves the irradiated spot in *lateral* direction, i.e., with

$$w_0 < v_0 \tau \tag{27}$$

lateral compression relaxes. The elastic problem is again quasi-static, but three-dimensional (3D). As long as Eq. (23) holds, heat conduction is still 1D, which allows one to obtain general results, *different* from the quasi-static unilateral expansion.

Finally, when the heat diffuses out of irradiated spot, i.e., with

$$w_0 < l_T \Rightarrow D\tau > w_0^2 \tag{28}$$

heat conduction becomes 3D. The result depends on the laser beam profile, and universal analytical solutions do not exist.

Note that with a *stationary* temperature distribution and a *semi-infinite* substrate the surface displacement is infinite. At big distances from the source the temperature rise decreases as *1/r*, and the surface displacement $l \propto \int_0^\infty T(z)dz$ diverges. Indeed, a stationary temperature distribution in the semi-infinite substrate requires an infinite time and energy. The stationary surface displacement for CW irradiation will be determined by the size of the specimen, and by the heat exchange with the surrounding. For example in ref. (Welsh, 1988), integral expressions for the surface displacement are given. These integrals, however, logarithmically diverge at any point.

3.2. General equations

The equations of *classical* isotropic thermoelasticity (Landau, 1986a; Sokolnikoff, 1956; Parkus, 1976) can be written in the form:

$$\begin{aligned} \rho\ddot{u} = {} & \frac{Y(1-\sigma)}{(1+\sigma)(1-2\sigma)} div\, grad\, \boldsymbol{u} + \frac{Y}{2(1+\sigma)(1-2\sigma)} rot\, rot\, \boldsymbol{u} - \\ & -\frac{\beta}{3}\frac{Y}{(1-2\sigma)} grad\, T. \end{aligned} \tag{29}$$

Here, ρ is the density of the material, $\boldsymbol{u}$ the displacement vector and T the temperature difference from the ambient temperature. Y is Young modulus,

σ the Poisson ratio, and β the coefficient of *volumetric* thermal expansion. It is three times larger than the coefficient of *linear* thermal expansion.

We employ the heat equation in the simplest form

$$c\rho\dot{T} = div(K\, grad\, T) + Q\,. \tag{30}$$

Here K is the thermal conductivity, c is the specific heat of the material, and Q is the source term. With very short pulses when dynamic (sound) terms are retained in Eq. (30), the assumption of infinite speed of heat propagation in Eq. (30) is not always valid, and equations (29) and (30) should be modified' (Tamma, 1997; Chandrasekharaiah, 1998). However, *ns* laser cleaning can be studied without these details. Likewise omitted are effects of high electronic temperature (Anisimov, 1995), shock wave generation, etc. This is justified by the non-destructive processing conditions required in laser cleaning.

We assume the following boundary conditions for equations (29)-(30). Firstly, there is no force at the free surface:

$$\sigma_{xz} = \sigma_{yz} = \sigma_{zz} = 0 \quad \text{at} \quad z = 0, \tag{31}$$

where stress σ_{ik} is related to strain via *generalized* Hook's law:

$$\sigma_{ik} = \frac{Y}{1+\sigma}\left(u_{ik} + \frac{\sigma}{1-2\sigma}u_{ll}\delta_{ik}\right) - \frac{\beta}{3}\frac{Y}{1-2\sigma}T\delta_{ik}, \tag{32}$$

$$u_{ik} = \frac{1}{2}\left(\frac{\partial u_i}{\partial x_k} + \frac{\partial u_k}{\partial x_i}\right)$$

Secondly, we assume no heat losses into the ambient

$$\frac{\partial T}{\partial z}\Big|_{z=0} = 0. \tag{33}$$

And finally, with semi-infinite substrate all quantities disappear inside the material at $z \to \infty$.

Though general solutions of such a problem are possible (Sokolnikoff, 1956; Parkus, 1976), they are not elucidating. For this reason we will discuss the most relevant cases.

3.3. Unilateral quasi-static expansion

In this case all quantities are independent of x, y and t, and only a z-component of displacement $\boldsymbol{u}$ is present. For brevity, we will denote it as u and use index z for differentiation with respect to z. We introduce the coefficient of *unilateral* thermal expansion

$$\beta_1 = \frac{\beta}{3}\frac{1+\sigma}{1-\sigma}. \tag{34}$$

The equations and the boundary conditions then reduce to

$$u_{zz} = \beta_1 T_z \ , \ u_z\big|_{z=0} = \beta_1 T \ , \ u, u_z\big|_{z\to\infty} = 0, \tag{35}$$

$$c\rho\dot{T} = (KT_z)_z - I_z \ , \ T_z\big|_{z=0} = 0 \ , \ T\big|_{z\to\infty} = 0. \tag{36}$$

Here I is intensity of laser light inside the material. Integrating Eq. (35) over z and using boundary conditions at infinity, we get:

$$u_z = \beta_1 T. \tag{37}$$

One can immediately write

$$u(0) = -\int_0^\infty \beta_1 T dz. \tag{38}$$

It is possible to avoid solving the heat equation, even with temperature-dependent parameters. We differentiate Eq. (37) with respect to time and substitute T from Eq. (36):

$$\dot{u}_z = \frac{\beta_1}{c\rho}(KT_z - I)_z. \tag{39}$$

Integrating over z and using the boundary conditions, we obtain for the displacement vector at the surface.

$$\dot{u}(0) = -\frac{\beta_1}{c\rho} I(0). \tag{40}$$

Finally, we introduce absorptivity A and the *transient* absorbed fluence,

$$\phi_a(t) = \int_0^t AI(t')dt'. \tag{41}$$

Then, integrating Eq. (40) over t, we obtain the surface displacement $u(0) < 0$ It is negative, as z was directed into the substrate. For the (positive) surface expansion l and for the expansion velocity that enter Eq. (15) we obtain:

$$l(t) \equiv -u(0) = \frac{1+\sigma}{3(1-\sigma)} \frac{\beta\phi_a(t)}{c\rho} \quad , \quad \dot{l} = \frac{1+\sigma}{3(1-\sigma)} \frac{\beta I_a(t)}{c\rho}. \tag{42}$$

Qualitatively, this result is as expected, as both the thermal expansion and the heat content within the material are proportional to the absorbed energy. It can be used even for temperature-dependent parameters. Indeed, the thermal conductivity and absorption coefficient, which change strongly for some materials, do not enter this expression. The absorption coefficient may be non-linear in intensity, etc. With temperature-dependent thermal expansion one has to substitute in the elastic equations, in particular in Eq. (37):

$$\beta_1 T \to \int_0^T \beta_1(T')dT'. \tag{43}$$

However, the time differentiation of Eq. (37) keeps Eq. (39) valid if $\beta_l(T)$ (and $\beta(T)$) is a *differential* coefficient as defined by the Eq. (43). The ratio $\beta_l/c\rho$ is approximately constant due to the Grüneisen relation (Gray, 1972; Ashcroft, 1976; Landau 1980) and the result (42) does not change. Temperature variations in the Poisson ratio and the Young modulus are usually insignificant (Landolt-Börnstein, 1982). As Young modulus does not enter the answer, Eq. (42) probably holds also if Y depends on temperature, as in the case of polymers.

Typical surface displacement, velocity and acceleration for an excimer laser pulse of the form

$$I(t) = I_0 \frac{t}{\tau} \exp\left(-\frac{t}{\tau}\right) \tag{44}$$

are shown in Fig. 3. With this definition, the laser fluence is given by $\phi = I_0\tau$ and the pulse duration at the full widths at half-maximum by $\tau_{FWHM} \approx 2.45\,\tau$.

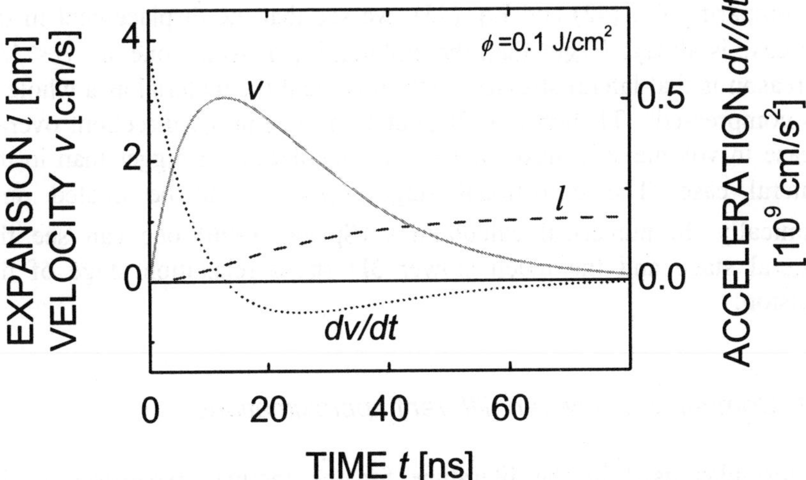

Fig. 3. Surface displacement l (dashed line), velocity v (solid line) and acceleration dv/dt (dotted line) for silicon substrate and typical temporal profile of the laser pulse given by Eq. (44). Laser fluence ϕ = 0.1 J/cm^2. Other parameters used in the calculations are listed in Table 2.

3.4. 3D quasi-static expansion for finite beams with 1D heat conduction

If the condition (27) is fulfilled, the stresses relax outside of the beam area, but the heat conduction can still be considered as 1D. The mathematical derivation is given in the Appendix A. Replacing the integral in Eq. (A.16) by the fluence, as in transition from the Eq. (38) to Eq. (42), we obtain:

$$l(t) = \frac{2(1+\sigma)}{3}\frac{\beta\phi_a(t)}{c\rho} \quad , \quad \dot{l} = \frac{2(1+\sigma)}{3}\frac{\beta I_a(t)}{c\rho}. \tag{45}$$

Although the laser spot size *does not* enter the formula *explicitly*, the consideration *implicitly* assumes that stress and displacement disappear at infinity. This requires 3D relaxation, which can be treated quasi-statically only if Eq. (27) holds.

This result *does not* depend on the spatial profile of the laser beam. Similar results in the particular case of epicentral displacement for the Gaussian beam with surface absorption can be obtained from ref. (Prokhorov, 1990), or found in ref. (Vicanek, 1994).

Comparing Eq. (42) and Eq. (45), we see that the displacement in the latter case is always *larger* than the unilateral quasi-static one, as $\sigma < 1/2$. The reason is that lateral stresses partly relax and the material as a whole is "less compressed". Though not all dilatation goes into z direction, overall increase in volume still makes the surface displacement bigger than in the unilateral case. The transitional stage $v_0 \tau \approx w_0$ can be treated only dynamically. In numerical calculations (Spicer, 1996) one can see the unilateral stage and the much slower 3D stress relaxation stage of the expansion.

3.4.1. Comparison between different approximations

The formulas used in the literature for the thermal expansion of the substrate differ in the dimensionless coefficient in expressions such as Eqs. (38), (42) and (45). The results are summarized in Table 1. The difference between various approximations can easily reach a factor of two. Clearly, one should not consider smaller effects, for example related to moderate temperature dependences of parameters, unless correct formula is used.

Table 1. Coefficient of proportionality between the surface displacement and absorbed fluence, that enters (42) and similar quasi-static expressions, and its value for representative values of Poisson coefficient.

Approximation	1D linear expansion	1D expansion	3D expansion 1D heat cond.	1D incomp expansion
Coefficient in Eqs. (42), (45)	$\frac{1}{3}$	$\frac{1+\sigma}{3(1-\sigma)}$	$\frac{2}{3}(1+\sigma)$	1
Poisson ratio σ				
-1	1/3	0	0	1
-1/2	1/3	1/9	1/3	1
0	1/3	1/3	2/3	1
1/4	1/3	5/9	5/6	1
1/2 (incomp.)	1/3	1	1	1

3.5. Particle influence on the expansion of the substrate

In some references (Lu, 1997a) for the absorbing particle, (Wu, 2000), for the absorbing substrate) the following expression describes the thermal stress at the surface.

$$\sigma_{zz} \sim Y\beta T\,. \tag{46}$$

This stress exists if the substrate (particle) *is not allowed* to expand. In reality, the expansion is restricted only by the elasticity and inertia of the particle. The latter is not that big due to the small particle size.

Let us consider an expanding substrate. The surface displacement during a nanosecond pulse is of the order of several nanometers (see Eq. (42) and experimental results (Dobler, 1999)). The radius of the particle is at least 10-100 times larger. If the particle does not move, expansion results in an indentation of depth $\sim l$ over an area $\sim a^2 \sim lr$ on the substrate. Such a deformation requires a force $\sim Y_s\,(lr)^{1/2} l \sim Y_s l^{3/2} r^{1/2}$ (see ref. (Landau, 1986a) or consider elastic part of the force in Eq. (5) with $h \sim l$). Such a force would have resulted in the following acceleration, velocity, and displacement of the particle at the end of the pulse:

$$\ddot{z} \sim \frac{Y_s r^{1/2} l^{3/2}}{\rho_p r^3} \sim v_0^2 \frac{l^{3/2}}{r^{5/2}}\,, \quad \dot{z} \sim \ddot{z}\tau \sim v_0^2 \frac{l^{3/2}}{r^{5/2}}\tau\,, \quad z \sim \ddot{z}\tau^2 \sim v_0^2 \frac{l^{3/2}}{r^{5/2}}\tau^2\,. \tag{47}$$

In the first estimation we assumed that the elastic constants of the substrate and particle are of the same order of magnitude. This hypothetical displacement $z >> l$ as long as

$$v_0^2 \frac{l^{3/2}}{r^{5/2}}\tau^2 >> l \Rightarrow \left(\frac{v_0 \tau}{r}\right)^2 \sqrt{\frac{l}{r}} >> 1\,. \tag{48}$$

For the typical numbers employed in ns laser cleaning (l~5 nm, r~1 μm) l.h.s. is about 10^3, i.e., condition (48) is practically always fulfilled in *ns* laser cleaning, especially with smaller particles. This means that the substrate is not appreciably slowed down by the particle. In other words, we can treat the influence of the particle on substrate expansion on the basis of the quasi-static Hertz problem. This *does not* mean that the substrate is not deformed. The total deformation in Eq. (15) contains particle and substrate parts $h = h_p + h_s$ (see Fig. 1), which are in relation (Landau, 1986a)

$$h_p : h_s = \frac{1-\sigma_p^2}{Y_p} : \frac{1-\sigma_s^2}{Y_s}. \tag{49}$$

Thus, "soft" substrates will have an indentation of the order of h, determined by Eq. (15), but not of the order of l. Note, that though we used a *geometrical* approximation for the contact radius a, Fig. 1 and equation (15), are based on a more realistic Hertzian picture. Within the geometrical approximation $h_p : h_s = r_s : r_p$ and the plane substrate with $r_s \to \infty$ is not deformed.

In *ps* cleaning one has to include the influence of the particle in the boundary conditions (31) for the substrate and the elastic problem becomes dynamic and essentially 3D even for wide beams.

3.6. Unilateral dynamic expansion

We give here for reference purposes the solution for *dynamic* expansion with *free* boundary in 1D case with constant parameters. The applicability of this solution to laser cleaning requires additional discussion, as the boundary not always can be considered as free. The derivation and consideration of sound related effects will be presented elsewhere. We introduce the longitudinal sound velocity

$$v_0^2 = \frac{Y(1-\sigma)}{\rho(1+\sigma)(1-2\sigma)} \tag{50}$$

and in the same notations as before rewrite Eqs. (29), (31), (32) for the 1D case in the form:

$$\ddot{u} = v_0^2(u_{zz} - \beta_1 T_z)\ ,\ u_z\big|_{z=0} = \beta_1 T\ ,\ u, u_z\big|_{z\to\infty} = 0\,. \tag{51}$$

Displacement l defined as in Eq. (42) is (see also (Maznev, 1997))

$$l(t) \equiv -u(0,t) = \beta_1 v_0 \int_0^t T(v_0(t-t_1), t_1)dt_1 = \beta_1 \int_0^{v_0 t} T(z, t - z/v_0)dz\,. \tag{52}$$

This is an obvious generalization of static expression (38). Indeed, with $v_0 \to \infty$ we recover static result.

3.7. Thermal expansion of absorbing particle

The temperature of a *small* particle with thermal diffusivity D_p is homogeneous if $D_p t >> r^2$, i.e., with $r \sim 10^{-5}$ cm and $D_p \sim 0.1$ cm^2/s, for $t >>$ 1 ns. If volumetric thermal expansion coefficient of the particle is β_p, the increase in volume V is given by:

$$\dot{V} = V\beta_p \dot{T}_p \Rightarrow \dot{r} = \frac{V}{S}\beta_p \dot{T}_p = \frac{\beta_p}{3} r\dot{T}_p. \tag{53}$$

Here, S is the surface area of the particle and the last equality assumes spherical shape. Temperature evolution of an *absorbing* particle with (quasi-static) heat contact with substrate can be approximated by

$$cm\dot{T}_p = \sigma_a I - 4K_s a(T_p - T_s). \tag{54}$$

where $\sigma_a \approx \pi r^2 A_p < \pi r^2$ is the total absorption cross section (for small particles diffraction effects should be considered and the expression for σ_a is quite different (Born, 1980)). Here, a is contact radius, and we used the formula (8.2.10) from (Carslaw, 1959) for the flux into the semi-infinite substrate from the uniformly heated disk. Indexes p and s refer to particle and substrate respectively, T_s being the temperature of the substrate "far away from the particle". Combining Eqs. (53) and (54) we obtain for the extreme case with no heat contact

$$\dot{r} = \frac{\beta_p r\sigma_a}{3c_p m} I \approx \frac{\beta_p}{4c_p\rho_p} I_a < \frac{\beta_p}{4c_p\rho_p} I, \tag{55}$$

which shows similarity with Eq. (42) for the substrate expansion. In the limiting case with poor thermal contact the material with the biggest thermal expansion coefficient provides the biggest contribution to thermally induced deformation and elastic forces. The case with the substrate/particle thermal contact deserves further consideration and will be presented elsewhere.

3.8. Transparent particle heated by the substrate

Another extreme case is the transparent particle, which does not disturb the absorption of light. Or, more accurately, it does not alter the temperature field. This requirement is less restrictive for small particles, as heat conduction smoothes small-scale intensity inhomogeneities during ns pulses. Let us assume that the particle/substrate contact is so good that their temperatures are equal. This will give us an upper limit for the particle temperature. Energetic estimation (7.5.8b) from (Bäuerle, 2000) yields for the surface temperature (constant parameters):

$$\begin{gathered}\phi_a \approx c_s \rho_s T_s (l_\alpha + l_T) \Rightarrow I_a \approx c_s \rho_s \dot{T}_s (l_\alpha + l_T) + c_s \rho_s T_s \dot{l}_T \Rightarrow \\ \dot{T}_s \approx \frac{I_a}{c_s \rho_s (l_\alpha + l_T)} - \frac{\phi_a \dot{l}_T}{c_s \rho_s (l_\alpha + l_T)^2}.\end{gathered} \tag{56}$$

Here l_α and l_T are absorption and thermal lengths. Substituting this into Eq. (53) we obtain

$$\dot{r} = \frac{\beta_p r}{3 c_s \rho_s (l_\alpha + l_T)} \left(I_a - \frac{\phi_a \dot{l}_T}{(l_\alpha + l_T)} \right) \le \left(\frac{r}{l_\alpha + l_T} \right) \frac{\beta_p I_a}{3 c_s \rho_s}. \tag{57}$$

This expression has a similar structure as Eq. (42). The ratio $r / (l_\alpha + l_T)$ is typically (much) less than one. Thus, particle expansion can be taken into account by a replacement $\beta_s \to \beta_s + C_l \beta_p$ in Eq. (42), which modifies coefficient C in Eq. (74).

3.9. Maximum velocity of ejected particles

Let us now estimate the *maximum* particle velocity v if the cleaning is based on thermal expansion and elasticity. Similar estimations, but not expressed in terms of laser fluence, were done in (Lu, 2000a). We neglect the initial adhesion energy, assume that the particle does not move during the pulse and that all elastic energy is later transformed into kinetic energy (big particles or short pulses). Then the energy balance yields:

$$\frac{2\overline{Y}}{5} r^{1/2} l^{5/2} = m\frac{\dot{z}^2}{2} \Rightarrow \dot{z} \sim \left(\frac{\overline{Y}\, l^{5/2}}{\rho\, r^{5/2}}\right)^{1/2} \sim v_0 \left(\frac{l}{r}\right)^{5/4}. \tag{58}$$

Here we assumed comparable material properties of particle and substrate. Including possible expansion of the particle Δr and using Eqs. (42), (55) and (74), we obtain with values from the Table 1 and $\phi_a \approx 1$ J/cm^2 an upper estimation:

$$\dot{z} \sim v_0 \left(\frac{l + \Delta r}{r}\right)^{5/4} \sim v_0 \left(C\frac{\beta\phi_a}{c\rho r}\right)^{5/4} \sim 10^4 \,\text{cm/s}. \tag{59}$$

As in most cases l, $\Delta r << r$, the velocity of the ejected particles is always rather small. It is even smaller for small particles, as they move as a whole during the expansion. Elastic mechanisms yield higher velocities. Indeed, the velocity of *non-deformable* particles cannot exceed that of the moving surface given by Eq. (42) or, better by Eq. (74):

$$\dot{z} \sim C\frac{\beta I_a}{c\rho} \sim 0.25\times 10^3 \,\text{cm/s}. \tag{60}$$

Results for intermediate particle sizes can be obtained by solving Eq. (15) as explained in sections 4,5. If measured particle velocities (Schrems, 2000) significantly exceed both Eqs. (59) and (60), this is a strong indication that other mechanisms are responsible for cleaning.

4. Cleaning threshold

4.1. General threshold conditions

Let us formulate the cleaning condition for the general law of particle movement:

$$\ddot{h} + \gamma\dot{h} + \frac{1}{m}\frac{\partial U}{\partial h} = \dot{v}. \tag{61}$$

Here $\dot{v}$ in the r.h.s. may include expansion of both substrate and particle (see Eq. (15)) or other cleaning forces, while U may include capillary effects, etc. We neglect initial and escape velocities, and define

$U_{ad} = U(t=\infty) - U_0$ (difference between escape and initial energies). Multiplying Eq. (61) by $\dot{h}$ and integrating we obtain the energy criterion

$$\int_0^\infty \dot{v}\dot{h}dt' > m^{-1}U_{ad} + \gamma \int_0^\infty \dot{h}^2 dt' . \tag{62}$$

The l.h.s. is the (specific) work of the cleaning force, while the second term on the r.h.s. is the dissipative loss caused by damping. To write the threshold condition in terms of fluence, one has to solve equation of motion (61). Some limiting cases allow more general consideration.

4.1.1. Short cleaning pulse

If the pulse is short i.e., $\tau << \tau_0, \gamma^{-1}$ cleaning force $m\dot{v}$ dominates *during the action of the pulse*. Neglecting damping and potential in Eq. (61) we obtain:

$$\dot{h} \approx v \Rightarrow h(\tau) \approx h_0 + \int_0^\tau v\,dt, \;\; \dot{h}(\tau) \approx 0 . \tag{63}$$

Thus, energy acquired at the end of the pulse is due to change in h (deformation). If damping is week ($\gamma << \omega_0$), cleaning will take place (after the pulse) if the accumulated (potential) energy is higher than the detachment energy. This is *elastic energy* cleaning regime. For the potential (5) with expansion of both substrate and particle taken into account, this results in

$$\begin{gathered} U(h_0 + \int_0^\tau v\,dt) > 0, \;\; \int_0^\tau v\,dt = l + \Delta r \Rightarrow \\ l + \Delta r > \left((5/2)^{2/3} - 1\right)h_0 \approx 0.84 h_0 \sim h_0 . \end{gathered} \tag{64}$$

Last approximation refers to a parabolic potential (9).

4.1.2. Long cleaning pulse

If the pulse is long, i.e., $\tau >> \tau_0$, one can solve Eq. (61) in a quasi-static approximation. Internal oscillations are weakly excited because there are no high harmonics in the spectrum of the cleaning force. As a result h is determined by the condition that cleaning force $m\dot{v}$ balances the force from the adhesion potential.

$$\frac{1}{m}\frac{\partial U}{\partial h} \approx \dot{v}. \qquad (65)$$

Thus, to clean, one has to overcome the biggest adhesion force *during the pulse*. This is *force (inertia)* cleaning regime. For the potential (5) the force is maximal with $h = 0$ and is positive in our notations. This results in:

$$(-m\dot{v})_{\max} = -m(\ddot{l} + \ddot{r})_{\max} > F_0. \qquad (66)$$

Detailed analysis shows that there exists a coefficient $C_l < 1$ in the r.h.s., which takes into account weak internal oscillations. Its value depends on the pulse shape and γ. The l.h.s. should be positive, i.e., for the mechanism based on thermal expansion, detachment occurs *in the deceleration* phase (Dobler, 1999) due to the inertia of the already accelerated particle.

4.1.3. Over-damped movement

If damping is strong ($\gamma >> \tau^{-1}$, ω_0), one can neglect "inertia" $\ddot{h}$ in Eq. (6). Then

$$\dot{h} \approx \gamma^{-1}\left(\dot{v} - \frac{1}{m}\frac{\partial U}{\partial h}\right). \qquad (67)$$

Thus, as with long pulses, one has to overcome the biggest adhesion force *during the pulse*.

4.1.4. Long pulses with steep fronts

Consider a long pulse ($\tau >> \tau_0$), which starts abruptly, so that v rises to v_f within time $t_f << \tau_0$. Then, from Eq. (63) particle "instantaneously" acquires "velocity" $\dot{h} \approx v_f$ towards the substrate. During the rest of the pulse the position h changes weakly. If the kinetic energy associated with $\dot{h}$ exceeds that of adhesion, the particle will detach. This is the *kinetic energy* cleaning regime with the criterion:

$$mv_f^2/2 > U_{ad}. \tag{68}$$

Similar consideration applies for the trailing edge of the pulse. In this case the particle acquires the velocity away from the substrate. In other words, to produce strong "force" the pulse should not necessarily be short. It is enough if it has sharp edges. The criterion (68) is often less restrictive than Eq. (66).

To obtain compact analytical results for the transitional stages we consider the following problem.

4.2. Single sinusoidal pulse in parabolic potential without damping

Let us neglect damping and use a parabolic approximation for the potential. For convenience we introduce $h_1 = h - h_0$ and count the potential energy from the bottom of the well where $h_1 = 0$. The equation of motion and initial conditions become:

$$\ddot{h}_1 + \omega_0^2 h_1 = \dot{v}, \tag{69}$$

$$\dot{h}_1(0) = h_1(0) = 0. \tag{70}$$

With these notations detachment occurs when $h_1 < -h_0$.

For the sake of generality and simplicity we consider the following "sinusoidal cleaning velocity":

$$v = \frac{l\omega}{2\pi}(1-\cos\omega t),\ 0 < t < \tau. \tag{71}$$

Here, l is the *total* displacement during the pulse, and $\tau \equiv 2\pi/\omega$ the *total* pulse duration with $\tau_{FWHM} = \pi/\omega$. Parameters l and τ are convenient characteristics of the expansion process. For laser cleaning v is proportional to the laser intensity and l to the laser fluence, see Eq. (42). Thus, Eq. (71) implies similar temporal profile of the laser pulse. Qualitative results are similar for any smooth pulse. This model problem retains important features of the original formulation. At the same time it allows complete theoretical exploration and compact formulas for the main relationships between the parameters.

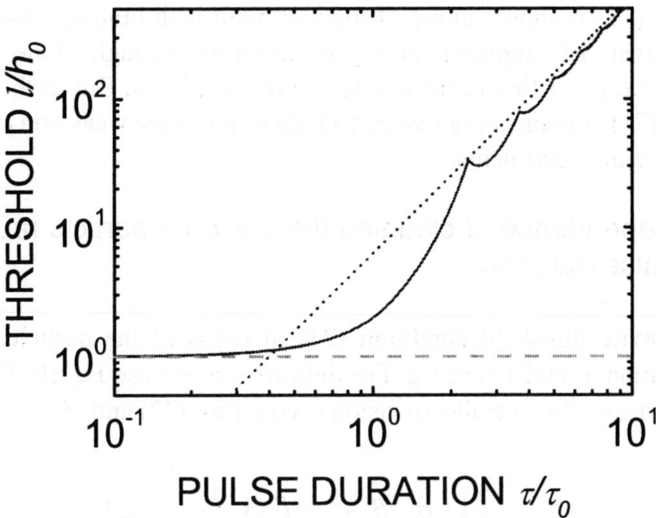

Fig. 4. Dimensionless threshold condition for a *cosinusoidal* pulse (71) and *parabolic* potential (9). Solid line - general threshold condition (73). One can see (week) resonance effects. Dashed line - short pulse limit, and dotted line - long pulse limit approximations from Eq. (73).

It is convenient to introduce the variable

$$y \equiv \frac{\omega_0}{\omega} = \frac{\tau}{\tau_0}, \tag{72}$$

where τ_0 is the resonant period of the oscillator. As shown in Appendix B, the threshold condition reads:

$$\frac{l}{h_0} > \begin{cases} \dfrac{\pi y(1-y^2)}{\sin \pi y} & \text{for } y<1 \text{ and } \approx 1 \text{ for } y \ll 1, \\[2ex] -\dfrac{2\pi y(y-1)}{\sin\left(\dfrac{2\pi}{y+1}\left[\dfrac{3}{4}(y+1)\right]\right)} & \text{for } y>1 \text{ and } \approx 2\pi y^2 \text{ for } y \gg 1. \end{cases} \tag{73}$$

Approximate expressions in Eq. (73) are convenient for fast estimations. The result (73) is shown in Fig. 4 together with both limiting cases. With $\tau \sim \tau_0$ neither of approximations is accurate enough. One can see characteristic kinks that occur due to resonance effects. But as the cleaning force Eq. (71) contains only *one* period, these kinks are weak and can hardly be observed in experiments.

4.3. Dependence of cleaning threshold on particle radius and pulse duration

Let us rewrite threshold condition (73) in terms of the particle radius r, pulse duration τ, and fluence ϕ. For definiteness we use Eq. (6) for h_0 and Eq. (10) for ω_0. The overall expansion is (see Eqs. (42) and (53))

$$l + \Delta r = \left(\frac{1+\sigma_s}{1-\sigma_s} \frac{\beta_s A_s}{3c_s \rho_s} + \frac{\beta_p A_p}{4c_p \rho_p} \right) \phi \approx C \frac{\beta A}{c\rho} \phi. \tag{74}$$

Here, the first term refers to the substrate and the second to the (absorbing) particle without thermal contact. For the transparent particle with thermal contact the second term will be of the order of Eq. (57). The last expression is a notation used for brevity. The contribution from the material with the biggest βA value dominates. C~0.25-1 is dimensionless coefficient. We neglect particle influence on the expansion of the substrate, i.e., assume that the heat conduction homogenizes the temperature near the particle and neglect field enhancement effects (Mosbacher, 2001; Leiderer, 2000; Luk'yanchuk, 2000, 2002). Rewriting the two limiting cases in Eq. (73) in dimensional quantities, we obtain the expressions for the threshold fluence:

$$\phi > C^{-1} \frac{c\rho}{\beta A} \begin{cases} \left(2\pi\varphi / \overline{Y}\right)^{2/3} r^{1/3} & \text{short pulses or small particles } \omega_0 \tau \ll 1, \text{ "elastic energy"}, \\ \dfrac{9\tau^2}{10\pi} \dfrac{\varphi}{\rho_p r^2} & \text{long pulses or small particles } \omega_0 \tau \gg 1, \text{ force (inertia)}. \end{cases} \tag{75}$$

Intermediate regimes for $\omega_0\tau \sim 1$ can be calculated numerically. The expression (73) recalculated into dimensional variables is shown in Fig. 7 by the dotted line, together with the results of more accurate calculations described in section 5.

The dependence on pulse duration is monotonic -- shorter pulses are more favorable for fixed fluence. With pulse durations shorter than the resonant period τ_0, a further decrease in pulse duration is *not* advantageous. With $\alpha v_0 \tau \lesssim 1$ one has to consider sound effects in the substrate, which will be discussed elsewhere.

The dependence on particle radius is less trivial. As ω_0 and h_0 depend on r, threshold dependence on r is non-monotonic. There exists an optimal radius for a given pulse duration. For this radius the resonant period τ_0 is close to the duration of the laser pulse τ.

With big radii, the native period τ_0 is long and the cleaning pulse is much shorter than one cycle of oscillations. Cleaning proceeds in the "elastic energy" regime. Heavy particles almost do not move during the pulse. The substrate surface moves much faster than the center of the particle. This leads to an increase in elastic energy (compression of substrate and particle). Detachment occurs after the pulse, in the first backward swing of the (internal) oscillation. This regime (for non-linear potential) is shown in Fig. 5 a. For a symmetric parabolic potential the elastic energy become positives in the compression stage if $l > h_0$, and this yields simplified detachment condition for such a potential. A more accurate estimation (64) for non-linear potential does not change the threshold significantly. This can be seen also from the comparison of the solid and dotted curves in Fig. 7. The $r^{1/3}$ increase in threshold with radius is due to bigger equilibrium value of h_0 and higher adhesion energy $|U_0|$ in Eq. (7) for bigger particles.

With smaller particles, the native period becomes shorter than the laser pulse duration. The response of the oscillator to the "low frequency" force is inefficient. Cleaning proceeds in the "quasi-static" regime, when fast and small internal oscillations in h are superimposed on the slow changes in h that obey Eq. (65). This regime is shown in Fig. 5b. Let us derive the threshold condition for this situation from the general expression (66). For the pulse (71) the largest deceleration occurs at $3\tau/4$, and in notations (74) the cleaning condition is:

$$mC\frac{\beta A}{c\rho}(-\ddot{I})_{\max} \equiv \left(\frac{4\pi r^3}{3}\rho_p\right)C\frac{\beta A\phi}{c\rho}\frac{2\pi}{\tau^2} > F_0 \equiv 2\pi r\varphi \Rightarrow$$

$$\phi > C^{-1}\frac{c\rho}{\beta A}\left(\frac{3\tau^2}{4\pi}\frac{\varphi}{\rho_p r^2}\right) \tag{76}$$

The results (75) and (76) slightly differ because the pull-out force F_0=(6/5)$2\pi r\varphi$ in parabolic approximation. The strong increase in the threshold fluence with smaller particles demonstrates the inefficiency of the "force (inertia)" cleaning regime.

Let us consider kinetic energy cleaning regime (68). The movement of the particle for a rectangular laser pulse is shown in Fig. 5c. If intensity change at the steep front is I_f,

$$\frac{m}{2}(\dot{l}+\dot{r})^2 \equiv \frac{4\pi r^3}{6}\rho_p\left(C\frac{\beta A I_f}{c\rho}\right)^2 > \left|U_0\right| \equiv \frac{3}{5}(2\pi\varphi)^{5/3}\overline{Y}^{-2/3}r^{4/3} \Rightarrow$$

$$I_f > C^{-1}\frac{c\rho}{\beta A}\left(\frac{3(2\pi)^{1/3}}{5^{1/2}}\frac{\varphi^{5/6}}{\rho_p^{1/2}\overline{Y}^{1/3}r^{5/6}}\right). \tag{77}$$

This can be *formally* written as a condition for fluence. For example assuming $I_f \approx \phi/\tau$ we get

$$\phi > C^{-1}\frac{c\rho}{\beta A}\left(\frac{3(2\pi)^{1/3}}{5^{1/2}}\frac{\varphi^{5/6}\tau}{\rho_p^{1/2}\overline{Y}^{1/3}r^{5/6}}\right). \tag{78}$$

But with steep fronts it is the *intensity* what is important. Note also much weaker dependence of threshold on r as compared with Eq. (76). This is crucial for small particles.

With $\tau \sim \tau_0$ no simple approximations for the threshold exists. This situation is shown in Fig. 6. Together with h and l, the movement of the particle center in the laboratory frame z - z_0 is shown. In the beginning, surface displacement l is faster than the particle movement (compression) and later the particle detaches with constant velocity.

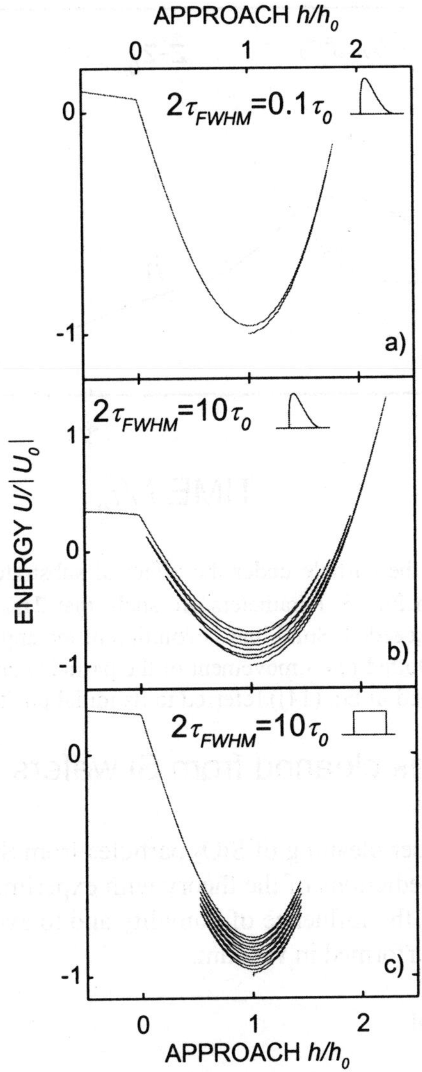

Fig. 5. Movement of the particle (evolution of the approach h in non-linear potential (5)) under the effect of substrate expansion somewhat above threshold. Temporal profile of the laser pulse is given by Eq. (44). a) Elastic energy cleaning regime for big particle. Laser pulse is much shorter than the period of oscillator $2\tau_{FWHM} = 0.1\,\tau_0$ b) Quasi-static force/inertia regime for small particle. Laser pulse is much longer than the period of oscillator $2\tau_{FWHM}=10\,\tau_0$. c) Kinetic energy regime for the pulse with steep fronts. Rectangular laser pulse is longer than the period of oscillator $2\tau_{FWHM} = 10\ \tau_0$, while rise/fall time of the fronts $t_f << \tau_0$.

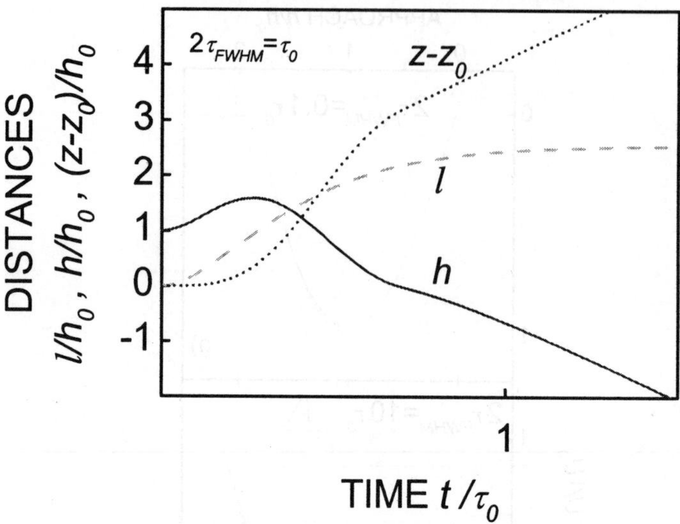

Fig. 6. Movement of the particle under the effect of substrate expansion. Potential and pulse shape as in Fig. 5. Parameters are such, that $2\tau_{FWHM} = \tau_0$ and cleaning threshold is slightly exceeded. Solid line - evolution of the approach h. Dashed line - surface expansion l. Dotted line - movement of the particle center z in the laboratory reference frame (defined in Eq. (14)) referred to its initial position z_0.

5. SiO_2 particles cleaned from Si wafers

Let us now study laser cleaning of SiO_2 particles from Si surfaces. The goal is to compare the predictions of the theory with experimental investigations. In order to diminish the influence of humidity and to avoid redeposition the experiments were performed in vacuum.

5.1. Experimental

The cleaning of SiO_2 particles (Bangs Laboratories, radii 200-2585 nm) from (100) Si wafers (Wacker Siltronic) was performed with a KrF excimer laser (Lambda Physik LPX 205, wavelength 248 nm, pulse duration 31 ns FWHM). The energy of the beam was controlled by an external attenuator (tilted quartz plate) and projected with a mask onto the target to a spot of 1 mm diameter. Such an imaging produces a uniform energy distribution within the irradiated area. The pressure within the chamber was ~ 4×10^{-5}

mbar. Particles are deposited onto the sample by spin-coating, which gives a high uniformity of the particle density. Optical microscopy and picture processing software, which can count the particles and measure their size, is used to evaluate the cleaning efficiency. Since the homogeneity of the samples is not perfect, a picture of the cleaned area is taken before irradiation and is compared with the picture after irradiation. With this technique it is possible to see the behaviour of clusters and redeposition of the particles.

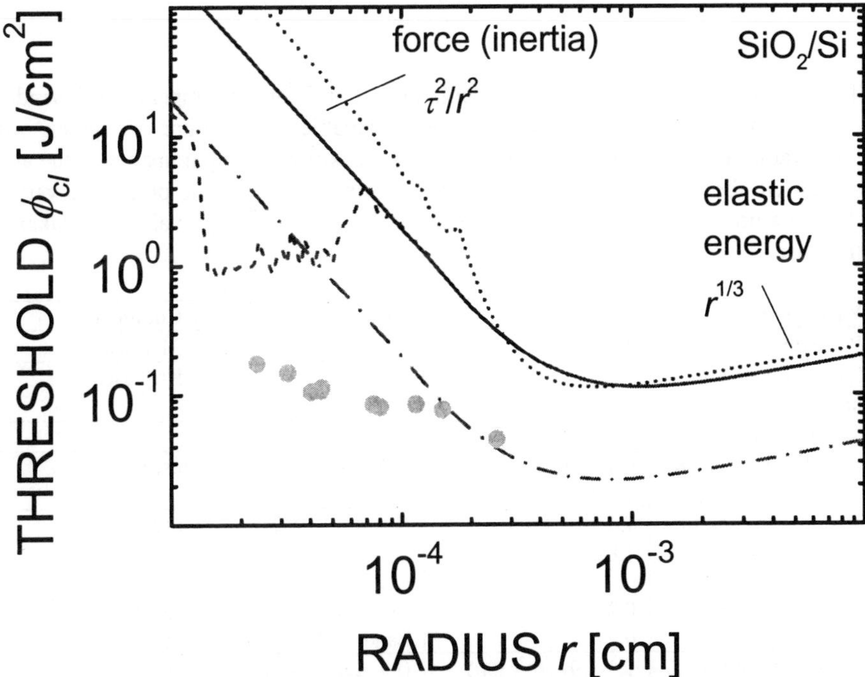

Fig. 7. Experimental and calculated cleaning fluence as a function of particle radius for SiO_2 particles on Si. Parameters used in the calculations are listed in Table 2. Only expansion of the substrate is taken into account. Circles – experimental points. Solid line - numerically calculated threshold. Dash-dotted line - numerically calculated threshold for ten times smaller adhesion. Resonance effects for the considered *anharmonic* potential (5) and *realistic* pulse shape (44) are absent. Dotted line - recalculated into dimensional quantities harmonic approximation (73) for the same τ_{FWHM}. Dashed line - numerically calculated threshold for "rigged" laser pulse (80).

5.2. Cleaning threshold vs. radius

Theoretical and experimental results are compared in Fig. 7. The parameters used in the calculations are listed in Table 2. Their choice is somewhat arbitrary. For example Si is rather anisotropic, vitreous SiO_2 does not accurately follow the Grüneisen relation, etc. Because it is more difficult to take into account thermal expansion of the particle (see section 3), only expansion of the substrate is included. This is justified, as expansion of *fused* silica is much smaller than that of silicon (Table 2) and should not significantly alter the results. Damping was ignored.

Table 2. Parameters used in the calculations. For strongly temperature-dependent parameters values for room and highest available temperature are given with temperature (in K) indicated in brackets. For weakly varying parameters value at or somewhat above room temperature is taken. Some of the elastic properties vary in the literature and for crystal may depend on direction. Average values are taken in this case.

Parameter	**Value(s)**	**Ref.**	**Comments**
Pulse duration τ ns	12.7		31 τ_{FWHM}
Laser wavelength λ nm	248		KrF laser
Substrate Si			
Specific heat c_s J/gK	0.72 (300) 1 (1500)	(Bäuerle, 2000)	used value
Volumetric thermal expansion coefficient β_s K^{-1}	7.7×10^{-6} (300) 13.2×10^{-6} (1400)	(Landolt, 1982)	used value
Poisson ratio σ_s	0.26-0.28 0.27	(Flina; Almaz; Landolt, 1982)	anisotropic used value
Young modulus Y_s dynes/cm^2	$1.3\text{-}1.9\times10^{12}$(300) $1.2\text{-}1.8\times10^{12}$(900) 1.6×10^{12}	(Landolt, 1982)	100-111 direction used value v_s=9.13×10^5 cm/s (110)
Density ρ_s g/cm^3	2.3	(Bäuerle, 2000)	used value
Absorption coefficient α_s cm^{-1}	1.67×10^6	(Bäuerle, 2000)	*weak $\alpha(T)$*
Absorptivity A_s	0.39	(Bäuerle, 2000)	*weak $A(T)$*
Melting temperature $T_{m\,s}$ K	1690	(Bäuerle, 2000)	

Table 2. (Continuation)

Parameter	Value(s)	Ref.	Comments
Particle SiO_2 (fused silica)			
Specific heat c_p J/gK	0.72 (300) 1.22 (1000) 1	(Bäuerle, 2000)	 used value
Volumetric thermal expansion coefficient β_p K^{-1}	1.65×10^{-6} (300) 1.2×10^{-6} (300) 1.8×10^{-6} (500) 1.4×10^{-6} (1100) 1.65×10^{-6}	(Prokhorov, 1990) (Gray, 1972)	(300-1300) used value
Poisson ratio σ_p	0.17	(Goodfellow; Weast, 1989)	used value
Young modulus Y_p dynes/cm^2	$0.7\text{-}0.75\times10^{12}$ 0.73×10^{12}	(Goodfellow; Weast, 1989)	 used value $v_s = 5.9\times10^5$ cm/s
Density ρ_p g/cm^3	2.2	(Bäuerle, 2000)	probably smaller (Bangslabs)
Absorption coefficient α_p cm^{-1}	1	(Bäuerle, 2000)	
Absorptivity A_p	0.94	(Bäuerle, 2000)	
Melting temperature $T_{m\,p}$ K	1873	(Bäuerle, 2000)	
Adhésion			
LVdW constant $\varsigma_{Si\text{-}Si}=(4\pi/3)H$, *eV* *H* - Hamaker constant	6.5-6.76 6.8-7.2 6.15 6.7	(Visser, 1976) (Bowling, 1989) (Dahneke, 1972)	 used value
LVdW constant $\varsigma_{SiO2\text{-}SiO2}=(4\pi/3)H$, *eV* *H* - Hamaker constant	1.7 1.9-12 2.33 1.32 2	(Bergström, 1997) (Visser, 1976) (Dahneke, 1972) (Heim, 1999)	assuming DMT (Derjaguin, 1975) used value
LVdW constant *eV* $\varsigma_{Si\text{-}SiO2} = (\varsigma_{Si-Si}\ \varsigma_{SiO2\text{-}SiO2})^{1/2}$	3.66		used value
Adhesion distance ε cm	4×10^{-8}	(Visser, 1976; Bowling, 1989)	rather universal
Work of adhesion φ erg/cm^2	140		used value

We used average parameters for Si and fused SiO_2 available in the literature and calculated the adhesion between them according to the formula (Visser, 1976)

$$\varphi_{12} = \sqrt{\varphi_{11}\varphi_{22}} \; . \tag{79}$$

But one has to have in mind, that

- In the presence of the native oxide the contact is essentially SiO_2-SiO_2, which reduces φ by about a factor of two.
- Several monolayers of moisture trapped in the particle-substrate interstice during the spin-on procedure can decrease adhesion by an order of magnitude due to screening of VdW interaction by water, which has a high dielectric constant.
- We *do not* discuss here influence of capillary effects on adhesion. One has to have in mind, however, that with elevated temperatures induced in laser cleaning, surface tension coefficient significantly decreases, making capillary forces less important.
- Likewise not discussed here is increase in adhesion due to possible charge of particles. This can be included into consideration, but is usually more important for bigger particles and/or in the post-detachment stage.
- Adhesion may increase with the storage time due to formation of covalent bonds and possible plastic deformation even for rather hard materials.
- Surface roughness of substrate *and/or* particle can further change adhesion.

Having this in mind, we present calculations also for the adhesion decreased by a factor of 10. The result, and dependence on other parameters, can be understood from expressions (75). Though it is more difficult to clean smaller particles, experimentally observed thresholds (circles) and the slope of $\phi_{cl}(r)$ dependence are much *lower* than theory predicts. Even small adhesion cannot explain these findings. What can be the reason for such a behavior?

5.3. Role of small oscillations in intensity

One possibility is the "bad quality" of the excimer laser pulse. Typical excimer pulses are too long so that the small particles are removed in the inefficient "force" regime. If the pulse contains high frequency components, they may significantly reduce the threshold for small particles, despite small fluence contained in each "spike". Numerically calculated threshold for the rigged pulse of the same total fluence is shown by the dashed curve. The rigged pulse had a temporal profile with harmonics characterized by j.

$$I(t)=\frac{2}{\{(j_{\max}-j_{\max})/dj\}+1}\left(\sum_{j=j_{\min},\ dj}^{j_{\max}}\cos^2\left(2\pi j\frac{t}{\tau}+j^3\right)\right)I_0\frac{t}{\tau}\exp\left(-\frac{t}{\tau}\right). \quad (80)$$

Here the integer part {} in the denominator makes the overall fluence independent on j_{min}, j_{max} and dj, which were taken as j_{min} = 1.1415, j_{max} = 5, and dj = 0.2718. The surface displacement, velocity, and acceleration are shown in Fig. 8. The displacement is virtually the same as for the smooth pulse (Fig.3). Velocity (which is proportional to intensity) and especially acceleration differ significantly. Short spikes in the intensity may resonantly decrease the threshold for small particles by two orders of magnitude.

Another reason, which seem to be more plausible (Leiderer, 2000; Lu, 2000b) may be field enhancement by the particles (Mosbacher, 2001; Luk'yanchuk, 2000, 2002; Zheng, 2001; Lu, 2000c) or explosive vaporization of residual moisture (Fourrier, 2001).

5.4. Suggestions for cleaning experiments

Oscillations of adhering particles may be used to increase efficiency of DLC. One can try to utilize possible resonance effects with the aim to remove smaller particles and to increase damage-free cleaning window. We discuss several possibilities.

Smooth excimer laser ns pulse is "too long" for sub-μm particles. If it is modulated with the frequency that matches internal "adhesion frequency" (10) for the particles of given size, one can expect resonance increase in oscillation amplitude. Calculations demonstrate that if the overall duration of the pulse stays constant, and the period of the oscillations is about one tenth of the overall pulse duration, cleaning threshold can decrease by 1-2 orders of magnitude. Due to the non-linearity of the potential (5), even without damping, at near threshold fluences resonance growth "saturates" after 5-10 oscillations. Detailed investigation of this effect will be presented

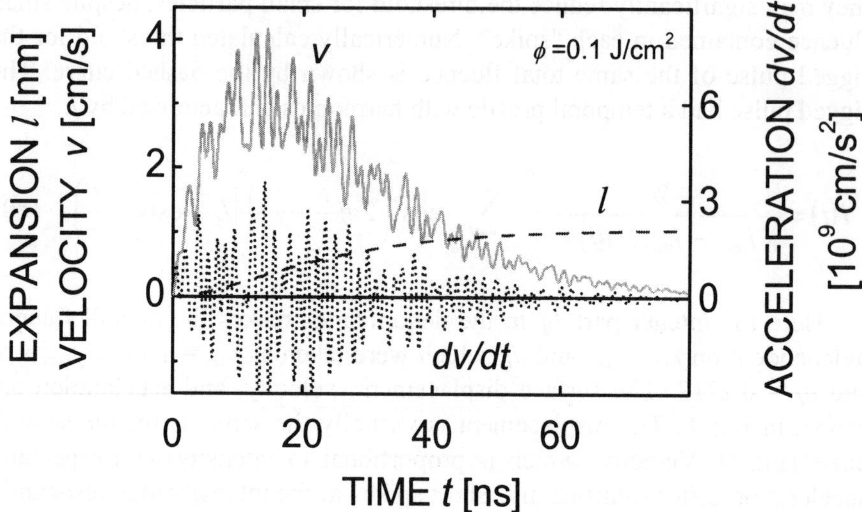

Fig. 8. Surface displacement l (dashed line), velocity v (solid line) and acceleration dv/dt (dotted line) for the temporal profile of rigged pulse (80) used for the calculations of dashed line in Fig. 7. All other parameters are as in Fig. 3. Note difference in scale for acceleration as compared to Fig. 3.

elsewhere (Arnold, 2002 a, b).

Without damping the cleaning effect of the pulse of duration $\tau = \tau_0$ (*single resonant* "push") and longer *modulated* pulse which includes n *resonant* "pushes" $\tau = n\ \tau_0$ is similar if they have *the same overall fluence*. But heating will be lower for the longer pulse, proportionally to $n^{-1/2}$ for surface absorption (Bäuerle, 2000). Thus, damage threshold will increase and the window for damage-free cleaning may widen.

"Infinitely short" pulse is the most efficient for the given fluence. It is more efficient than the modulated pulse of arbitrary duration. But with short pulses damage threshold is determined by l_α and is much lower than for ns pulses (Bäuerle, 2000). One can replace one short pulse with several pulses with the fixed delay between them. Mode locked lasers are natural candidates for such experiments. If the delay matches internal frequency of the oscillations, the *cleaning* effect will be the same. *Damage* threshold will be determined rather by the *overall* duration τ of the pulse train, provided that $l_T \sim (D_s\tau)^{1/2} >> l_\alpha$. Note, that the description of ps laser cleaning requires consideration of sound related effects that become important with $\alpha v_0 \tau < 1$ or $v_0 \tau / r < 1$.

6. Conclusions

In this article we have provided a theoretical analysis of dry laser cleaning using ns pulses. Expressions for the thermal expansion of the substrate are derived and discussed for different situations. The formula for the 1D quasi-static thermal expansion of the substrate does not require solution of the heat equation and is valid over a broad range of temperature-dependent material parameters. The expansion of absorbing and transparent particles is discussed as well.

A simple approximation for a combined elastic-VdW potential has been suggested. The laser cleaning process is formulated as an escape problem from the non-linear potential under the action of cleaning force produced by thermal expansion. Expansion of the substrate and the particle are treated on a unified basis. Possible damping mechanisms are discussed.

Two parameters characterize the adhesion potential -- the period of oscillations near the bottom of the potential well τ_0 and the equilibrium deformation (approach) h_0. They serve as natural temporal and spatial scales. Their analytical dependence on particle size r and material properties is provided. Laser pulse duration τ should be compared with τ_0 and overall thermal expansion $l+\Delta r$ with h_0.

Formulas for the cleaning fluence ϕ_{cl} in different regimes are derived and compared with numerical calculations. In particular, with $\tau < \tau_0$ (large particles) cleaning proceeds in the "elastic energy regime" which reduces to the condition $l+\Delta r > h_0$. As a result, $\phi_{cl} \propto r^{1/3}$. With $\tau > \tau_0$ (small particles) cleaning proceeds in the inefficient "force regime" and $\phi_{cl} \propto \tau^2/r^2$, which favors shorter laser pulses. With $\tau > \tau_0$, but steep edges of the pulse $t_f << \tau_0$, cleaning requires that particle kinetic energy exceeds that of adhesion $m(\dot{l} + \Delta\dot{r})_f^2 / 2 > |U_0|$. This leads to $\phi_{cl} \propto \tau / r^{5/6}$ for the "kinetic energy" regime.

Comparison with experimental $\phi_{cl}(r)$ dependence for SiO_2 particles on Si surface shows that commonly assumed mechanisms of dry laser cleaning do not explain experimental findings. Experimentally observed thresholds are *too low*. Among possible explanations are fast spatial-temporal variations in intensity of excimer (KrF) laser pulse and field enhancement effects suggested earlier by other authors.

Utilization of resonance effects either by modulation of ns laser pulse or employing the train of ps pulses with delay equal to $\tau_0(r)$ is suggested. Developed approach can be applied to the cases when other adhesion forces (capillary, electrostatic, chemical bonding, etc.) may dominate.

Acknowledgments

I want to thank DI M. Mosbacher, DI G. Schrems, Dr. S. Pleasants and Prof. B. Luk'yanchuk for useful comments, and Dr. J. Durrell for the careful reading of the manuscript. My special thanks to Prof. D. Bäuerle, who attracted my attention to the physics of laser cleaning. His numerous scientific advises and tireless support of collaboration between theory and experiment made this work possible. The work was financed by the Fonds zur Förderung der wissenschaftlichen Forschung in Österreich, project P14700-TPH and by the EU within the frame of the TMR project Laser Cleaning, contract #ERBFMRXCT98 0188.

Appendix A. Quasi-static 3D thermal expansion

Here we derive the surface displacement for the semi-infinite substrate $z > 0$. The idea is to write equations and boundary conditions for u_z and $div\, \boldsymbol{u}$ only (more accurately for some function f introduced below), and to solve them by Fourier transform in x-y plane. Henceforth ∂_z denotes derivative with respect to z, while index z refers to the component of a vector. Stationary equation (29) for z component can be written as:

$$\Delta u_z + \frac{1}{1-2\sigma}\partial_z div\, \boldsymbol{u} - \frac{\beta}{3}\frac{2(1+\sigma)}{1-2\sigma}\partial_z T = 0 . \qquad \text{(A.1)}$$

At the same time, applying div to the stationary equation (29) we get

$$\Delta(div\, \boldsymbol{u} - \beta_1 T) = 0 . \qquad \text{(A.2)}$$

Boundary conditions (31) also can be written in terms of *div* ***u*** and u_z only. They are valid in *x-y* plane $z = 0$ and can be differentiated in this plane. The following combination does not contain u_x and u_y separately, and can be used as a boundary condition for u_z

$$\partial_x \sigma_{xz} + \partial_y \sigma_{yz} = \frac{Y}{2(1+\sigma)} \left(\partial_{xx} u_z + \partial_{yy} u_z - \partial_{zz} u_z + \partial_z div\, \boldsymbol{u} \right) = 0 \text{ at } z = 0 . \quad \text{(A.3)}$$

To deal only with the first order boundary conditions, we exclude $\partial_{zz} u_z$ using Eq. (A.1).

$$\Delta_\perp u_z + \frac{1-\sigma}{1-2\sigma} \partial_z div\, \boldsymbol{u} - \frac{\beta}{3} \frac{1+\sigma}{1-2\sigma} \partial_z T = 0 \text{ at } z = 0 . \quad \text{(A.4)}$$

Here $\Delta_\perp$ stands for 2D Laplacian in *x-y* plane. Second boundary condition is Eq. (31) for normal stress σ_{zz} with definition Eq. (32).

$$\sigma_{zz} = 0 \Rightarrow \partial_z u_z + \frac{\sigma}{1-2\sigma} div\, \boldsymbol{u} - \frac{\beta}{3} \frac{1+\sigma}{1-2\sigma} T = 0 \; at \; z = 0 . \quad \text{(A.5)}$$

The variable more convenient than *div* ***u*** is *f* defined as

$$f = div\, \boldsymbol{u} - \beta_1 T . \quad \text{(A.6)}$$

Rewriting equations (A.2) and (A.1) in terms of *f* we get:

$$\Delta f = 0, \qquad \Delta u_z + \frac{\partial_z f}{1-2\sigma} - \beta_1 \partial_z T = 0. \quad \text{(A.7)}$$

And for the boundary conditions (A.4) and (A.5) at $z = 0$ and at infinity we obtain

$$\Delta_{\perp} u_z + \frac{1-\sigma}{1-2\sigma}\partial_z f = 0 \quad \text{at } z = 0,$$

$$\partial_z u_z + \frac{\sigma}{1-2\sigma} f - \beta_1 T = 0 \text{ at } z = 0, \tag{A.8}$$

$$u_z\,,\ f\,,\ T \to 0 \qquad\qquad \text{at } z \to \infty.$$

These coupled equations are solved by Fourier transform in *x-y* plane. There exist subtle requirements, that all quantities (including *f*, which is a difference of two "good" functions, see Eq. (A.6)) disappear at infinity and can be Fourier transformed. The former property holds in physically admissible situations. The latter is more restrictive. It is *not* satisfied for the stationary temperature distribution in the semi-infinite substrate induced by a permanent finite source. In this case $T \sim 1/r$ at large distances and Fourier transform *of displacement* u_z does not exist. This is the mathematical reason why one cannot obtain unilateral expansion (42) as a limiting case of the formulas from this appendix. With time dependent temperature distributions induced by spatially finite sources present results should be used.

The Fourier transform of Eq. (A.7) (taking into account conditions at infinity) with wave vector $\boldsymbol{k}$ (length *k*), results in trivial equation for *f*, which can be immediately solved, and in differential equation for u_z

$$\tilde{f} = \tilde{f}_0 e^{-kz},$$
$$\partial_{zz}\tilde{u}_z - k^2\tilde{u}_z - \frac{k\tilde{f}_0 e^{-kz}}{1-2\sigma} - \beta_1 \partial_z \tilde{T} = 0. \tag{A.9}$$

Here, tilde denotes Fourier image and $f_0 \equiv f\ (z{=}0)$. Transformed boundary conditions (A.8) look like

$$k^2\tilde{u}_z + \frac{1-\sigma}{1-2\sigma} k\tilde{f}_0 = 0 \qquad \text{at } z = 0,$$
$$\partial_z \tilde{u}_z + \frac{\sigma}{1-2\sigma}\tilde{f}_0 - \beta_1 \tilde{T} = 0 \text{ at } z = 0, \tag{A.10}$$
$$\tilde{u}_z \to 0 \qquad\qquad \text{at } z \to \infty.$$

We find $\widetilde{f}_0$ from the first equation and exclude it from the remaining boundary condition *and the equation* for u_z. Introducing $\widetilde{u}_z(0) \equiv \widetilde{u}_z(z=0)$ we get the equation for the Fourier image of u_z only.

$$\begin{aligned} &\partial_{zz}\widetilde{u}_z - k^2\widetilde{u}_z + \frac{k^2\widetilde{u}_z(0)}{1-\sigma}e^{-kz} - \beta_1\partial_z\widetilde{T} = 0, \\ &\partial_z\widetilde{u}_z(0) - \frac{\sigma k\widetilde{u}_z(0)}{1-\sigma} - \beta_1\widetilde{T}(0) = 0\ , \quad \widetilde{u}_z(\infty) \to 0. \end{aligned} \tag{A.11}$$

This linear equation can be solved in the general case. Expression that satisfies condition at infinity is:

$$\begin{aligned} \widetilde{u}_z = &\frac{1+2kz}{4(1-\sigma)}\widetilde{u}_z(0)e^{-kz} + \frac{\beta_1}{2k}e^{kz}\int_\infty^z e^{-kz_1}\widetilde{T}(z_1)dz_1 - \\ &-\frac{\beta_1}{2k}e^{-kz}\int_0^z e^{kz_1}\widetilde{T}(z_1)dz_1 + c_1e^{-kz}. \end{aligned} \tag{A.12}$$

Coefficient c_1 is from the solution of homogeneous equation (decaying at $z\to\infty$). This solution should be self-consistent, i.e., $\widetilde{u}_z(z=0) = \widetilde{u}_z(0)$, and it should satisfy boundary condition in Eq. (A.11). This results in:

$$\begin{aligned} &\frac{1}{4(1-\sigma)}\widetilde{u}_z(0) + \frac{\beta_1}{2k}\int_\infty^0 e^{-kz_1}\widetilde{T}(z_1)dz_1 + c_1 = \widetilde{u}_z(0), \\ &k\left(-c_1 + \frac{1-4\sigma}{4(1-\sigma)}\widetilde{u}_z(0)\right) + \frac{\beta_1}{2}\left(\int_\infty^0 e^{-kz_1}\partial_z\widetilde{T}(z_1)dz_1 - 2\widetilde{T}(0)\right) = 0. \end{aligned} \tag{A.13}$$

Resolving this couple of equations for c_1 and $\widetilde{u}_z(0)$, substituting β_1 from Eq. (34) and performing integration by parts we find:

$$\widetilde{u}_z(0) = \frac{2\beta(1+\sigma)}{3k}\left(\int_\infty^0 e^{-kz_1}\partial_z\widetilde{T}(z_1)dz_1 - \widetilde{T}(0)\right) = \frac{2\beta(1+\sigma)}{3}\int_\infty^0 e^{-kz_1}\widetilde{T}(z_1)dz_1\,. \tag{A.14}$$

$$c_1 = \frac{\beta(1+\sigma)}{3k(1-\sigma)}\left((1-2\sigma)\int_\infty^0 e^{-kz_1}\partial_z \tilde{T}(z_1)dz_1 - \frac{3-4\sigma}{2}\tilde{T}(0)\right) =$$
$$= \beta_1\left((1-2\sigma)\int_\infty^0 e^{-kz_1}\tilde{T}(z_1)dz_1 - \frac{\tilde{T}(0)}{2k}\right).$$

Let us give for reference purposes the resulting compact expression for $\tilde{u}_z$

$$\tilde{u}_z = \beta_1\left[\left(\frac{3-4\sigma}{2} + kz\right)e^{-kz}\int_\infty^0 e^{-kz_1}\tilde{T}(z_1)dz_1 + \right.$$
$$\left. + \frac{1}{2}\left(e^{-kz}\int_0^z e^{kz_1}\tilde{T}(z_1)dz_1 + e^{kz}\int_\infty^z e^{-kz_1}\tilde{T}(z_1)dz_1\right)\right]. \tag{A.15}$$

These expressions can be useful for dry laser cleaning problem with tightly focused beams, or with local field enhancement under the particle. In both cases the source term in the heat equation and temperature distribution are 3D, but *elasticity* can be considered quasi-statically. The approach developed here may prove useful also for the time-dependent elasticity. Similar considerations are known in acoustic studies (Dubois, 1994) with more complicated problems, but there numerical calculations were heavily involved. Our presentation provides compact, closed form results, especially for the measurable surface displacement.

One can simplify the results for given temperature distribution, or write the equation for the Fourier image of temperature from the heat equation and relate the displacement directly to the source term (or its Fourier image). This will be considered elsewhere. General result for *non-transformed* quantities can be obtained if temperature distribution is almost 1D, but is nevertheless limited in *x-y* directions. In this case stresses, strains, *and displacements* disappear at $x,\ y \to \infty$ and Fourier transforms of *all* quantities (in particular displacements) exist. In the last expression for $\tilde{u}_z(0)$ in Eq. (A.14) $\tilde{T}(k, z_1)$ significantly differs from zero only at small k. In this region exp ($-k\,z$) ~ 1, and Fourier transform can be inverted:

$$l \equiv -u_z(0) = \frac{2\beta(1+\sigma)}{3}\int_0^\infty T(z)dz. \tag{A.16}$$

This result for surface displacement is valid *independently* on the spatial profile of the laser beam in 3D static elasticity if all stresses relax at infinity and heat conduction is 1D in the sense that spatial temperature distribution in z-direction is much smaller than in x-y direction. It is *not equal* to pure 1D case, when there is *no* stress relaxation at infinity.

Appendix B. Cleaning threshold with the single sinusoidal pulse

The solution of the problem Eqs. (69)-(71) *during the pulse* is

$$h_1 = \frac{l\omega^2(\omega \sin\omega_0 t - \omega_0 \sin\omega t)}{2\pi\omega_0(\omega^2 - \omega_0^2)}, \quad \dot{h}_1 = \frac{l\omega^3(\cos\omega_0 t - \cos\omega t)}{2\pi(\omega^2 - \omega_0^2)}. \tag{B.1}$$

If particle detaches *after* the pulse, the total energy at the pulse end should be bigger than adhesion energy. In other words

$$\dot{h}_1^2(\tau) + \omega_0^2 h_1^2(\tau) > \omega_0^2 h_0^2 \Rightarrow \left| \frac{l\omega^3 \sin\dfrac{\omega_0\tau}{2}}{\pi(\omega^2 - \omega_0^2)} \right| > \omega_0 h_0 \Rightarrow \frac{l}{h_0} > \left| \frac{\pi y(1-y^2)}{\sin\pi y} \right|. \tag{B.2}$$

The situation is more complicated if the detachment occurs *during* the pulse. Turning points for h_1 are given by the condition

$$\dot{h}_1 = 0 \Rightarrow \cos y\omega t = \cos\omega t \Rightarrow y\omega t = 2\pi n \pm \omega t \Rightarrow \omega t = \frac{2\pi n}{y \mp 1}. \tag{B.3}$$

where n is integer number. The value of h_1 at these turning points is:

$$h_1(\dot{h}_1 = 0) = \frac{l}{2\pi}\frac{\sin y\omega t - y\sin\omega t}{y(1-y^2)} = \frac{l}{2\pi}\frac{(\pm 1 - y)\sin\omega t}{y(1-y^2)} = \frac{l}{2\pi}\frac{\sin\left(\dfrac{2\pi n}{y\mp 1}\right)}{y(y\pm 1)}. \tag{B.4}$$

Detachment during the pulse occurs if this expression is smaller than -h_0 for some n, in other words

$$\frac{l}{h_0} > \min_{n,\pm}\left(-\frac{2\pi y(y\pm 1)}{\sin\left(\dfrac{2\pi n}{y\mp 1}\right)}\right). \qquad \text{(B.5)}$$

The argument of the sine function (which is always smaller than 2π as long as we are within the pulse) should be as close to $3\pi/2$ as possible (then the value of sine is close to - 1). It can also be shown, that lower sign should be always preferred (numerator is smaller for " - " sign, while denominator is always close to "- 1"), This results in the condition

$$n = \left[\frac{3}{4}(y+1)\right], \qquad \text{(B.6)}$$

where [] denotes the closest integer number. Finally, it can be verified that with $y < 1$ Eq. (B.2) is always smaller than Eq. (B.5), which means that with $y < 1$ particle always detaches after the pulse. For $y>1$ the situation is the opposite and particle always detaches during the pulse, at $\omega t \approx 3\pi/2$ or $t/\tau \approx 3/4$, i.e., in the second half of the (symmetric) pulse. Combining these two cases, we arrive at the expression (73) in the text.

References

Almaz, *http://www.almazoptics.com/homepage/Si.htm*

Anisimov S. I., Khokhlov V. A., *Instabilities in Laser-Matter Interaction* (CRC Press, Boca Raton, 1995)

Arnold N., Schrems G., Mühlberger T., Bertsch M., Mosbacher M., Leiderer P., Bäuerle D., *Dynamic particle removal by ns dry laser cleaning: Theory*, Proc. SPIE, vol. **4426**, pp. 340-346 (2002a)

Arnold N., *Resonance and steep fronts effects in nanosecond dry laser cleaning*, To appear in: Appl. Surf. Sci. (2002b)

Ashcroft N. W., Mermin N. D., *Solid State Physics* (Holt, Rinehart & Winston, Philadelphia, 1976)

Assendel'ft E. Y., Beklemyshev V. I., Makhonin I. I., Petrov Y. N., Prokhorov A. M., Pustovoi V. I., *Optoacoustic effect on the desorption of microscopic particles from a solid surface into a liquid*, Sov. Tech. Phys. Lett., **14**(6), pp. 444-445 (1988)

Bangslabs, *http://www.bangslabs.com/technote/104.pdf*

Bäuerle D., *Laser Processing and Chemistry*, 3^{d} ed. (Springer, Berlin, 2000)

Bergström L., *Hamaker constants of inorganic materials*, Adv. Colloid Interface Sci., **70**, pp. 125-169 (1997)

Born M., Wolf E., *Principles of Optics* (Pergamon Press, Oxford, 1980)

Bowling R. A., *A theoretical review of particle adhesion*, In Mittal K. L. (Ed.), Particles on Surfaces, v.1, pp. 129-142 (Plenum Press, New York, 1988)

Bradley R. S., *The cohesive force between solid surfaces and the surface energy of solids*, Philos. Mag. **13**, pp. 853-862 (1932)

Carslaw H. S., Jaeger J. C., *Conduction of Heat in Solids* (Oxford Univ. Press, New York, 1959)

Chandrasekharaiah D. S. *Thermoelasticity with second sound: A review*. Appl. Mech. Rev., **39**, pp. 355-376 (1986)

Chandrasekharaiah D. S., *Hyperbolic thermoelasticity: A review of recent literature*, Appl. Mech. Rev., **51**(12), pp. 705-729 (1998)

Dahneke B., *The influence of flattening on the adhesion of particles*, J. Colloid Interface Sci., **40**(1), pp. 1-13 (1972)

Derjaguin B. V., Muller V. M., Toropov Yu. P., *Effect of contact deformations on the adhesion of particles*, J. Colloid Interface Sci., **53**(2), pp. 314-326 (1975)

Dobler V., Oltra R., Boquillon J. P., Mosbacher M., Boneberg J., Leiderer P., *Surface acceleration during dry laser cleaning of silicon*, Appl. Phys. A. **69**, pp. 335-337 (1999)

Dubois M., Enguehard F., Bertrand L., Choquet M., Monchalin J. P., *Modeling of laser thermoelastic generation of ultrasound in an orthotropic medium*, Appl. Phys. Lett., **64**(5), pp. 554-556 (1994)

Flina, *http://flina.com/silicon.html*

Fourrier T., Schrems G., Mühlberger T., Heitz J., Mosbacher M., Boneberg J., Leiderer P., Arnold N., Bäuerle D., *Laser cleaning of polymer surfaces*, Appl. Phys. A., **72,** pp. 1-6 (2001)

Goodfellow, *http://www.goodfellow.com/static/e/si61.html*

Gray D. E. (Ed.), *American Institute of Physics Handbook*, (McGraw-Hill, New York, 1972)

Greenwood J. A., Johnson K. L., *An alternative to the Maugis model of adhesion between elastic spheres*, J Phys. D: Appl. Phys., **31**, pp. 3279-3290, (1998)

Halfpenny R. D., Kane D. M., *A quantitative analysis of single pulse ultraviolet dry laser cleaning*, J. Appl. Phys., **86**(12), pp. 6641-6646 (1999)

Heim L. O., Blum J., Preuss M., Butt H. J., *Adhesion and friction forces between spherical micrometer-sized particles,* Phys. Rev. Lett., **83**(16), pp. 3328-3331 (1999) and ref.[26] therein

Heroux J. B, Boughaba S., Sacher E., Meunier M., *CO_2 laser-assisted particle removal from silicon surfaces*, Can. J. Phys. (Suppl.), **74**(1), pp. 95-99, (1996)

Hertz H., *Über die Berührung fester elastischer Körper*, J. Reine Angew. Math., **92**, pp. 156-171 (1881)

Johnson K. L, Kendall K., Roberts A. D., *Surface energy and the contact of elastic solids*, Proc. R. Soc. London Ser. A., **324**, pp. 301-313 (1971)

Kern W., *The evolution of silicon wafer cleaning technology*, J. Electrochemical Soc., **137**(6), pp. 1887-1991 (1990)

Kern W., in *Handbook of Semiconductor Wafer Cleaning Technology*, Ed. Kern W, (Noyes Publication, New Jersey, 1993)

Kolomenskii A. A., Schuessler H. A., Mikhalevich V. G., Maznev A. A., *Interaction of laser-generated surface acoustic pulses with fine particles: Surface cleaning and adhesion studies*, J. Appl. Phys., **84**(5), pp. 2404-2410 (1998)

Landau L. D., Lifshitz E. M., *Course of Theoretical Physics, v. V, Statistical Physics Part I*, (Pergamon Press, New York, 1980)

Landau L. D., Lifshitz E. M., *Course of Theoretical Physics, v. VII, Theory of elasticity*, (Pergamon Press, New York, 1986a), §§6,7

Landau L. D., Lifshitz E. M., *Course of Theoretical Physics, v. VII, Theory of Elasticity*, (Pergamon Press, New York, 1986b), §22, problem 3, Eq. (3)

Landau L. D., Lifshitz E. M., *Course of Theoretical Physics, v. VII, Theory of Elasticity*, (Pergamon Press, New York, 1986c), §35, Eq. (35.4)

Landau L. D., Lifshitz E. M., *Course of Theoretical Physics, v. VI, Fluid Mechanics*, (Pergamon Press, New York, 1987), §74, problem 1, p. 286, see also problems 3,4

Landold-Börnstein, *New Series III 17a Semiconductors* (Springer, Berlin, 1982)

Lee S. J., Imen K., Allen S. D., *Laser-assisted micron scale particle removal*, Appl. Phys. Lett., **58**(2), pp. 203-205 (1991)

Lee S. J., Imen K., Allen S. D., *Shock wave analysis of laser assisted particle removal*, J. Appl. Phys., **74**(12), pp. 7044-7047 (1993)

Leiderer P., Boneberg J., Mosbacher M., Schilling A., Yavas O., *Laser cleaning of silicon surfaces*, Proc. SPIE, vol. **3274**, pp. 68-78 (1998)

Leiderer P., Boneberg J., Dobler V., Mosbacher M., Münzer H. -J., Chaoui N., Siegel J., Solis J., Afonso C. N., Fourrier T., Schrems G., Bäuerle D., *Laser-induced particle removal from silicon wafers*, Proc. SPIE, **4065**, pp. 249-259 (2000)

Lifshitz E. M., Pitaevskii L. P., *Course of Theoretical Physics, v. X, Physical Kinetics*, (Pergamon Press, New York, 1981), §12, p. 41

Lu Y. F., Song W. D., Hong M. H., Teo B. S., Chong T. C., Low T. S., *Laser removal of particles from magnetic head sliders*, J. Appl. Phys. **80**(1), pp. 499-504 (1996)

Lu Y. F., Song W. D., Ang B. W., Hong M. H., Chan D. S. H., Low T. S., *A theoretical model for laser removal of particles from solid surfaces*, Appl. Phys. A, **65**(1), pp. 9-13 (1997a)

Lu Y. F., Song W. D., Ye K. D., Lee Y. D., Chan D. S. H., Low T. S., *A cleaning model for removal of particles due to laser-induced thermal expansion of substrate surface*, Jpn. J. Appl. Phys. **36**(2, 10A), L1304-L1306 (1997b)

Lu Y. F., Lee Y. P., Zhou M. S., *Laser cleaning of etch-induced polymers from via holes*, J. Appl. Phys., **83**(3), pp. 1677-1684 (1998a)

Lu Y. F., Song W. D., Tee C. K., Chan D. S. H., Low T. S., *Wavelength effects in laser cleaning process*, Jpn. J. Appl. Phys. **37**(1, 3a), pp. 840-844, (1998b)

Lu Y. F., Zheng Y. W., Song W. D., *An energy approach to the modelling of particle removal by pulsed laser irradiation*, Appl. Phys. A, **68**(5), pp. 569-572 (1999)

Lu Y. F., Zheng Y. W., Song W. D., *Characterization of ejected particles during laser cleaning*, J. Appl. Phys., **87**(1), pp. 549-552 (2000a)

Lu Y. F., Zheng Y. W., Song W. D., *Laser induced removal of spherical particles from silicon wafers*, J. Appl. Phys., **87**(1), 1534-1539 (2000b)

Lu Y. F., Zhang L., Song W. D., Zheng Y. W., Luk'yanchuk B. S., *Laser writing of sub-wavelength structure on silicon (100) surfaces with particle enhanced optical irradiation*, JETP Lett., **72**(9), pp. 457-459 (2000c)

Luk'yanchuk B. S., Zheng Y. W., Lu Y. F., *Laser cleaning of solid surface: Optical resonance and near-field effects*, Proc. SPIE, vol. **4065**, pp. 576-587 (2000)

Luk'yanchuk B. S., Zheng Y. W., Lu Y. F., *A new mechanism of laser dry cleaning*, Proc. SPIE, vol. **4423**, pp. 115-126 (2001)

Luk'yanchuk B. S., Zheng Y. W., Lu Y. F., *Basic physical problems related to dry laser cleaning*, RIKEN Review, No. **43**, pp. 28-34 (2002)

Maugis D., *Adhesion of spheres: the JKR-DMT transition using a Dugdale model*, J. Colloid Interface Sci., **150**(1), pp. 243-269, (1992)

Maznev A. A., Hohlfeld J., Güdde J., *Surface thermal expansion of metal under femtosecond laser irradiation*, J. Appl. Phys., **82**(10), pp. 5082-5085 (1997)

Mittal K. L., *Particles on Surfaces*, vol.1, (Plenum Press, New York, 1988)

Mittal K. L., *Particles on Surfaces*, vol. 4 (Marcel Dekker, Inc., New York, 1995)

Mosbacher M., Chaoui N., Siegel J., Dobler V., Solis J., Boneberg J., Afonso C. N., Leiderer P., *A comparison of ns and ps steam laser cleaning of Si surfaces*, Appl. Phys. A. **69** (Suppl.), pp. 331-334 (1999)

Mosbacher M., Dobler V., Boneberg J., Leiderer P., *Universal threshold for the steam laser cleaning of submicron spherical particles from silicon*, Appl. Phys. A, **70**(6), pp. 669-672 (2000)

Mosbacher M., Münzer H. –J., Zimmermann J., Solis J., Boneberg J., Leiderer P., *Optical field enhancement effects in laser-assisted particle removal*, Appl. Phys. A, **72**(1), pp. 41-44 (2001)

Muller V. M., Yushchenko V. S., Derjaguin B. V., *General theoretical consideration of the influence of surface forces on contact deformations and the reciprocal adhesion of elastic spherical particles*, J. Colloid Interface Sci., **92**(1), pp. 92-101, (1983)

Park H. K., Grigoropoulos C. P., Leung W. P., Tam A. C., *A practical excimer laser-based cleaning tool for removal of surface contaminants,* IEEE Trans Comp., Pack., and Manuf. Tech. A, **17**(4), pp. 631-644 (1994)

Parkus H., *Thermoelasticity* (Springer-Verlag, New York, 1976)

Prokhorov A. M., Konov V. I., Ursu I., Mihailescu I. N., *Laser Heating of Metals* (Adam Hilger, Bristol, 1990)

Schrems G., *Laser Cleaning von Polymeroberflächen*, Diploma Thesis (Linz, 2000)

She M., Kim D., Grigoropoulos C. P., *Liquid-assisted pulsed laser cleaning using near-infrared and ultraviolet radiation, J. Appl. Phys.*, **86**(11), pp. 6519-6524 (1999)

(SIA 1994), *The national Technology Roadmap for Semiconductors* (Semiconductor Industry Association), San Jose, CA, 1994, p. 116

Sokolnikoff I. S., *Mathematical theory of elasticity*, 2nd ed. (McGraw Hill, New York, 1956), §99

Spicer J. B., Hurley D. H., *Epicentral and near epicenter surface displacements on pulsed laser irradiated metallic surfaces*, Appl. Phys. Lett., **68**, pp. 3561-3563 (1996)

Tam A. C., Leung W. P., Zapka W., Ziemlich W., *Laser-cleaning techniques for removal of surface particulates*, J. Appl. Phys. **71**, pp. 3515-3523 (1992)

Tamma K. K., Namburu R. R., *Computational approaches with applications to non-classical and classical thermomechanical problems*, Appl. Mech. Rev., **50**(9), pp. 514-551 (1997)

Teutsch C. W., Miller C. F., Fournier C., In: Mittal K. L., *Particles on Surfaces*, vol. 3, (Plenum Press, New York, 1990), p. 173

Vereecke G., Röhr E., Heyns M. M., *Laser-assisted removal of particles on silicon wafers*, J. Appl. Phys., **85**(7), pp. 3837-3843 (1999)

Vicanek M., Rosch A., Piron F., Simon G., *Thermal deformation of a solid surface under laser irradiation*, Appl. Phys. A, **59**, pp. 407-412 (1994)

Visser J., *Adhesion of colloidal particles*. Chapter 1 in: Matijevich E. (Ed.), *Surface and Colloid Science*, vol. 8, (Wiley, New York, 1976), pp. 3-84

Weast R. C. (Ed.), *CRC Handbook of Chemistry and Physics*, 70th Ed., (CRC Press, Inc., Boca Raton, Florida, 1989)

Welsh L. P., Tuchman J. A., Herman I. P., *The importance of thermal stresses and strains induced in laser processing with focused Gaussian beams*, J. Appl. Phys., **64**(11), pp. 6274-6286 (1988)

Wu X., Sacher E., Meunier M., *The effects of hydrogen bonds on the adhesion of inorganic oxide particles on hydrophilic silicon surfaces*, J. Appl. Phys., **86**(3), pp. 1744-1748 (1999)

Wu X., Sacher E., Meunier M., *The modeling of excimer laser particle removal from hydrophilic silicon surfaces*, J. Appl. Phys., **87**(8), pp. 3618-3627 (2000) and refs. therein

Zapka W., Ziemlich W., Tam A. C., *Efficient pulsed laser removal of 0.2 μm sized particles from a solid surface*, Appl. Phys. Lett. **58**(20), pp. 2217-2219 (1991)

Zheng Y. W., Luk'yanchuk B. S., Lu Y. F., Song W. D., Mai Z. H., *Dry laser cleaning of particles from solid substrates: experiments and theory*, J. Appl. Phys., **90**(5), pp. 2135-2142 (2001)

Chapter 3

OPTICAL RESONANCE AND NEAR-FIELD EFFECTS IN DRY LASER CLEANING

B. S. Luk'yanchuk, M. Mosbacher, Y. W. Zheng, H. - J. Münzer, S. M. Huang, M. Bertsch, W. D. Song, Z. B. Wang, Y. F. Lu, O. Dubbers, J. Boneberg, P. Leiderer, M. H. Hong, T. C. Chong

Optical problems, related to the particle on the surface, i.e. optical resonance and near-field effects in laser cleaning are discussed. It is shown that the small transparent particle with size by the order of the wavelength may work as a lens in the near-field region. This permits to focus laser radiation into the area with the sizes, smaller than the radiation wavelength. It leads to 3D effects in surface heating and thermal deformation, which influences the mechanisms of the particle removal.

Keywords: Near-field effect, Optical resonance, Modeling, Dry laser cleaning, Threshold, Oscillations, SiO_2 particles, Si.

PACS: 42.62.Cf, 81.65.Cf, 81.07.Wx

1. Introduction

Up to now there is no completely satisfactory industrial solution for the surface cleaning involving submicrometer particles. The technique of laser cleaning, utilizing short-pulsed laser irradiation, sometimes in conjunction with the deposition of a liquid-film on the surface, has been studied and utilized since the late 1980's in several Institutes and R&D Centers (Beklemyshev, 1987; Assendel'ft, 1988 a, b; Kolomenskii, 1991; Zapka, 1991 a; Imen, 1991 a). Around 1990's a few successful examples for microelectronics and data-storage industrial applications were demonstrated (see patents of Zapka, 1989; Boykov, 1991; Imen, 1991 b).

There are two basic mechanisms of laser cleaning: the mechanism, based on evaporation of liquid film is called a "steam" laser cleaning, while laser cleaning without the usage of liquid film is called "dry" laser cleaning. It is considered that thermal expansion effects cause dry laser cleaning.

Previous examinations of the dry laser cleaning were investigated for two particular cases: 1) expansion of absorbing particle on the transparent substrate and 2) expansion (thermal deformation) of the absorbing substrate with non-absorbing particle (Bäuerle, 2000). The mutual influence of the particle and substrate was ignored in this approach. Nevertheless, the latest researches (Luk`yanchuk, 2000; Mosbacher, 2001) show that these feedback effects play an important role and suggest the understanding of dry laser cleaning in somewhat different terms than traditional approach. For example, the material optical properties influence the distribution of absorbed and scattered energy, which can be rather complex. In free space these distributions for the spherical particle can be found from the Mie theory (Born & Wolf, 1999; Barber & Hill, 1990).

As it was shown in the recent publications (Lu, 2000 a, b, c; Leiderer, 2000; Luk'yanchuk, 2000, 2001, 2002 a, b; Mosbacher, 2001, 2002; Zheng, 2001; Münzer, 2001, 2002) the small transparent particle can work as a focusing lens even at the particle size (radius a) comparable with radiation wavelength, λ. If one considers the particle as a perfect sphere, then in the Mie theory a size parameter $q = 2\pi a/\lambda$ appears. As this parameter changes, extinction and other scattering characteristics of the particle show oscillations, caused by optical resonance (Born & Wolf, 1999; Kerker, 1989). The optical resonance in microcavity is a subject of big interest, related to fluorescence and lasing in microspheres (Fields, 2000).

A small transparent contaminant particle on the surface works as a lens in the near-field region. It leads to 3D thermal expansion of the substrate, which is strongly different from the 1D thermal expansion model. Meanwhile this 1D model was considered for the last ten years as a basic model of dry laser cleaning.

The strong enhancement (about few tens) of laser intensity can be obtained within the region with size < 100 nm on the substrate under the particle. Naturally, this effect is very important for optical lithography and many other applications (Lu, 2000 c; Mosbacher, 2001, Huang, 2002 a).

The substrate strongly influenced the distribution of laser intensity in the near field region due to secondary scattering of reflected radiation (see in Fig. 1). The necessity to take this secondary scattering effect into account in laser cleaning has been declared many times (Donovan, 1988), nevertheless exact analytic solution of this problem is surprisingly rigorous (Bobbert & Vlieger, 1986 a). Nevertheless, a practical example of calculations with this solution for laser cleaning problem was done just recently (Luk'yanchuk, 2000).

Another way is direct numerical solution of the Maxwell equation by finite difference method (Wojcik, 1987; Mishenko, 2000). Although this method is universal, it can be applied to particles of different shapes, etc.; it needs powerful computers and is not flexible for the analysis of numerous experimental situations.

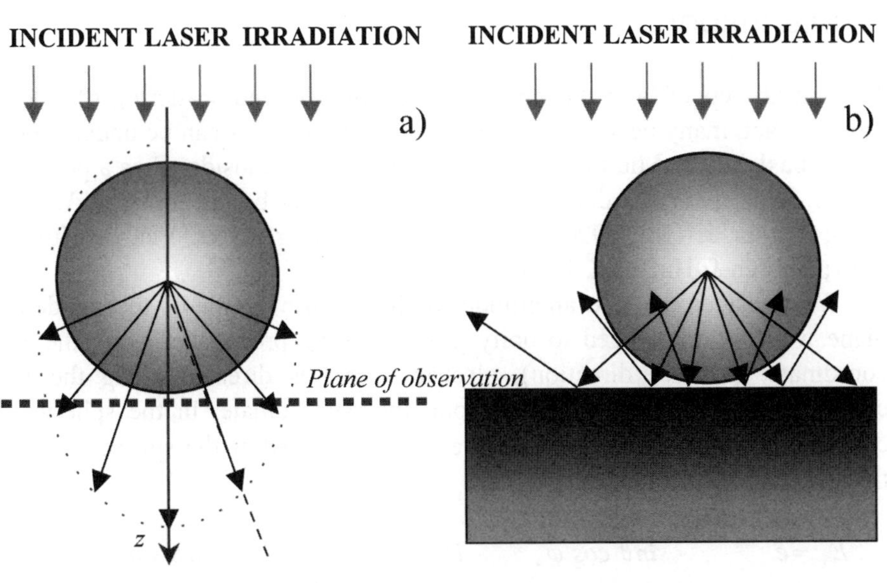

Fig. 1. Schematic for the particle scattering within the Mie theory, where the distribution of field is studied in arbitrary points, for example, in some observation plane (a). At typical consideration of the Mie theory (Born & Wolf, 1999) the incident plane wave propagates along the z-coordinate, and electric vector is directed along the x-coordinate. Particle on the surface (b) – the scattered radiation reflects from the surface and participates in the secondary scattering.

The methods of "intermediate power" like semianalytical field calculations using the Multiple Multipole (MMP) technique (Hafner, 1990) and ring multipoles (Zheng, 1990) should be mentioned. An example of calculations with these methods was done just recently (Münzer, 2002).

A careful calculation of intensity distribution permits to use more realistic "input" numbers for the further solution of the heat and thermal expansion problems. The goal of the given paper is to present the results of an examination of the mentioned optical problems and to discuss the basic influence of these effects on the efficiency of dry laser cleaning.

2. Optical resonance and near-field effects within the Mie theory

The initial step in laser cleaning is the absorption and scattering of laser light. In fact, many peculiarities of the scattering process can be understood on the basis of the Mie theory, where the particle is considered as a perfect sphere. This theory is discussed in detail in many books (Born & Wolf, 1999; Barber & Hill, 1990; Stratton, 1941; Kerker, 1969; Van de Hulst, 1981; Bohren & Huffman, 1983).

We consider that the amplitude of the electric vector of the incident plane wave is normalized to unity, and the wave propagates along the z-coordinate (positive direction), electric vector is directed along the x-coordinate, and magnetic vector along the y-coordinate. In the spherical coordinate system $\{r, \theta, \varphi\}$ with the origin, situated at the sphere center, this plane wave can be expressed as:

$$
\begin{aligned}
E_r &= e^{i k_m r \cos\theta} \sin\theta \cos\varphi, & H_r &= \sqrt{\varepsilon_m}\, e^{i k_m r \cos\theta} \sin\theta \sin\varphi, \\
E_\theta &= e^{i k_m r \cos\theta} \cos\theta \cos\varphi, & H_\theta &= \sqrt{\varepsilon_m}\, e^{i k_m r \cos\theta} \cos\theta \sin\varphi, \\
E_\varphi &= -e^{i k_m r \cos\theta} \sin\varphi, & H_\varphi &= \sqrt{\varepsilon_m}\, e^{i k_m r \cos\theta} \cos\varphi.
\end{aligned} \tag{1}
$$

Here ε_m is a complex dielectric permittivity of media, $\sqrt{\varepsilon_m} = n_m + i\kappa_m$. The wave vector of radiation in the media is indicated by $k_m = 2\pi\sqrt{\varepsilon_m}/\lambda$, where λ is the wavelength of radiation. In a similar way we shall indicate the corresponding values for the particle as $\sqrt{\varepsilon_p} = n_p + i\kappa_p$ and $k_p = 2\pi\sqrt{\varepsilon_p} / \lambda$. The wave vector for vacuum is indicated by $k_0 = 2\pi / \lambda$.

In terms of the spherical waves the fields (1) are expressed as following (index "*i*" indicates the incident wave):

$$E_r^{(i)} = \frac{\cos\varphi}{(k_m r)^2} \sum_{\ell=1}^{\infty} i^{\ell-1} (2\ell+1)\, \psi_\ell(k_m r) P_\ell^{(1)}(\cos\theta),$$

$$E_\theta^{(i)} = -\frac{\cos\varphi}{k_m r} \sum_{\ell=1}^{\infty} i^{\ell-1} \frac{2\ell+1}{\ell(\ell+1)} \left[\psi'_\ell(k_m r) P_\ell^{(1)'}(\cos\theta)\sin\theta - i\,\psi_\ell(k_m r) \frac{P_\ell^{(1)}(\cos\theta)}{\sin\theta} \right],$$

$$E_\varphi^{(i)} = -\frac{\sin\varphi}{k_m r} \sum_{\ell=1}^{\infty} i^{\ell-1} \frac{2\ell+1}{\ell(\ell+1)} \left[\psi'_\ell(k_m r) \frac{P_\ell^{(1)}(\cos\theta)}{\sin\theta} - i\,\psi_\ell(k_m r) P_\ell^{(1)'}(\cos\theta)\sin\theta \right],$$

$$H_r^{(i)} = \frac{\sqrt{\varepsilon_m}\,\sin\varphi}{(k_m r)^2} \sum_{\ell=1}^{\infty} i^{\ell-1} (2\ell+1)\, \psi_\ell(k_m r) P_\ell^{(1)}(\cos\theta), \qquad (2)$$

$$H_\theta^{(i)} = i\frac{\sin\varphi}{k_0 r} \sum_{\ell=1}^{\infty} i^{\ell-1} \frac{2\ell+1}{\ell(\ell+1)} \left[\psi_\ell(k_m r) \frac{P_\ell^{(1)}(\cos\theta)}{\sin\theta} + i\,\psi'_\ell(k_m r) P_\ell^{(1)'}(\cos\theta)\sin\theta \right],$$

$$H_\varphi^{(i)} = -i\frac{\cos\varphi}{k_0 r} \sum_{\ell=1}^{\infty} i^{\ell-1} \frac{2\ell+1}{\ell(\ell+1)} \left[\psi_\ell(k_m r) P_\ell^{(1)'}(\cos\theta)\sin\theta + i\,\psi'_\ell(k_m r) \frac{P_\ell^{(1)}(\cos\theta)}{\sin\theta} \right],$$

where the radial dependence is expressed through the Bessel function (regular at ρ = 0) and prime indicates differentiation

$$\psi_\ell(\rho) = \sqrt{\frac{\pi\rho}{2}}\, J_{\ell+\frac{1}{2}}(\rho), \quad \psi'_\ell(\rho) = \frac{\partial\psi_\ell(\rho)}{\partial\rho}. \qquad (3)$$

The angular dependence in (2) is related to spherical functions, where $P_n^{(m)}(x)$ are associated Legendre polynomials. There is a well-known problem important for numerical calculations and related to the cutting of sums in (2) by value $\ell \le \ell_{max}$ (Barber & Hill, 1990). The recommended values are given by $\ell_{\max} \approx q + 4q^{1/3} + 1$, where q is the corresponding size parameter. The scattered field for the non-magnetic particle ($\mu = 1$) immersed in vacuum is presented by (index "s" stands for indication of the scattered wave):

$$E_r^{(s)} = \frac{\cos\varphi}{(k_m r)^2} \sum_{\ell=1}^{\infty} \ell(\ell+1)\, {}^eB_\ell\, \zeta_\ell(k_m r)\, P_\ell^{(1)}(\cos\theta),$$

$$E_\theta^{(s)} = -\frac{\cos\varphi}{k_m r} \sum_{\ell=1}^{\infty} \left[{}^eB_\ell\, \zeta_\ell'(k_m r)\, P_\ell^{(1)'}(\cos\theta)\sin\theta - i\, {}^mB_\ell\, \zeta_\ell(k_m r) \frac{P_\ell^{(1)}(\cos\theta)}{\sin\theta} \right],$$

$$E_\varphi^{(s)} = -\frac{\sin\varphi}{k_m r} \sum_{\ell=1}^{\infty} \left[{}^eB_\ell\, \zeta_\ell'(k_m r) \frac{P_\ell^{(1)}(\cos\theta)}{\sin\theta} - i\, {}^mB_\ell\, \zeta_\ell(k_m r)\, P_\ell^{(1)'}(\cos\theta)\sin\theta \right], \tag{4}$$

$$H_r^{(s)} = \frac{\sqrt{\varepsilon_m}\,\sin\varphi}{(k_m r)^2} \sum_{\ell=1}^{\infty} \ell(\ell+1)\, {}^mB_\ell\, \zeta_\ell(k_m r)\, P_\ell^{(1)}(\cos\theta),$$

$$H_\theta^{(s)} = i\frac{\sin\varphi}{k_0 r} \sum_{\ell=1}^{\infty} \left[{}^eB_\ell\, \zeta_\ell(k_m r) \frac{P_\ell^{(1)}(\cos\theta)}{\sin\theta} + i\, {}^mB_\ell\, \zeta_\ell'(k_m r)\, P_\ell^{(1)'}(\cos\theta)\sin\theta \right],$$

$$H_\varphi^{(s)} = -i\frac{\cos\varphi}{k_0 r} \sum_{\ell=1}^{\infty} \left[{}^eB_\ell\, \zeta_\ell(k_m r)\, P_\ell^{(1)'}(\cos\theta)\sin\theta + i\, {}^mB_\ell\, \zeta_\ell'(k_m r) \frac{P_\ell^{(1)}(\cos\theta)}{\sin\theta} \right],$$

where

$$\zeta_\ell(\rho) = \rho\, h_\ell^{(1)}(\rho), \qquad \zeta_\ell'(\rho) = \frac{\partial \zeta_\ell(\rho)}{\partial \rho}. \tag{5}$$

Here $h_\ell^{(1)}$ is related to the Hankel function, i.e. the Bessel function of the third kind, which vanished at infinity

$$h_\ell^{(1)}(\rho) = H_{\ell+\frac{1}{2}}^{(1)}(\rho) = J_{\ell+\frac{1}{2}}(\rho) + iN_{\ell+\frac{1}{2}}(\rho), \tag{6}$$

where $N_\ell(\rho)$ is the Neumann function (designation $Y_\ell(\rho)$ is used in some books for this function).

Coefficients ${}^eB_\ell$ and ${}^mB_\ell$ in formulae (4) are given by

$$ {}^eB_\ell = i^{\ell+1}\frac{2\ell+1}{\ell(\ell+1)}a_\ell, \qquad {}^mB_\ell = i^{\ell+1}\frac{2\ell+1}{\ell(\ell+1)}b_\ell, \tag{7} $$

where a_ℓ and b_ℓ are defined as

$$ a_\ell = \frac{q_p\psi'_\ell(q_m)\psi_\ell(q_p)-q_m\psi_\ell(q_m)\psi'_\ell(q_p)}{q_p\varsigma'_\ell(q_m)\psi_\ell(q_p)-q_m\psi'_\ell(q_p)\varsigma_\ell(q_m)}, \quad q_m = k_m a, $$

$$ b_\ell = \frac{q_p\psi'_\ell(q_p)\psi_\ell(q_m)-q_m\psi_\ell(q_p)\psi'_\ell(q_m)}{q_p\psi'_\ell(q_p)\varsigma_\ell(q_m)-q_m\psi_\ell(q_p)\varsigma'_\ell(q_m)}, \quad q_p = k_p a. \tag{8} $$

The internal fields (indicated by index "a") inside the particle are given by

$$ E_r^{(a)} = \frac{\cos\varphi}{(k_p r)^2}\sum_{\ell=1}^{\infty}\ell(\ell+1)\,{}^eA_\ell\,\psi_\ell(k_p r)P_\ell^{(1)}(\cos\theta), $$

$$ E_\theta^{(a)} = -\frac{\cos\varphi}{k_p r}\sum_{\ell=1}^{\infty}\left[{}^eA_\ell\,\psi'_\ell(k_p r)P_\ell^{(1)'}(\cos\theta)\sin\theta - i\,{}^mA_\ell\,\psi_\ell(k_p r)\frac{P_\ell^{(1)}(\cos\theta)}{\sin\theta}\right], $$

$$ E_\varphi^{(a)} = -\frac{\sin\varphi}{k_p r}\sum_{\ell=1}^{\infty}\left[{}^eA_\ell\,\psi'_\ell(k_p r)\frac{P_\ell^{(1)}(\cos\theta)}{\sin\theta} - i\,{}^mA_\ell\,\psi_\ell(k_p r)P_\ell^{(1)'}(\cos\theta)\sin\theta\right], $$

$$ H_r^{(a)} = \frac{\sqrt{\varepsilon_p}\,\sin\varphi}{(k_p r)^2}\sum_{\ell=1}^{\infty}\ell(\ell+1)\,{}^mA_\ell\,\psi_\ell(k_p r)P_\ell^{(1)}(\cos\theta), \tag{9} $$

$$ H_\theta^{(a)} = i\frac{\sin\varphi}{k_0 r}\sum_{\ell=1}^{\infty}\left[{}^eA_\ell\,\psi_\ell(k_p r)\frac{P_\ell^{(1)}(\cos\theta)}{\sin\theta} + i\,{}^mA_\ell\,\psi'_\ell(k_p r)P_\ell^{(1)'}(\cos\theta)\sin\theta\right], $$

$$ H_\varphi^{(a)} = -i\frac{\cos\varphi}{k_0 r}\sum_{\ell=1}^{\infty}\left[{}^eA_\ell\,\psi_\ell(k_p r)P_\ell^{(1)'}(\cos\theta)\sin\theta + i\,{}^mA_\ell\,\psi'_\ell(k_p r)\frac{P_\ell^{(1)}(\cos\theta)}{\sin\theta}\right]. $$

Coefficients ${}^{e}A_{\ell}$ and ${}^{m}A_{\ell}$ in formulae (9) are given by

$$ {}^{e}A_{\ell} = i^{\ell+1}\frac{2\ell+1}{\ell(\ell+1)}c_{\ell}, \qquad {}^{m}A_{\ell} = i^{\ell+1}\frac{2\ell+1}{\ell(\ell+1)}d_{\ell}, \tag{10} $$

where c_{ℓ} and d_{ℓ} are defined as

$$ \begin{aligned} c_{\ell} &= -\frac{q_p\left[\varsigma_{\ell}'(q_m)\psi_{\ell}(q_m)-\varsigma_{\ell}(q_m)\psi_{\ell}'(q_m)\right]}{q_p\varsigma_{\ell}'(q_m)\psi_{\ell}(q_p)-q_m\psi_{\ell}'(q_p)\varsigma_{\ell}(q_m)}, \\ d_{\ell} &= \frac{q_p\left[\varsigma_{\ell}'(q_m)\psi_{\ell}(q_m)-\varsigma_{\ell}(q_m)\psi_{\ell}'(q_m)\right]}{q_p\psi_{\ell}'(q_p)\varsigma_{\ell}(q_m)-q_m\psi_{\ell}(q_p)\varsigma_{\ell}'(q_m)}. \end{aligned} \tag{11} $$

The time-averaged Poynting vector gives the power per unit of area carried by the wave; see e.g. (Stratton, 1941):

$$ \mathbf{S} = \frac{1}{2}\mathrm{Re}\left\{\mathbf{E}\times\mathbf{H}^{\bullet}\right\}. \tag{12} $$

The z-component of this vector for the plane wave (1) is given by $S_z = \left\langle \cos^2 \omega t \right\rangle = 1/2$. This value characterizing the homogeneous light intensity falls normally to $\{x, y\}$ plane. In some books (Barber & Hill, 1990) the light intensity is defined as

$$ I = \mathbf{E}\cdot\mathbf{E}^{*} \equiv \left|\mathbf{E}\right|^{2}. \tag{13} $$

Definitions $I = S_z$ and $I = \left|\mathbf{E}^2\right|$ yield the same *time-averaged* value for a purely transversal electromagnetic wave, e.g. for the plane wave (1). For the near-field region (with the longitudinal field components) these two intensities are different.

As it was mentioned a small transparent particle can work as a focusing lens even at the particle size, $2a$, comparable with radiation wavelength, λ. The enhanced laser intensity arises near the particle surface at the distances, which are small compared to λ (near-field effects). This behavior can be clearly seen from Fig. 2, where the intensity distributions are shown in $\{x, z\}$ and $\{y, z\}$ planes.

For practical applications it is important to understand how these distributions vary with radiation wavelengths. As an example Fig. 3 shows the distribution of laser intensity along the z - axis of the particle with radius $a = 0.5$ μm for radiation with $\lambda = 1064$, 532, 266 and 157 nm. The z - axis coincides with the direction of the wave vector for incident radiation. Particle is nonabsorbing ($\kappa = 0$) with refractive index $n = 1.6$. The background media is vacuum. Intensity is understood as a square of the electric vector. One can see from the Fig. 3 that both maximal intensities (inside and outside the particle) increase with a decrease of the radiation wavelength. Oscillations inside the particle (standing wave pattern) are resulting from interference between the refracted and internally reflected field components, while outside the particle they are caused by interference of incident and scattered radiation (Barber & Hill, 1990).

The maximal intensity out of the particle (see in Fig. 3 a) can be by the order of magnitude higher than the incident intensity. With $a \approx (2-3)\lambda$ this intensity may exceed the incident intensity by two orders of magnitude (see in Fig. 3 c, d). This maximal intensity is situated exactly on the surface (Fig. 3a, b, c) or below the particle (Fig. 3 d).

The distribution of laser intensity within the tangential $\{x, y\}$ - plane under the particle is shown in Fig. 4. One can see the high localization of laser intensity. It is clear that this effect can be used for optical lithography and nanopatterning of the surface (Lu, 2000c, 2002; Mosbacher, 2001; Huang, 2002 a, b) as well as for near-field microscopy (Münzer, 2001).

The extinction, absorption and scattering cross sections are given by $\sigma_{ext} = \pi a^2 Q_{ext}$, $\sigma_{abs} = \pi a^2 Q_{abs}$, $\sigma_{sca} = \pi a^2 Q_{sca}$, where related efficiencies Q for polarized and non-polarized light are presented by (Born & Wolf, 1999):

$$Q_{ext} = \frac{2}{q^2}\sum_{\ell=1}^{\infty}(2\ell+1)Re(a_\ell + b_\ell), \quad Q_{sca} = \frac{2}{q^2}\sum_{\ell=1}^{\infty}(2\ell+1)\left\{|a_\ell|^2 + |b_\ell|^2\right\}, \tag{14}$$

$$\overline{cos\,\theta}\cdot Q_{sca} = \frac{4}{q^2}\sum_{\ell=1}^{\infty}\frac{\ell(\ell+2)}{\ell+1}Re\left(a_\ell a_{\ell+1}^* + b_\ell b_{\ell+1}^*\right) + \frac{4}{q^2}\sum_{\ell=1}^{\infty}\frac{2\ell+1}{\ell(\ell+1)}Re\left(a_\ell b_\ell^*\right),$$

$$Q_{abs} = Q_{ext} - Q_{sca}, \quad Q_{pr} = Q_{ext} - \overline{\cos\theta}\cdot Q_{sca},$$

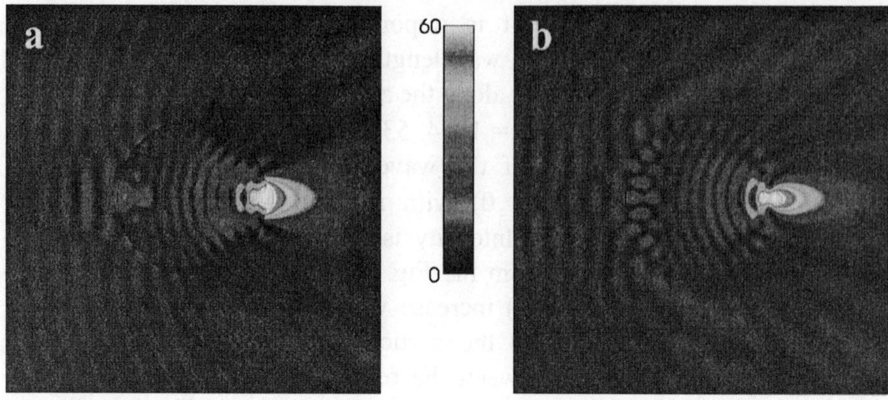

Fig. 2. Intensity distribution, $I = |E|^2$, inside and outside the 1 μm SiO_2 particle, illuminated by radiation with λ = 266 nm, and polarization parallel (a) and perpendicular (b) to image plane. The maximum intensity enhancement in calculations is about 60 for both regions.

where the star indicates a complex conjugation, and size parameter is given by $q = k_0 a = 2\pi a/\lambda$. Q_{pr} describes the effects of radiation pressure (van de Hulst, 1981).

The extinction versus size parameter, q, demonstrates the "low frequency" transition oscillations, and "high frequency" modulation, which can be seen in Fig. 5, where the dependence $Q_{ext}(q)$ is shown for sphere with n = 1.6. At a very big size parameter $q \to \infty$ extinction tends to value $Q_{ext} = 2$; this is the so-called "extinction paradox". The oscillations are related to excitation of partial **E** and **H** resonance modes (Born & Wolf, 1999). The extinction is the integral characteristic caused by far-field scattering. This scattering diagram in *x-y* plane is given by modes (Born & Wolf, 1999):

$$I_{\parallel}^{(far)} = \left(\frac{\lambda}{2\pi r}\right)^2 \left| \sum_{\ell=1}^{\infty} (-i)^{\ell} \left[{}^{e}B_{\ell} P_{\ell}^{(1)'}(\cos\theta)\sin\theta - {}^{m}B_{\ell} \frac{P_{\ell}^{(1)}(\cos\theta)}{\sin\theta} \right] \right|^2 , \tag{15}$$

$$I_{\perp}^{(far)} = \left(\frac{\lambda}{2\pi r}\right)^2 \left| \sum_{\ell=1}^{\infty} (-i)^{\ell} \left[{}^{e}B_{\ell} \frac{P_{\ell}^{(1)}(\cos\theta)}{\sin\theta} - {}^{m}B_{\ell} P_{\ell}^{(1)'}(\cos\theta)\sin\theta \right] \right|^2 .$$

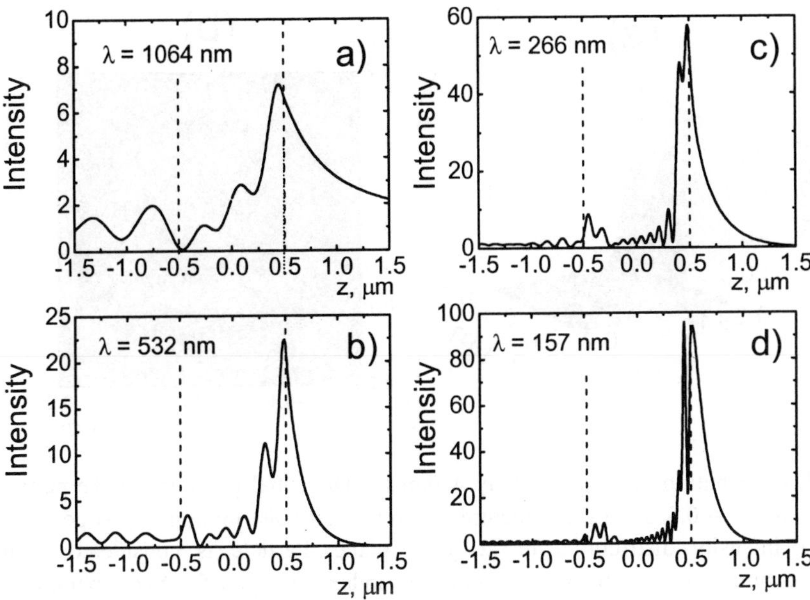

Fig. 3. Distribution of laser intensity $I = \left|\mathbf{E}^2\right|$ inside and outside the particle with radius a = 0.5 μm for different radiation wavelength λ. Particle is considered to be nonabsorbing (κ = 0) with refractive index n = 1.6. Background media is vacuum. Intensity is understood as a square of the electric vector.

Formulae (15) follow from the asymptotic expansion of the electric field in the far-field region $r \gg \lambda$ (we use indexes $^{(far)}$ and $^{(nf)}$ for indication far field and near field distributions). Within the near field region, where $r \sim \lambda$, instead of (15) one should write the exact formulae for the field components:

$$I_{\parallel}^{(nf)} = \left(\frac{\lambda}{2\pi r}\right)^2 \left| \sum_{\ell=1}^{\infty} (-i)^{\ell} \left[{}^{e}B_{\ell}\zeta_{\ell}'(kr) P_{\ell}^{(1)'}(\cos\theta)\sin\theta - i\, {}^{m}B_{\ell}\zeta_{\ell}(kr)\frac{P_{\ell}^{(1)}(\cos\theta)}{\sin\theta} \right] \right|^2$$

$$I_{\perp}^{(nf)} = \left(\frac{\lambda}{2\pi r}\right)^2 \left| \sum_{\ell=1}^{\infty} (-i)^{\ell} \left[{}^{e}B_{\ell}\zeta_{\ell}'(kr)\frac{P_{\ell}^{(1)}(\cos\theta)}{\sin\theta} - i\, {}^{m}B_{\ell}\zeta_{\ell}(kr) P_{\ell}^{(1)'}(\cos\theta)\sin\theta \right] \right|^2 \tag{16}$$

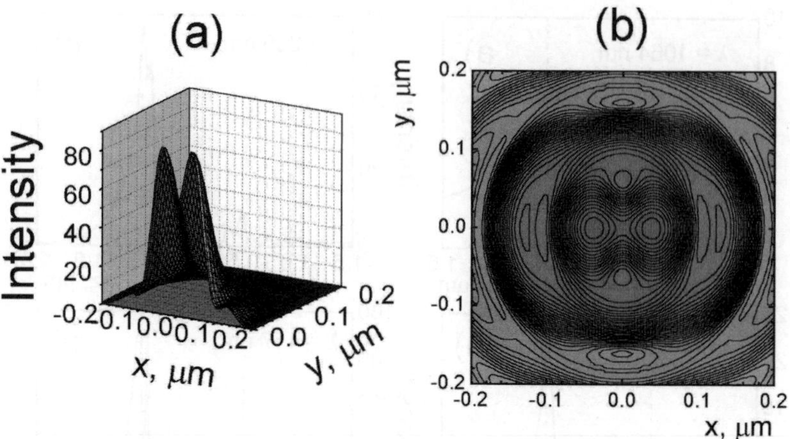

Fig. 4. Distribution of laser intensity within the tangential plane under the particle with radius a = 0.5 μm. (a) 3D picture of the intensity distribution. (b) Topography of the intensity distribution in $\{x, y\}$ - plane. Particle is considered to be nonabsorbing ($\kappa = 0$) with refractive index $n = 1.5$. Size parameter: $q = 2\pi\ a/\lambda = 20$, i.e. $\lambda \approx 157$ nm. Intensity is understood as a square of the electric vector.

These diagrams are shown in Fig. 6. One can see a big difference in the far field and near field intensity distributions, e.g. the main directional lobe in the near field region is narrower in the far-field region, new lobes appear in back scattering diagram, etc.

According to this diagram one would expect that optical resonance strongly influenced the near-field scattering characteristics compared to far-field characteristics. As an example, in Fig. 7 the laser radiation intensity under the transparent particle is shown. The calculation is performed on the basis of the Mie theory. This intensity is clearly a near-field characteristic. One can see variations of this intensity are by the order of magnitude higher than in the extinction (far-field characteristic) shown in Fig. 5. Big variations of intensity inside the particle (see in Fig. 3) are important for nonlinear optics in microspheres (Fields, 2000).

The concept of the optical resonance suggests high variations in cleaning efficiency with a small change in size parameter. It is different from conventional point of view (Lu, 2000 a; Mosbacher, 2001), which suggests the dry laser cleaning efficiency varies monotonously with particle size. A similar comment can be done with respect to increase of cleaning

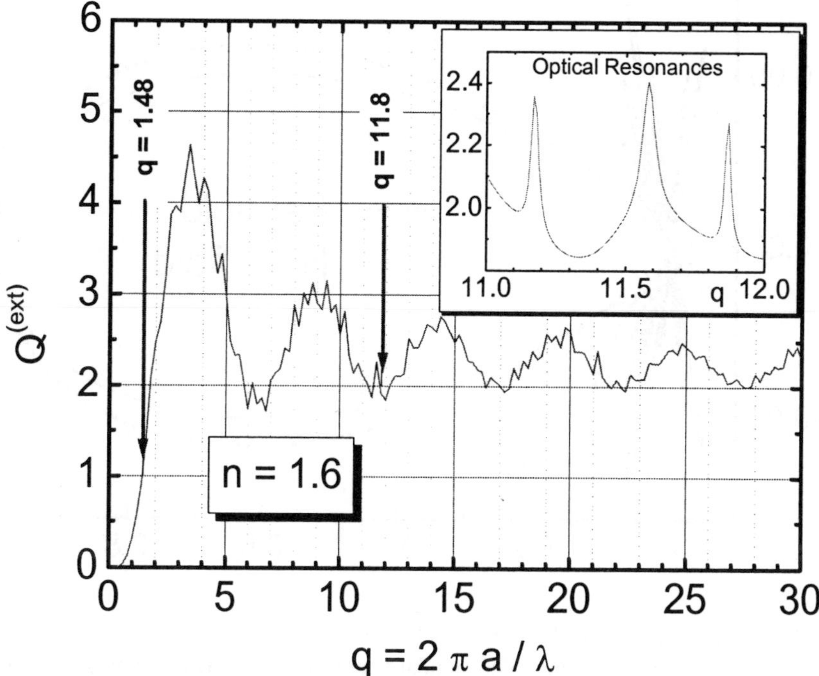

Fig. 5. The extinction coefficient $Q^{(ext)}$ versus size parameter q is calculated with the help of (14) for a spherical silica particle with $n = 1.6$. The arrows indicate particular values, examined in the paper of Lu, 2000 a: $q = 1.48$ ($a = 0.25$ μm, $\lambda = 1.064$ μm) and $q = 11.81$ ($a = 0.5$ μm, $\lambda = 0.266$ μm). The insertion shows the resonance structure within the range $11 \le q \le 12$.

efficiency for shorter wavelength radiation (Lu, 2000). Fig. 8 shows a big effect in the intensity distribution within the range of size parameter variation between the two nearest optical resonances (maximal intensity varies twice, when the q varies just of 2 % !). Once again, we should remind that under the approximation of the Mie theory one ignores the secondary scattering effects, produced by radiation reflected from the substrate surface, see in Fig. 1.

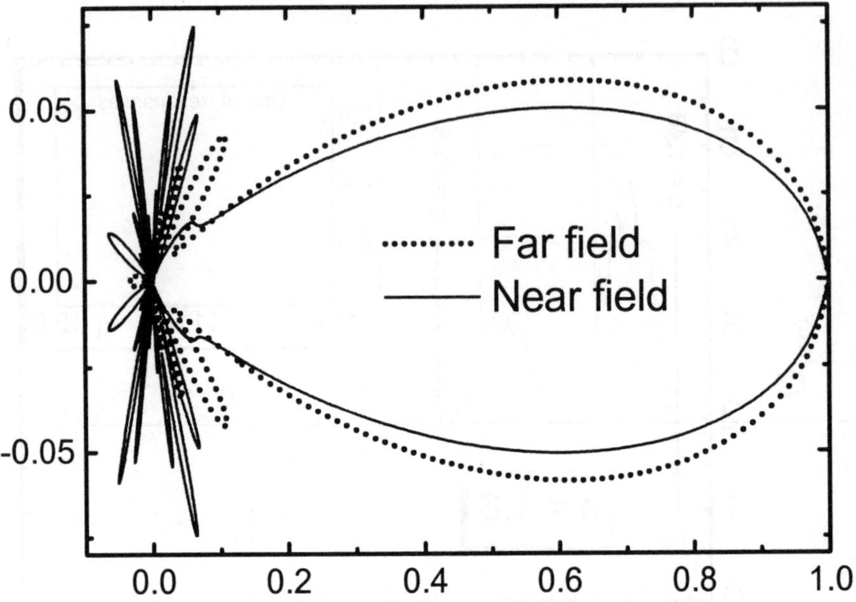

Fig. 6. The polar diagrams for the far field ($r \gg \lambda$) and near field ($r = a \sim \lambda$, $a = 0.5$ µm, $\lambda = 0.266$ µm) scattering. The standard representation (Born & Wolf, 1999) is used, i.e. the polar angle corresponds to θ, while the radius vector in the plots is proportional to the perpendicular intensity $I_{\perp}$. The diagrams are normalized to unity in the direction of direct scattering, i.e. at $\theta = 0$.

3. Particle on the surface. Beyond the Mie theory

The effect of secondary scattering can be qualitatively understood under the following simplification. Let us consider that reflected radiation is presented by a plane wave (in reality it is a spherical wave). This yields:

$$I = \frac{I_m(0)}{1 - R_0\, I_m(\pi)}, \tag{17}$$

where R_0 is reflection coefficient, and $I_m(\theta)$ is a distribution of intensity versus polar angle from the Mie theory.

Although formula (17) is applicable just for $R_0 I_m(\pi) < 1$ and exceeds the true value of intensity, it shows by a correct way the main effect of reflection, i.e. a fast increase of field enhancement at resonant points with increase of size parameter.

One can see in Fig. 7 important consequences of the substrate reflection onto the optical resonance effect. Oscillations of the intensity versus size parameter become more pronounced with a higher surface reflection coefficient. The effect can be seen even with a small reflection $R_0 = 0.02$. The surface played the role of a resonator mirror for a spherical cavity. This, in turn, leads to a sharpening in intensity distribution. Inhomogeneity in laser intensity leads to temperature distribution inhomogeneity, producing a "hot point" under the particle. It results in 3D-thermoelastic deformations, which are quite different from conventional 1D thermal expansion model (Kelly, 1993; Lu, 1997). The important limitation of 1D model is that it does not permit a fast backward motion of the substrate surface. As a result, 1D model predicts threshold fluences for laser cleaning, which exceed the experimental values by the order of magnitude (Luk'yanchuk, 2001, 2002c; Zheng, 2001; Arnold, 2002).

Formula (17) is quite crude and valid just for qualitative consideration. For quantitative analysis one can use the exact solution of the problem "particle on the surface". Bobbert & Vliger, 1986 a, found this solution. Although this solution is rigorous, the idea of the solution is rather simple.

Let a wave $\mathbf{V}^{in}$ (e.g. a plane wave) be incident on this system. If the sphere was absent we could satisfy the boundary conditions at the interface between the ambient and the substrate by adding a wave $\mathbf{V}^{R}$ (just Fresnel reflection in the case of a plane wave). In the presence of the sphere there will be an additional scattered wave $\mathbf{W}^{S}$ as a result of the currents flowing inside the sphere. But this wave will also be reflected by the substrate – i.e. induce currents flowing inside the substrate and will give rise to a secondary reflected scattered wave $\mathbf{V}^{SR}$. The fields $\mathbf{V}^{SR}$ and $\mathbf{W}^{S}$, once again, should be linearly related by some matrix $\hat{\mathbf{A}}$, characterizing the reflection of spherical waves by the substrate:

$$\mathbf{V}^{SR} = \hat{\mathbf{A}} \cdot \mathbf{W}^{S}. \qquad (18)$$

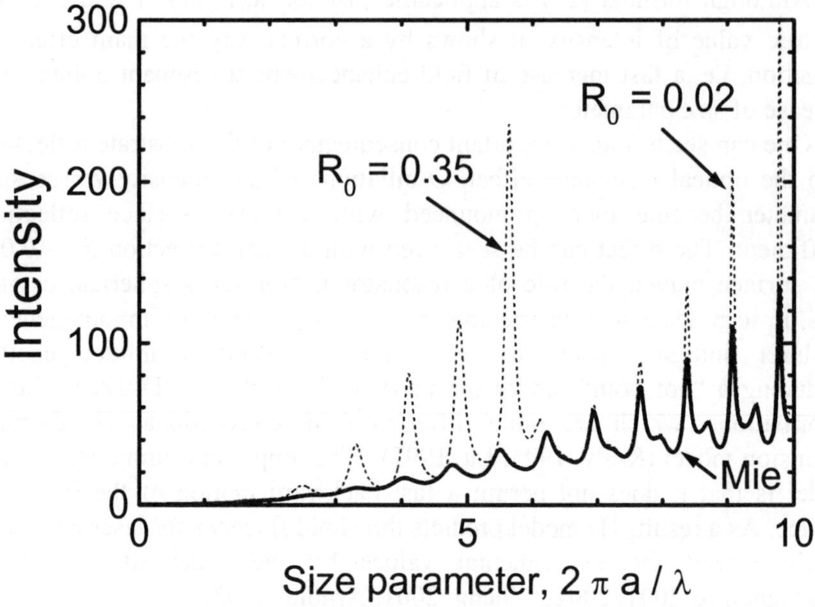

Fig. 7. The intensity distribution under the particle with refractive index n = 1.6. Solid line presents solution from the Mie theory. Dashed lines show variation in intensity, caused by reflection of substrate. These lines are according to approximating formula (17) for small (R_0 = 0.02) and high (R_0 = 0.35) substrate reflection coefficients.

Consider (18) together with overall equation

$$\mathbf{W}^S = \hat{\mathbf{B}} \cdot \left(\mathbf{V}^{in} + \mathbf{V}^R + \mathbf{V}^{SR}\right). \tag{19}$$

one can easily find the formal solution in terms of $\mathbf{V}^{in}$ and $\mathbf{V}^{\mathrm{R}}$ vectors:

$$\mathbf{W}^{\mathrm{S}} = (\hat{1} - \hat{\mathbf{B}}\hat{\mathbf{A}})^{-1} \cdot \hat{\mathbf{B}} \cdot (\mathbf{V}^{\mathrm{in}} + \mathbf{V}^{\mathrm{R}})\,. \tag{20}$$

Thus, the technical problem is related to the calculation of "reflection matrix", $\hat{\mathbf{A}}$, and the inverse matrix $\left(\hat{\mathbf{1}} - \hat{\mathbf{B}}\hat{\mathbf{A}}\right)^{-1}$ in the above equation. The

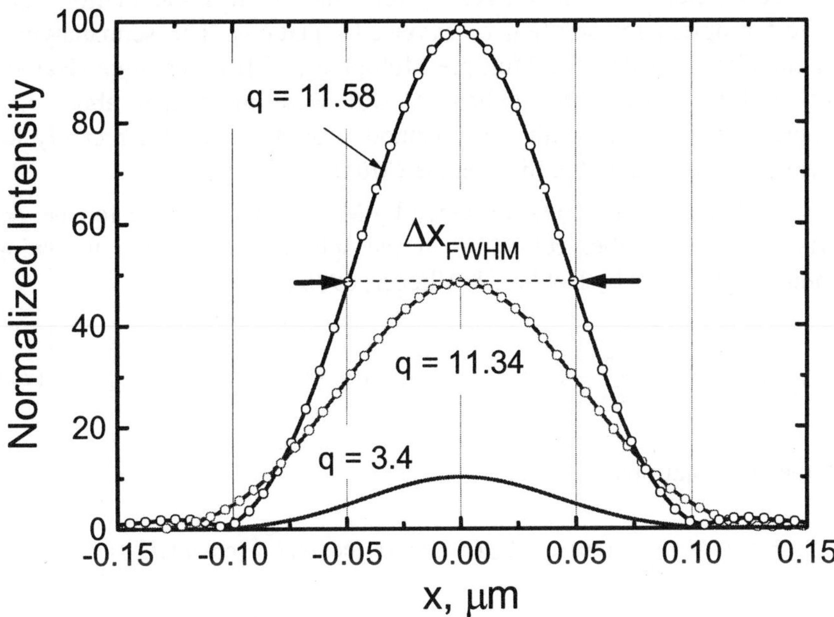

Fig. 8. The intensity distributions (z-component of the Poynting vector) along the x-coordinate for different size parameters q. Values $q = 11.58$ and $q = 11.34$ correspond to the nearest maximal and minimal values of $Q^{(ext)}$ within particular optical resonance (see insertion in Fig. 5). We use in calculations $\lambda = 0.266$ μm and the variation of the size parameter was due to a variation in the particle size. For two upper curves particle sizes were, $2a = 0.98$ and 0.96 μm, respectively.

numerical calculations with exact formula (20) were presented by (Luk'yanchuk, 2000) for the SiO_2 particles on the silicon substrate. In papers of (Bobbert, 1986 a, b) authors used the expansion of inverse matrix. Some particular cases were also analyzed: perfectly conducting substrate or far-away scattered field. Examples of practical calculations for Si particles on the silicon substrate were presented by Wojcik, 1987. They were done with the help of discrete numerical solutions of Maxwell's equations. This direct way needs a powerful computer or even supercomputer (calculations were performed with CRAY 2 supercomputer). Although during the last decade there was big progress with these computations (Mishenko, 2000), the exact solutions are still interesting for both, practical calculations, and as a test problem for solution of Maxwell equations. The semianalytical field calculations using the Multiple Multipole (MMP) technique (Hafner, 1990) and ring multipoles (Zheng, 1990) were mentioned above. An example of calculations with these methods was demonstrated recently for polystyrene particles on the Si substrate (Münzer, 2002).

The $\hat{\mathbf{B}}$-matrix in (20) is presented by Mie formulae, while for the $\hat{\mathbf{A}}$-matrix (which describes reflection of the spherical wave) the following formulae were found by Bobbert & Vlieger, 1986 a:

$$A_{\ell',m';\ell,m} = i^{\ell'-1}\sqrt{\frac{2\ell'+1}{\ell'(\ell'+1)}}\,(-1)^{m-1}\,\delta_{mm'}\int_0^{\pi/2-i\infty} d\alpha\,\sin\alpha\; e^{2iq\cos\alpha}\, a^m_{\ell',f';\ell,f}\;, \quad (21)$$

with the abbreviations

$$a^m_{\ell',e;\ell,e} = r_p(\cos\alpha)\,\tilde{V}^m_\ell(\cos\alpha)\,d^{\ell'}_{m,-}(\pi-\alpha) + r_s(\cos\alpha)\,\tilde{U}^m_\ell(\cos\alpha)\,d^{\ell'}_{m,+}(\pi-\alpha),$$

$$a^m_{\ell',h;\ell,e} = -i\,\{r_p(\cos\alpha)\,\tilde{V}^m_\ell(\cos\alpha)\,d^{\ell'}_{m,+}(\pi-\alpha) + r_s(\cos\alpha)\,\tilde{U}^m_\ell(\cos\alpha)\,d^{\ell'}_{m,-}(\pi-\alpha)\},$$

$$a^m_{\ell',e;\ell,h} = i\,\{r_p(\cos\alpha)\,\tilde{U}^m_\ell(\cos\alpha)\,d^{\ell'}_{m,-}(\pi-\alpha) + r_s(\cos\alpha)\,\tilde{V}^m_\ell(\cos\alpha)\,d^{\ell'}_{m,+}(\pi-\alpha)\},$$

$$a^m_{\ell',h;\ell,h} = r_p(\cos\alpha)\,\tilde{U}^m_\ell(\cos\alpha)\,d^{\ell'}_{m,+}(\pi-\alpha) + r_s(\cos\alpha)\,\tilde{V}^m_\ell(\cos\alpha)\,d^{\ell'}_{m,-}(\pi-\alpha), \quad (22)$$

$$\tilde{U}^m_\ell = i^{1-\ell}\,\frac{\ell-1}{2}\left\{\sqrt{\frac{(\ell+m-1)(\ell+m)}{(2\ell-1)(2\ell+1)}}\,\tilde{P}^{m-1}_{\ell-1} - \sqrt{\frac{(\ell-m-1)(\ell-m)}{(2\ell+1)(2\ell-1)}}\,\tilde{P}^{m+1}_{\ell+1}\right\}$$

$$-\,i^{-1-\ell}\,\frac{\ell+2}{2}\left\{\sqrt{\frac{(\ell-m+1)(\ell+m+2)}{(2\ell+3)(2\ell+1)}}\,\tilde{P}^{m-1}_{\ell-1} + \sqrt{\frac{(\ell-m+1)(\ell+m+2)}{(2\ell+3)(2\ell+1)}}\,\tilde{P}^{m+1}_{\ell+1}\right\}, \quad (23)$$

$$\widetilde{V}_\ell^{\,m} = \frac{1}{2} i^{1-\ell} \left\{ \sqrt{(\ell - m + 1)(\ell + m)}\ \widetilde{P}_\ell^{\,m-1} - \sqrt{(\ell + m + 1)(\ell - m)}\ \widetilde{P}_\ell^{\,m+1} \right\},$$

where tilda indicates the normalized Legendre functions $\widetilde{P}_\ell^m(\cos\alpha) = \sqrt{(2\ell+1)\dfrac{(\ell-m)!}{(\ell+m)!}}\ P_\ell^m(\cos\alpha)$.

The function $d^{\,\ell}_{m,m'}(\alpha)$ is defined as:

$$d^{\,\ell}_{m,m'}(\alpha) \equiv \sqrt{(\ell+m')!(\ell-m')!(\ell+m)!(\ell-m)!}$$
$$\sum_k (-1)^k \frac{(\cos\frac{\alpha}{2})^{2\ell+m'-m-2k}(\sin\frac{\alpha}{2})^{m-m'+2k}}{(\ell-m-k)!(\ell+m'-k)!(k+m-m')!k!} \tag{24}$$

and indexes plus and minus indicate the symmetrical and antisymmetrical combinations

$$\begin{aligned} d^{\,\ell'}_{m,-}(\pi-\alpha) &= \frac{1}{2}\left\{d^{\,\ell'}_{m,1}(\pi-\alpha) - d^{\,\ell'}_{m,-1}(\pi-\alpha)\}\right\}, \\ d^{\,\ell'}_{m,+}(\pi-\alpha) &= \frac{1}{2}\left\{d^{\,\ell'}_{m,1}(\pi-\alpha) + d^{\,\ell'}_{m,-1}(\pi-\alpha)\right\}. \end{aligned} \tag{25}$$

The $r_p(\cos\alpha)$ and $r_s(\cos\alpha)$ functions in (22) present the Fresnel reflection coefficients for parallel and perpendicular polarization. Coefficients d arise due to rotation matrix, for each plane wave propagating in another direction than the z-direction we need to express the basic Mie solution in rotated coordinate system (r, θ', φ') in terms of those in the coordinate system (r, θ, φ), i.e. transform the spherical harmonics. This transform can be done with the help of addition theorem for Legender polynomials, which yields (Rose, 1957):

$$Y_\ell^{m'}(\theta', \varphi') = \sum_{m=-\ell}^{\ell} (-1)^{m-m'} D^{\,\ell}_{m,m'}(\beta, \alpha, \gamma)\, Y_\ell^m(\theta, \varphi),$$
$$D^{\,\ell}_{m,m'}(\beta, \alpha, \gamma) = e^{-im\beta}\, d^{\,\ell}_{m,m'}(\alpha)\, e^{-im'\gamma}. \tag{26}$$

Here rotation matrix D is expressed through the Euler angles (α,β,γ) arising during rotation of (r,θ,φ) system to obtain the (r,θ',φ') system. The argument $\pi-\alpha$ in formulae (22) and (25) expresses the angle of reflected light. Indexes f and f' in (21) indicate transitions between the electrical, $f=e$, and magnetic, $f=h$, components.

Suppose the incident wave vector $\mathbf{k}$ to be in the x-z plane of the coordinate system $\mathbf{k}=k\,(\sin\theta_i,\,0,\,\cos\theta_i)$, k is the absolute value of the wave-vector, θ_i is the incident angle, and amplitude $|\mathbf{E}|=1$. If the plane wave is p-polarized (electric vector in the plane of incidence i.e. x-z plane), $\mathbf{V}^{in}$ and $\mathbf{V}^{\mathrm{R}}$ have components, see in Bobbert & Vlieger, 1986 a:

$$
\begin{aligned}
{}^{e}v_{\ell}^{m\,\mathrm{in}} &= \frac{i^{\ell-1}}{k}\sqrt{\frac{2\ell+1}{\ell(\ell+1)}}\,(-1)^{m-1}\,d^{\ell}_{m,-}(\theta_i),\\
{}^{h}v_{\ell}^{m\,\mathrm{in}} &= -\frac{i^{\ell}}{k}\sqrt{\frac{2\ell+1}{\ell(\ell+1)}}\,(-1)^{m-1}\,d^{\ell}_{m,+}(\theta_i)\\
{}^{e}v_{\ell}^{m\,\mathrm{R}} &= \frac{i^{\ell-1}}{k}\,e^{2iq\cos\theta_i}\,r_p\,(\cos\theta_i)\sqrt{\frac{2\ell+1}{\ell(\ell+1)}}\,(-1)^{\ell}\,d^{\ell}_{m,-}(\theta_i),\\
{}^{h}v_{\ell}^{m\,\mathrm{R}} &= -\frac{i^{\ell}}{k}\,e^{2iq\cos\theta_i}\,r_p\,(\cos\theta_i)\sqrt{\frac{2\ell+1}{\ell(\ell+1)}}\,(-1)^{\ell-1}\,d^{\ell}_{m,+}(\theta_i).
\end{aligned}
\tag{27}
$$

If it is s-polarized (electric vector normal to the plane of incidence) $\mathbf{V}^{\mathrm{in}}$ and $\mathbf{V}^{\mathrm{R}}$ have following components:

$$
\begin{aligned}
{}^{e}v_{\ell}^{m\,\mathrm{in}} &= -\frac{i^{\ell}}{k}\sqrt{\frac{2\ell+1}{\ell(\ell+1)}}\,(-1)^{m-1}\,d^{\ell}_{m,+}(\theta_i),\\
{}^{h}v_{\ell}^{m\,\mathrm{in}} &= \frac{i^{\ell+1}}{k}\sqrt{\frac{2\ell+1}{\ell(\ell+1)}}\,(-1)^{m-1}\,d^{\ell}_{m,-}(\theta_i),\\
{}^{e}v_{\ell}^{m\,\mathrm{R}} &= -\frac{i^{\ell}}{k}\,e^{2iq\cos\theta_i}\,r_s(\cos\theta_i)\sqrt{\frac{2\ell+1}{\ell(\ell+1)}}(-1)^{\ell-1}d^{\ell}_{m,+}(\theta_i),\\
{}^{h}v_{\ell}^{m\,\mathrm{R}} &= \frac{i^{\ell+1}}{k}\,e^{2iq\cos\theta_i}\,r_s(\cos\theta_i)\sqrt{\frac{2\ell+1}{\ell(\ell+1)}}\,(-1)^{\ell}\,d^{\ell}_{m,-}(\theta_i).
\end{aligned}
\tag{28}
$$

With the incident, reflected and scattered waves available, the Debye potential with respect to the *total* electromagnetic field (which is necessary for calculation of Poynting vector), can be expressed as:

$$
\begin{aligned}
{}^{e}D(r) &= \sum_{\ell=1}^{\infty} \sum_{m=-\ell}^{\ell} \left\{ \left({}^{e}v_{\ell}^{m\,\mathrm{in}} + {}^{e}v_{\ell}^{m\,\mathrm{R}} + {}^{e}v_{\ell}^{m\,\mathrm{SR}} \right) \Psi_{\ell}^{m}(r) + {}^{e}w_{\ell}^{m\,\mathrm{S}}\, \Pi_{\ell}^{m}(r) \right\}, \\
{}^{h}D(r) &= \sum_{\ell=1}^{\infty} \sum_{m=-\ell}^{\ell} \left\{ \left({}^{h}v_{\ell}^{m\,\mathrm{in}} + {}^{h}v_{\ell}^{m\,\mathrm{R}} + {}^{h}v_{\ell}^{m\,\mathrm{SR}} \right) \Psi_{\ell}^{m}(r) + {}^{h}w_{\ell}^{m\,\mathrm{S}}\, \Pi_{\ell}^{m}(r) \right\},
\end{aligned}
\tag{29}
$$

where ${}^{e}v_{\ell}^{m\,\mathrm{SR}}$ and ${}^{h}v_{\ell}^{m\,\mathrm{SR}}$ mean the corresponding elements of the vector $\mathbf{V}^{\mathrm{SR}}$, as given by (18), and scattering coefficients ${}^{e}w_{\ell}^{m\,\mathrm{S}}$ and ${}^{h}w_{\ell}^{m\,\mathrm{S}}$ are derived from (20). Functions Π_{l}^{m} and ψ_{l}^{m} are solutions of scalar Helmholtz equation:

$$
\Pi_{\ell}^{m} = h_{\ell}^{(1)}(kr)\, Y_{\ell}^{m}(\theta,\phi), \quad \Psi_{\ell}^{m} = j_{\ell}(kr)\, Y_{\ell}^{m}(\theta,\phi), \tag{30}
$$

where the spherical harmonics are defined by

$$
Y_{\ell}^{m}(\theta,\phi) = [(2\ell+1)\frac{(\ell-m)!}{(\ell+m)!}]^{1/2}\, P_{\ell}^{m}(\cos\theta)\, e^{im\phi}. \tag{31}
$$

Spherical Hankel and Bessel functions in (30) are given by formulae (6) and (3):

$$
h_{\ell}^{(1)}(\rho) \equiv \sqrt{\frac{\pi}{2\rho}}\, H_{\ell+1/2}^{(1)}(\rho), \qquad j_{\ell}(\rho) \equiv \sqrt{\frac{\pi}{2\rho}}\, J_{\ell+1/2}(\rho). \tag{32}
$$

The electric and magnetic fields can be derived from the Debye potentials (29) by a well-known formulae, see e.g. Born & Wolf, 1999:

$$
\mathbf{E} = \nabla \times \nabla \times \left(\mathbf{r}\, {}^{e}D\right) + ik\, \nabla \times (\mathbf{r}\, {}^{h}D),
$$

$$\mathbf{H}=\sqrt{\varepsilon_m}\left[\nabla\times\nabla\times\left(\mathbf{r}\,{}^hD\right)-ik\,\nabla\times(\mathbf{r}\,{}^eD)\right]. \tag{33}$$

Thus, formula (29) presents the final result of calculations, all quantaties within these formulae were defined above. Probably some of the written formulae were not explicitly derived in Bobbert 1986 a, b papers, but they follow from these papers after some simple transformations.

We use "Mathematica-4" software (Wolfram, 1999) for calculations. First of all we examine the calculation accuracy and different particular cases, some remarks should be done. First, the integrand in equation (21) contains a highly oscillating function. This is the so-called Weyl type integrals (Stratton, 1941; Morse & Feshbach, 1953; Bobbert & Vlieger, 1986 a) defined by

$$\Pi_\ell^m(r)=\frac{i^{-\ell}}{2\pi}\int_0^{2\pi}d\beta\int_0^{\pi/2-i\infty}\sin\alpha\, e^{ikr\cos\gamma}\,Y_\ell^m(\alpha,\beta)d\alpha\,, \tag{34}$$

where $Y_\ell^m(\alpha,\beta)$ are the spherical harmonics, related to associated Legendre polynomials (Wolfram, 1999).

For typical calculations it was insufficient to use the Laguerre polynomial approximation method, which was recommended in Bobbert, 1986 b. Thus, everywhere we use the "honest" integration in (34). The integral (34) is defined in the plane of complex variable, α, along some special path. We construct the integration contour as

$$\alpha=\frac{\pi}{2}\frac{cx}{1+cx}+i\,cx\,, \tag{35}$$

where c is any positive real value, and x varies from zero to infinity. Along this integration contour we found that, at the cost of longer calculation time, a stable value can be achieved. The matrix elements of $\hat{\mathbf{A}}$ have to be calculated only for non-negative m values, the negative parts can be mirrored as:

$$a_{l',e;l,e}^{-m}=a_{l',e;l,e}^{m},\quad a_{l',h;l,e}^{-m}=-a_{l',h;l,e}^{m},\quad a_{l',e;l,h}^{-m}=-a_{l',e;l,h}^{m},\quad a_{l',h;l,h}^{-m}=a_{l',h;l,h}^{m}\,. \tag{36}$$

We found that the cut-off parameter $\ell_{\max}$ within the sum (29) should be *two units greater* than that recommended by Barber & Hill, 1990, i.e. $\ell_{\max} \approx q + 4q^{1/3} + 3$. We calculate not the field directly, but the Debye potentials, which should be further differentiated according to (33). This implies that higher accuracy can be achieved with this truncating level. Verification was made by setting the same refractive indexes for the substrate and the medium (the reflecting matrix is zero). The result was compared with the Mie scattering. Another verification was done by setting the reflex index of the particle equal to the medium (the scattering matrix is zero), the result was compared with the Fresnel reflection. The numerical deviation within all the calculated quantaties (**E**, **H**, Pointing vector) was less than 5% withing the range $r \le a$.

The results of calculations are presented in Figs. 9-11. These pictures illustrate the main peculiarities of "particle on the surface" solution compared with the conventional Mie theory.

From Fig. 9 one can see that the true solution demonstrates 1.5 times higher intensity in the center than the Mie solution. Here we set solution slightly below the substrate to see the absorbed radiation. This result was expected, because the multi-reflection of radiation between the particle and the substrate results in more energy flowing into the substate.

The FWHM for the intensity distribution is even smaller than for the Mie solution. It means that the near-field sharpening effect shown in Fig. 8 for Mie solution will not diffuse due to secondary scattering. This result can be understood from the overall energy flux conservation within the range of the particle size.

One can see in Fig. 9 new oscillations (compared to Mie solution) in the intensity distribution. In the limit of geometrical optics such oscillations present the well-know effect, Newton rings, which arise due to interference of scattered and reflected radiation. Position for the dark Newton rings can be written as (Born & Wolf, 1999):

$$\frac{m\lambda}{2} = a - \sqrt{a^2 - r_m^2}, \quad m = 1,2,... \tag{37}$$

Formally, applying this formula for a = 0.5 μm and λ = 0.266 μm one can expect the appearance of three Newton rings at r = 0.34, 0.44 and 0.49 μm.

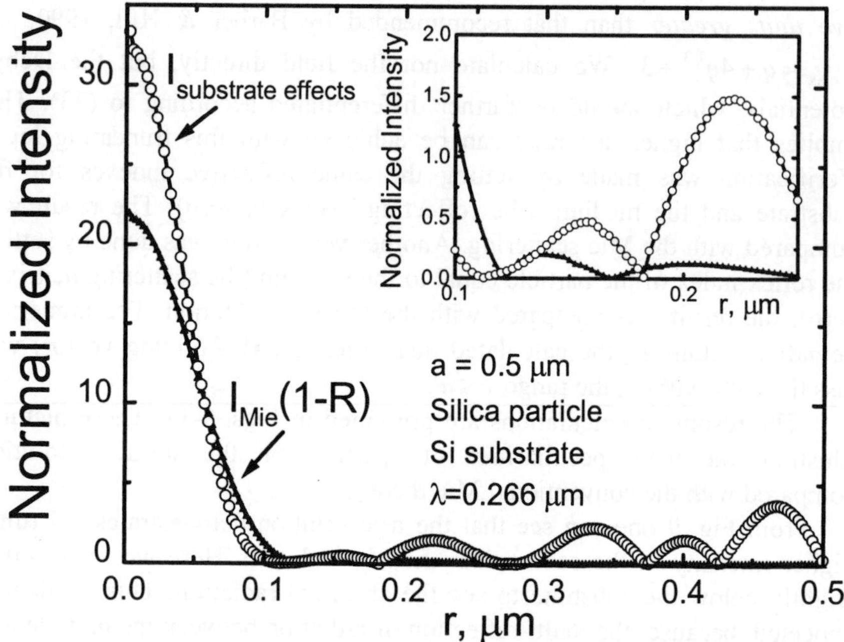

Fig. 9. Near-field light intensity along the radial axis with $\varphi = 45\,^{\circ}$, calculated with the Mie theory (triangular) and from reflective $\hat{\mathbf{A}}$ matrix (circles). The enhancement by the reflection of the substrate is observed near the contacting point. The ripple structure near the pedestal is shown in the insertion in the range between 0.1 μm and 0.25 μm (first two "rings").

Nevertheless, the necessary condition, $a \gg \lambda$, for formula (37) does not fulfill and this simplified consideration is not valid. Although the number of the ripples corresponds approximately to (37) some additional maximums arise. Some of these maximums exist with Mie solution (see insertion in Fig. 9) and they enhanced due to secondary scattering effects.

To analyze the origin of the sidelobe structure, we lift the particle above the substrate with a small distance δ, and examine the intensity profile on the substrate. The Newton rings should shrink inwards with increasing δ. The result of this examination is shown in Fig. 10. It shows that sidelobes firstly (up to the $\delta = 3$ nm) move outwards with increasing δ, which implies

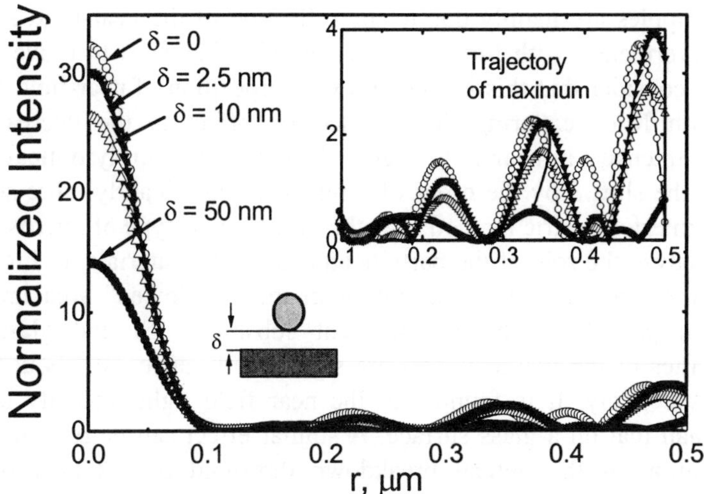

Fig. 10. Scattered light intensity on the substrate versus radial coordinate for different lifting distances δ between particle and the substrate. Light intensity drops drastically with increasing of lifting distance. The side-lobe structure shifts firstly outward with the distance, then inward. The insertion shows the trajectory of the maximum for the second "ring" is in between 0.1 μm and 0.5 μm.

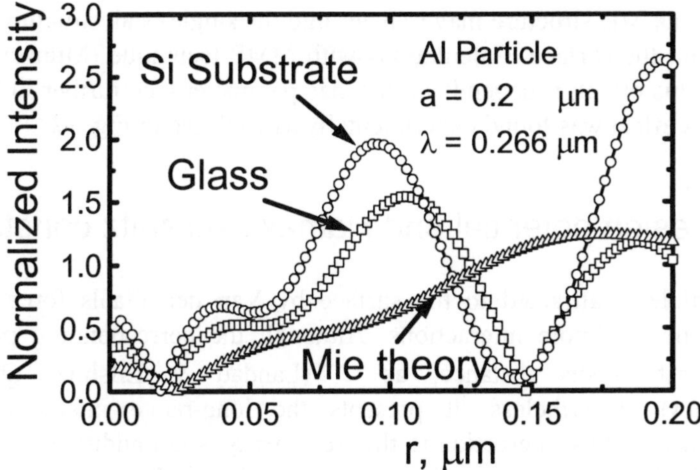

Fig. 11. Scattered light intensity in the "shadowed" area on the surface of Si and glass substrates. Higher light intensity is achieved on Si substrate due to its high reflectivity. Distribution of intensity, calculated from the Mie theory is also shown.

that the ripples are mainly due to the reflection of the near-field scattered light. Nevertheless with further increasing of δ they start to move inward. Thus we consider that this is a complex superposition of near-field Newton rings with Mie scattering. The noticeable effect is the decreasing of maximal intensity in near field region. It permits by purely optical way to measure the shift δ by the order of 1 nm! It is significantly more sensitive than the interferometric method; it is the idea for new type of microscopy.

To analyze the role of the particle and substrate materials, in Fig. 11 we present the light intensity distribution in the "shadowed" area under the aluminum particle, with two different substrates (Si and glass). The reflectivities of the two subsrates are significantly different, about 0.7 and 0.08, respectively. It is found that the near field light intensity on Si is higher than that on a glass surface. A similar effect can be responsible for the variation of the optical breakdown threshold of gas near different surfaces (Prokhorov, 1990).

One of the pecularities of the Mie solution is the typical double-peak structure in the intensity distribution for small particles, see in Fig. 4. The substrate reflection and secondary scattering can qualitatively change this picture. Due to the sharpening effect the amplification of intensity in the center is higher than at its periphery. Thus, at some range of parameters double-peak Mie structure may transfer into the single-peak structure for the particle on the surface. Calculations with MMP technique (Münzer, 2002) confirm this effect; it depends on the size parameter (see further in Section 7.1). This effect was found experimentally as well, see in Fig. 12.

4. Adhesion potential and Hamaker-Lifshitz constant

The particle is attracted to the surface by Van der Waals force, which occurs due to dipole interactions. Although the corresponding potential varies fast versus distance, as r^{-6} (Landau & Lifshitz, *Quantum Mechanics*), nevertheless, it presents the long-range interaction, e.g. contribution of this interaction to the free energy is not additive, it depends on the body shape and configuration (Lifshitz & Pitaevsky, *Statistical Physics*, Part 2). If one considers the particle as a deformed sphere, see in Fig. 13, then, according to Hamaker, 1937 the attraction force is given by

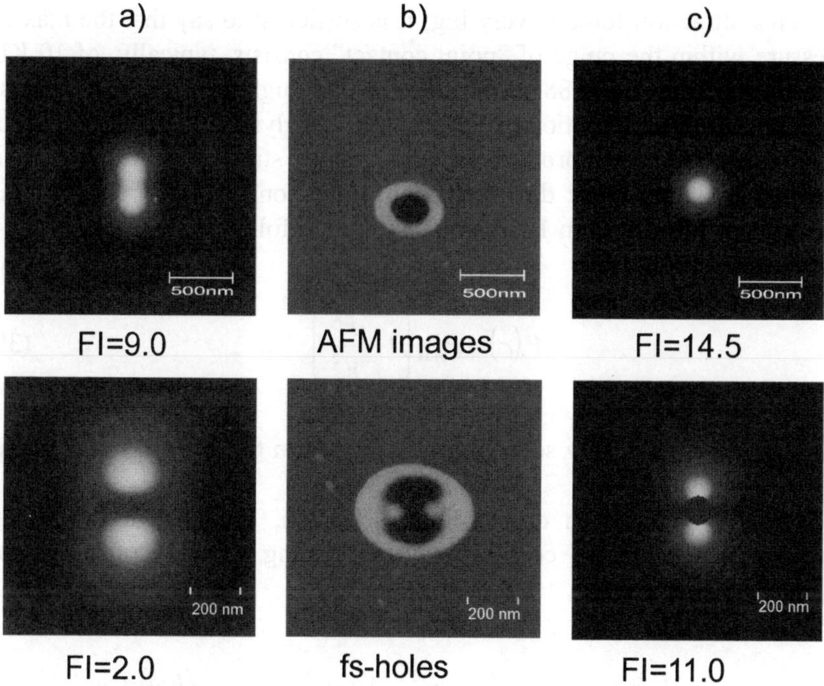

Fig. 12. Field intensity enhancement and ablation pattern underneath colloidal particles with diameters of 800 nm (top) and 320 nm (bottom) on a silicon substrate. The direction of the electrical field is orientated in the vertical direction. Calculated field enhancement by neglecting the influence of the substrate (a). Ablation pattern resulting from illumination by a fs laser pulse (experiment) (b). Calculation including the influence of the substrate (c). From Münzer, 2002.

$$F = \frac{\langle \hbar\omega \rangle a}{8\pi h^2} + \frac{\langle \hbar\omega \rangle r_c^{\,2}}{8\pi h^3}, \tag{38}$$

where a is radius of the particle, h is separation distance ($h \approx 4$ Å), r_c is radius of contact, The Lifshitz constant $\langle \hbar\omega \rangle$ is related to Hamaker constant A by $A = \frac{3}{4\pi}\langle \hbar\omega \rangle$. The Hamaker constant depends on the properties of the particle, substrate and medium.

This attraction force is very big; it is sufficient to say that the maximal pressure within the range of "point contact" consists, typically, of 10 Kbar and higher (Bowing, 1988). It is clear that this high loading leads to elastic or even plastic deformation of the material. Analysis of these deformations as well as the general problem of adherence is still under discussion; see, e.g. Rimai, 1995. Hertz did the first examination of pressure distribution within the contact area in 1882; this distribution follows to parabolic law:

$$P(r) = P_{max}\left(1 - \frac{r^2}{r_c^2}\right)^{1/2}, \tag{39}$$

see analysis of the Hertz solution, for example, in § 9 in Landau & Lifshitz, *Theory of Elasticity*.

Assuming Hertzian distribution, Derjaguin, 1934 found the relation between the radius of contact, r_c, and loading force, P_ℓ, for spherical particle:

$$r_c^3 = \frac{3}{4}\frac{P_\ell a}{E^*}, \quad \frac{1}{E^*} = \left(\frac{1-\sigma_1^2}{E_1} + \frac{1-\sigma_2^2}{E_2}\right), \quad P_0 = \frac{\langle \hbar\omega \rangle a}{8\pi h^2}. \tag{40}$$

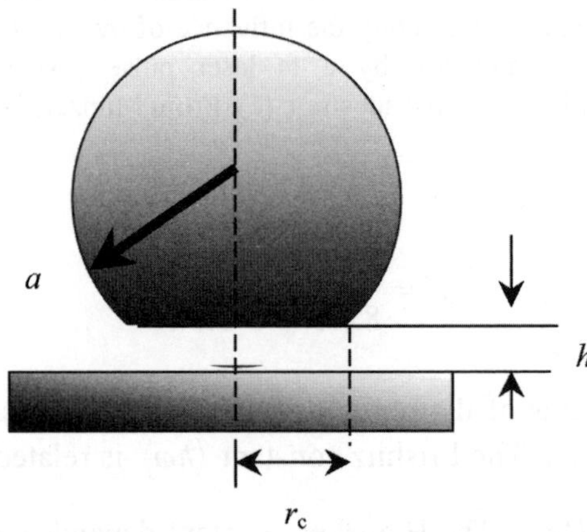

Fig. 13. Schematic for the particle on the surface. The attracting force in (38) is the result of integration with $h \ll a$ and $r_c \ll a$.

where $\sigma_{1,2}$ and $E_{1,2}$ are the Poisson coefficients and Young's modulus for the particle and substrate, without the external loading force, P_{ℓ}, is presented by the first term in (38). We denote this force as P_0. The adhesion-induced deformations are quite complex, and some other factors (adhesion forces outside the area of contact, etc.) should be taken into account to describe the experimental data well. At present, two models of adhesion are commonly acceptable: Derjaguin-Muller-Toropov (DMT) model for "hard" materials (Derjaguin, 1975, 1980; Muller, 1980, 1983) and Johnson-Kendall-Roberts (JKR) model for "soft" materials (Johnson, 1971, 1976). The transition between the two models was also discussed (Maugis, 1992, 1995). Without discussing the details, we want to pay attention just to the phenomena of instability, important for understanding laser cleaning. Namely, under the action of external force, P, the size of the contact in JKR-theory varies as (Johnson, 1976):

$$\left(\frac{r_c}{a_c}\right)^3=\frac{P}{P_c}+2\pm 2\sqrt{1+\frac{P}{P_c}}\,,\quad a_c=\left(\frac{9\pi\,\gamma\,a^2}{4E^*}\right)^{1/3},\quad P_c=3\pi\gamma a\,, \qquad (41)$$

where γ is a surface energy per unit area (work in separating the surfaces), it includes loading due to adhesion. Sign "plus" in (41) corresponds to stable, and "minus" to unstable brunches of the solution.

One can see in Fig. 14 that applying the negative force (tensile) with the critical load $P = - P_c$ the jump-like disconnection of the particle arises. On the contrary, one can approach particle to the surface without load, but at the moment, when particle touches the surface the jump-like adhesion force arises.

This hysteresis can be seen well when the load of the particle is performed with the help of atomic force microscope (AFM), which is shown in Fig. 15. Such an experiments are very popular now to estimate the Hamaker constants, see, e.g. (Shaefer, 1995; Mizes, 1995).

The Hamaker constant is used as phenomenological parameter in the DMT and JRK theories. This constant meanwhile can be calculated from the "first principles". In the macroscopic theory, the Van der Waals interaction in a material medium is regarded as brought about through a long-wavelength electromagnetic field; this concept suggested by Lifshitz, 1955 (see also Landau & Lifshitz, *Electrodynamics of Continuous Media*)

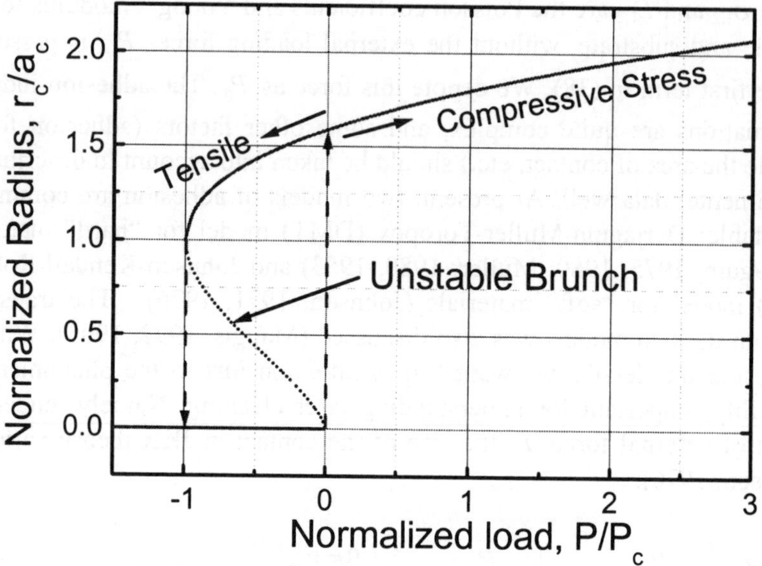

Fig. 14. Contact size-load, according to Eq. (41). In the paper of Johnson, 1976 the experimental points are shown additionally. They cover well the stable branch.

includes not only thermal fluctuations but also the zero-point oscillations of the field.

The Lifshitz theory is based on the theory of electromagnetic fluctuations developed by (Leontovich & Rytov, 1952; Rytov, 1953). Later, the Lifshitz formula was proved from the microscopic point of view, using methods of the quantum theory of field (Dzyaloshinski, 1959; 1960 a, b). The force F acting on a unit area of two bodies ("1" and "2") separated by a gap of width h, filled with a liquid (or some other substance "3") in the frame of microscopic theory is expressed through the complex dielectric constants $\varepsilon_i(\omega)=\varepsilon_i'(\omega)+i\varepsilon_i''(\omega)$, $i=1, 2, 3$ of three materials, see e.g. monograph (Abrikosov, 1965):

$$F=\frac{A\,a}{6h^2}=\frac{\langle\hbar\omega\rangle a}{8\pi\,h^2}=\frac{k_B T}{\pi\,c^3}\sum_{n=0}^{\infty}\varepsilon_3^{3/2}\omega_n^3\int_1^{\infty}p^2\left\{[L_1-1]^{-1}+[L_2-1]^{-1}\right\}dp\,,\quad(42)$$

where

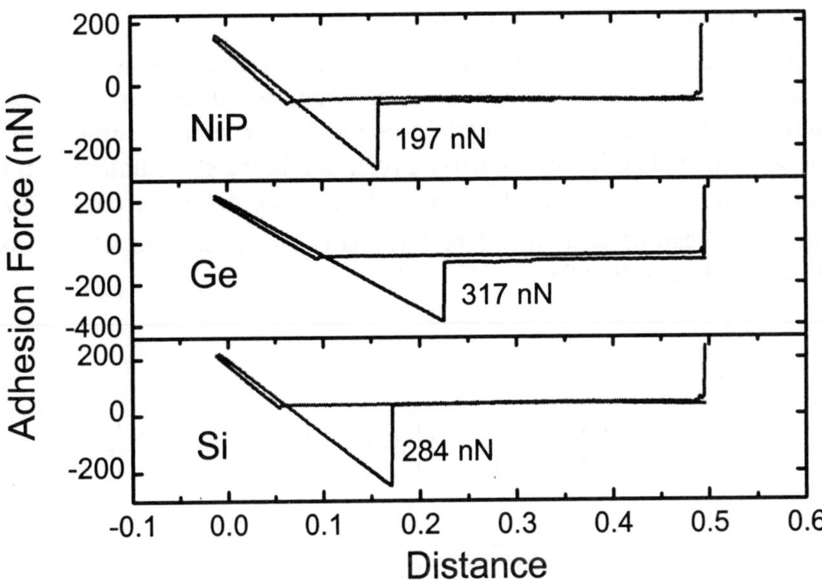

Fig. 15. AFM point-contact mode measuring the pull-off force of tip from NiP, Ge and Si substrates. The pulling force reflects different Hamaker constants of these substrates. The relative ratio of jumps is equal approximately to the ratio of corresponding Hamaker constants. Hamaker constants for Si and Ge have been reported in literature, data for NiP is original from Zheng, 2000.

$$L_1 = \frac{s_1 + p}{s_1 - p}\frac{s_2 + p}{s_2 - p}\exp\left[2p\sqrt{\varepsilon_3}\frac{\omega_n h}{c}\right],$$

$$L_2 = \frac{s_1 + p\dfrac{\varepsilon_1}{\varepsilon_3}}{s_1 - p\dfrac{\varepsilon_1}{\varepsilon_3}}\;\frac{s_2 + p\dfrac{\varepsilon_2}{\varepsilon_3}}{s_2 - p\dfrac{\varepsilon_2}{\varepsilon_3}}\exp\left[2p\sqrt{\varepsilon_3}\frac{\omega_n h}{c}\right], \qquad (43)$$

$$s_1 = \sqrt{\frac{\varepsilon_1}{\varepsilon_3} - 1 + p^2}\,,\quad s_2 = \sqrt{\frac{\varepsilon_2}{\varepsilon_3} - 1 + p^2}\,,\quad \omega_n = 2\pi n\frac{k_B T}{\hbar}.$$

During the summation in (42) the term with $n = 0$ should be taken with weight $1/2$, and the values of ε_1, ε_2 and ε_3 in (43) are considered as functions of the *imaginary frequency* $i\omega_n$, i.e. $\varepsilon = \varepsilon(i\omega_n)$. When the bodies are separated by vacuum, $\varepsilon_3 = 1$.

Formula (42) can be simplified, when the temperature T is sufficiently small (change from summation to integration, see details in Lifshitz & Pitaevsky, *Statistical Physics*, Part 2). Some further simplifications can be done with the help of Kramers-Kronig relation. In the static limit, when the permittivities ε_i are replaced by their electrostatic dielectric constants ε_{i0} formula (42) can be written as (see formula (82.4) in Lifshitz & Pitaevsky, *Statistical Physics*, Part 2):

$$F = \frac{\hbar c}{32\pi^2 h^4}$$

$$\int_0^\infty dx \int_1^\infty dp \frac{x^3}{p^2}\left\{\left[\frac{(s_{10}+p)(s_{20}+p)}{(s_{10}-p)(s_{20}-p)}e^x - 1\right]^{-1} + \left[\frac{(s_{10}+p\varepsilon_{10})(s_{20}+p\varepsilon_{20})}{(s_{10}-p\varepsilon_{10})(s_{20}-p\varepsilon_{20})}e^x - 1\right]^{-1}\right\}, \tag{44}$$

$$s_{10} = \sqrt{\varepsilon_{01} - 1 + p^2}, \quad s_{20} = \sqrt{\varepsilon_{02} - 1 + p^2}\ .$$

This formula was used for calculations in Lifshitz & Pitaevsky, *Statistical Physics*, Part 2. In the limit of two good metals, $\varepsilon_{10} = \varepsilon_{20} \to \infty$, the highest attraction force, F_∞, is reached (numerical number is presented for $h = 4$ Å)

$$F_\infty = \frac{\pi^2}{240}\frac{\hbar c}{h^4} \approx 0.5\,\text{Mbar}\,. \tag{45}$$

In Fig. 15 we present the dependence of attraction force for two identical insulators ($\varepsilon_{10} = \varepsilon_{20} = \varepsilon_0$) versus dielectric constant ε_0 in the same units as Fig. 18 in Lifshitz & Pitaevsky, *Statistical Physics*, Part 2:

$$F = F_\infty \left(\frac{\varepsilon_0 - 1}{\varepsilon_0 + 1}\right)^2 \Phi(\varepsilon_0). \tag{46}$$

One can see in Fig. 16 that the function $\Phi(\varepsilon_0)$ varies from 1 (at $\varepsilon_0 \to \infty$) to 0.35 (at $\varepsilon_0 \approx 1$), as it was found in Lifshitz & Pitaevsky,

Statistical Physics, Part 2. This function has a shallow minimum at $\varepsilon_0 \approx 1.6$. The structure of this minimum is shown in the insertion of Fig. 16.

Another approximation for the Lifshitz constant $\langle \hbar\omega \rangle$ (or for Hamaker constant A) can be derived from (42), taking into account the first terms in the sum. For the identical materials in vacuum the Lifshitz constant is given by (see, e. g. Lee, 1990; Israelachvili, 1991):

$$\langle \hbar\omega \rangle = \hbar \int_0^\infty \left[\frac{\varepsilon'(i\zeta)-1}{\varepsilon'(i\zeta)+1} \right]^2 d\zeta \,. \tag{47}$$

Here ε' is the real part of the complex dielectric function, it can be expressed through the imaginary part, ε'', with the help of Kramers-Kronig relation

$$\varepsilon'(i\zeta) = 1 + \int_0^\infty \frac{\omega\,\varepsilon''(\omega)}{\omega^2 + \zeta^2} d\omega \,. \tag{48}$$

The Hamaker constant was calculated from formula (47) using spectral data for many metals (see, e. g. Osborne-Lee, 1988). The simple formula for Hamaker constant follows under the approximation of the Drude theory, where the dielectric permittivity is given by

$$\varepsilon(\omega) = 1 - \frac{\omega_p^2}{\omega^2 + \gamma^2} + i\frac{\gamma}{\omega}\frac{\omega_p^2}{\omega^2 + \gamma^2} \,. \tag{49}$$

In the Eq. (49) ω_p is plasma frequency and γ is frequency of collisions. Substituting (49) into (48) and performing integration in (47) one can easily find

$$\langle \hbar\omega \rangle = \hbar\,\omega_p f(\gamma/\omega_p), \tag{50}$$

where

$$f(x) = \frac{\pi^2}{4}\frac{1}{\pi - x^2}\left\{ \frac{1}{\sqrt{\pi - x^2}}\left[\frac{\pi}{2} - ArcTan\left(\frac{x}{\sqrt{\pi - x^2}} \right) \right] - \frac{x}{\pi} \right\}. \tag{51}$$

Because for metals, typically, $\gamma \ll \omega_p$, one can use the asymptotic expansion

$$f(x) \approx \frac{\pi^{3/2}}{8} - \frac{x}{2} + \frac{3\sqrt{\pi}\,x^2}{16}, \quad \text{at } x \ll 1. \tag{52}$$

Some examples of calculations with formula (47) and (50) are shown in Table 1. The Drude parameters and optical spectra for three metals were taken from Shiles, 1980; Palik, 1985; Ordal, 1985; Qiu, 1995.

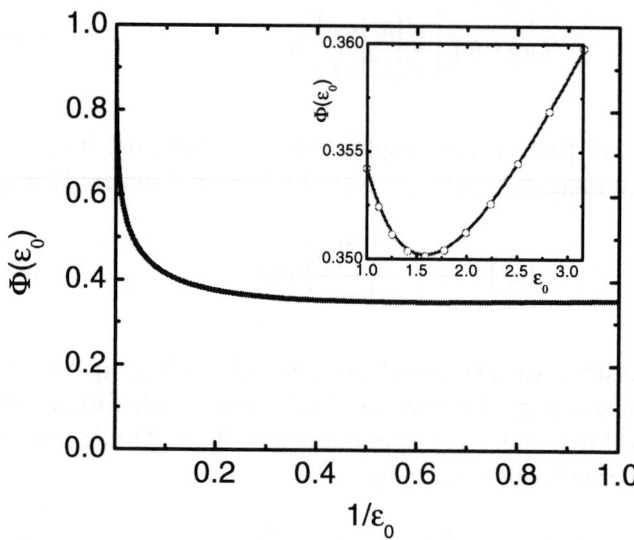

Fig. 16. Function Φ, which presents the attractive force according to Eq. (46) between two identical insulators ($\varepsilon_{10} = \varepsilon_{20} = \varepsilon_0$) versus dielectric constant ε_0.

Table 1. Calculations of the Lifshitz constant for pairs metal-metal.

Parameters	Al	Cu	Au
Plasma frequency, ω_p [s^{-1}]	$2.24\ 10^{16}$	$1.12\ 10^{16}$	$7.81\ 10^{15}$
Collision frequency, γ [s^{-1}]	$1.24\ 10^{14}$	$1.38\ 10^{13}$	$6.6\ 10^{13}$
Lifshitz constant, $\langle\hbar\omega\rangle$ [eV] from formula (50)	8.11	5.15	5.37
Lifshitz constant, $\langle\hbar\omega\rangle$ [eV] from the whole spectrum integration	8.59	10.7	12.3

One can see from Table 1 that for Al, the Lifshitz constant calculated from the approximation based on Drude theory is close to the number from the integration of the whole spectrum. However, for Cu and Au, a big difference exists between the results that are calculated from the two methods. This difference reveals the fact that for Cu and Au, the free-electron approximation of Drude theory is not suitable for their spectrum curve, since the 3S electron of these metals exerts strong effect outside the kernel shell. Therefore to Au and Cu, the "honest" integration through the whole spectrum is preferred for calculating Lifshitz - Hamaker constant. Klimchitskaya, 2000, did the generalization of the Lifshitz formalism for the case when the spatial dispersion can be important, in addition to the frequency dependence.

With very small particles the pressure under the particle within the contact region, $P = F / \pi r_c^2$ is growing up, from equation (38) follows:

$$P = \frac{\langle \hbar\omega \rangle}{8\pi^2 h^3}\left(1 + \frac{ah}{r_c^2}\right). \tag{53}$$

Radius of contact depends on the particle radius: $r_c \propto a^{\mu}$. The value $\mu = 2/3$ follows from the DMT - model, see Eq. (40). This exponent is typical for elastic deformations. Other exponents were also discussed in the literature (Rimai, 1995). Exponent $\mu = 1/2$ is typical for plastic deformations. For elastic deformation from (53) follows the growth of the pressure with a decrease of particle size. We mentioned above that the pressure, developed under the small particle varies typically from a few tens to a few hundreds Kbar. One can expect noticeable variations in pressure, when $ah/r_c^2 \geq 0.05$. On the other hand dielectric constant ε varies with pressure, one can consider, for simplicity, a linear dependence

$$\varepsilon(P) = \varepsilon_0 + \frac{\partial \varepsilon}{\partial P} P, \tag{54}$$

where $\partial\varepsilon / \partial P$ can be tried as constant.

Variation in the Lifshitz constant versus ε (or dependence $P = P(\varepsilon)$ between the flat surfaces) can be estimated, e.g. from (46). What is important is that the curve $P = P(\varepsilon)$ has the "Arrhenius-like" shape. The roots of the equation are given by points where two dependences (46) and (53) are crossed. This situation is typical for Semenov diagram in the theory of combustion (see, e.g. Karlov, 2000). Depending on the value of $\partial\varepsilon/\partial P$ coefficient equation may have one or three solutions. With higher loading pressure we have a higher ε and higher attraction force. This positive feedback may lead even to bistability in optical parameters (or in Lifshitz-Hamaker constant, or in attraction force) versus particle size.

The bistability versus particle size (or versus external loading force) looks like optical phase transition what is very attractive to use for a high-density information recording. This topic needs special discussion, which is out of the frame of the present paper.

5. Temperature under the particle

The temperature rise developed under the particle plays a decisive role for further analysis. A growth of temperature leads to thermal expansion of material, i.e. thermal deformations and stresses. Thus, we have to discuss temperature distribution within the substrate, $T = T(x, y, z, t)$, in more detail. This temperature distribution can be found from the heat equation:

$$c_s\rho_s\dot{T} = div[\kappa_s \; grad\, T] + \alpha A_0 I(x, y, t)e^{-\alpha z},$$
$$T\Big|_{z=\infty} = T\Big|_{x,y=\pm\infty} = T\Big|_{t=0} = 0 \qquad (55)$$

where surface intensity is understood as

$$I(x, y, t) = S_z(x, y) I_0(t). \qquad (56)$$

Here S_z is a z-component of the Poynting vector (12) above the surface, for the unit incoming intensity and function $I_0(t)$ describes the temporal pulse shape. A_0 is the substrate absorptivity, α is the absorption coefficient and κ_s is thermal conductivity of the substrate. The problem is that distribution $S_z(x, y)$ can be sufficiently complex due to near-field focusing (see e.g. examples shown in Sections 2 and 3). One can significantly simplify the problem, considering the near-field light intensity in the symmetric Gaussian form

$$S_z(x,y)=S_0 e^{-r^2/r_0^2}, \qquad (57)$$

where r is radial coordinate, r_0 is the radius of Gaussian beam, S_0 is the field enhancement factor.

This distribution fits well the "true" distribution of near-field light intensity when latter has "one-peak" distribution. The example in Fig. 17 demonstrates the main lobe of the "true" field is quite close to the Gaussian. The contribution of intensity oscillations on the periphery (side lobes) can be ignored. The special cases where these side lobes can be important are discussed further.

The "two-peak" distribution, similar to shown in Fig. 4 can be approximated by "two-peak" Gaussian profile:

$$S_z(x,y)=S_0\left[e^{-(x-x_1)^2/x_0^2}+e^{-(x+x_1)^2/x_0^2}\right]e^{-y^2/y_0^2}. \qquad (58)$$

For a linear heat equation the necessary solution can be presented through the Green function. For a nonlinear case, when functions c_s, κ_s, A_0 and α depend on the temperature, it needs numerical calculations and great calculation time. In the case of Gaussian profile (57) solution of linear heat equation is presented by a well-known formula:

$$T(r,z,t)=S_0\frac{(1-R)\alpha\chi r_0^2}{2\kappa_s}\int_0^t dt_1\, I_0(t-t_1)\frac{e^{-\frac{r^2}{r_0^2+4\chi t_1}}}{r_0^2+4\chi t_1}F(z,t_1), \qquad (59)$$

where $\chi=\kappa_s/c_s\rho_s$ is the thermal diffusivity, S_0 field enhancement factor (dimensionless) and F - function is given by

$$F(z,t)=e^{\alpha^2\chi t}\left\{e^{\alpha z}\,erfc\left[\alpha\sqrt{\chi t}+\frac{z}{2\sqrt{\chi t}}\right]+e^{-\alpha z}\,erfc\left[\alpha\sqrt{\chi t}-\frac{z}{2\sqrt{\chi t}}\right]\right\}. \qquad (60)$$

The smooth pulse shape, $I_0(t)$, for excimer laser can be described by

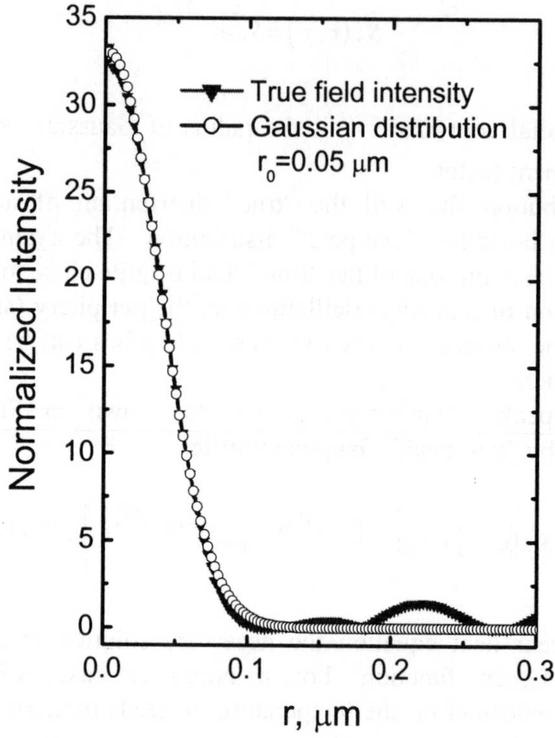

Fig. 17. The "true" intensity profile and its fitting by Gaussian beam. The "true" intensity (see in Fig. 9) presents the result of solution of the problem "particle on the surface" for 0.5 μm (radius) SiO_2 particle on the surface of Si substrate along $\varphi = 45^0$ (this presents non-polarized radiation). Radiation wavelength $\lambda = 266$ nm.

$$I_0(t) = \frac{\Phi t}{t_\ell^2} \exp\left[-\frac{t}{t_\ell}\right], \tag{61}$$

where $t_\ell = 0.409\, t_{FWHM}$ (the duration of the pulse defined at the full width at half maximum), the laser pulse energy is given by $E_\ell = \pi R^2 \Phi$, and Φ is an averaged fluence (input fluence).

Having done as many simplifications, as possible, we shall estimate the field enhancement factor S_0 from the Mie theory and then estimate r_0 value from the overall energy conservation condition. The particle geometrical cross-section as πR^2, area of the main lobe of scattered light can be expressed as πr_0^2. The area of all the side lobes (with the same efficient "brightness" of scattering) is expressed as πr_1^2; here r_1 is effective radius. One can defines r_1 by such a way that the ratio of corresponding cross-sections will give the field-enhanced factor $S_0 = R^2/\left(r_0^2 + r_1^2\right)$. Typically, the variation of r_1 is within the limits: $0 < r_1 < r_0$. This consideration yields the following estimation for radius r_0

$$R / S_0^{1/2} < r_0 < R / 2S_0^{1/2}. \qquad (62)$$

Although it is not a strong relation, it is sufficient for estimations. In Fig. 17 one can see S_0 = 33 for r_0 = 0.05 μm. From (62) follows 0.044 < r_0 < 0.087 μm. The width of near-field focusing intensity distribution is shown in Fig. 18. It is clear that r_0 value oscillates versus the particle size. This width is typically in between 50 and 100 nm, which is very attractive for many applications.

Using values S_0 and r_0 one can estimate the temperature rise from (59). This estimation shows that maximal temperature versus particle size oscillates (see in Fig. 19). Solving linear heat equation we used parameters of Si at T = 300 K, which are presented in Table 2. In reality, the "true" temperature is higher than in Fig. 19, because the optical and thermophysical parameters of Si strongly vary with temperature. The role of temperature dependent parameters can be seen in Fig. 20, where the solution of nonlinear heat equation is presented.

Oscillations in the temperature can be more pronounced than in Fig. 19 due to secondary scattering effect, which leads to higher intensity variations, see in Fig. 7. Figure 20 b shows behaviors of 1D and 3D temperature distributions for the field enhancement, where two distributions yield approximately the same maximal temperatures. Nevertheless one can see that 3D distribution produces a faster heating and cooling.

Table 2. Parameters of Si at room temperature (300 K).

ρ_s, g/cm^3	c_s, J/(g K)	κ_s, W/(cm K)	λ, nm	α, cm^{-1}	R_0
2.3	0.72	1.23	248	1.7 10^6	0.61

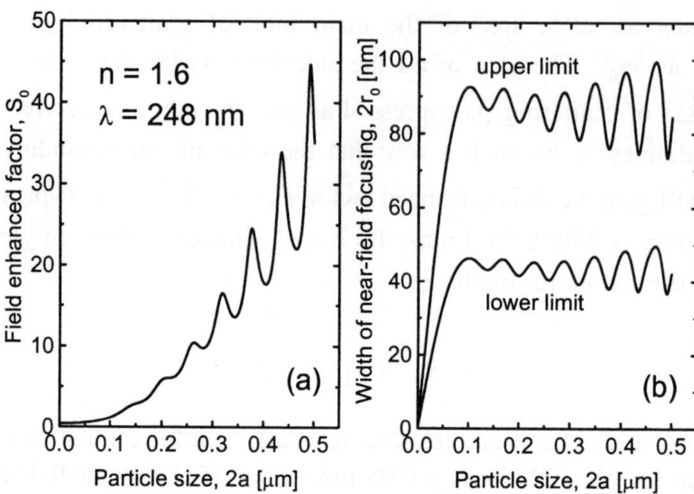

Fig. 18. Field enhancement factor, S_0 calculated from the Mie theory (a) and the width of near-field focusing (b) estimated from formula (62) for upper and lower limits. The refractive index $n = 1.6$ and the radiation wavelength $\lambda = 248$ nm.

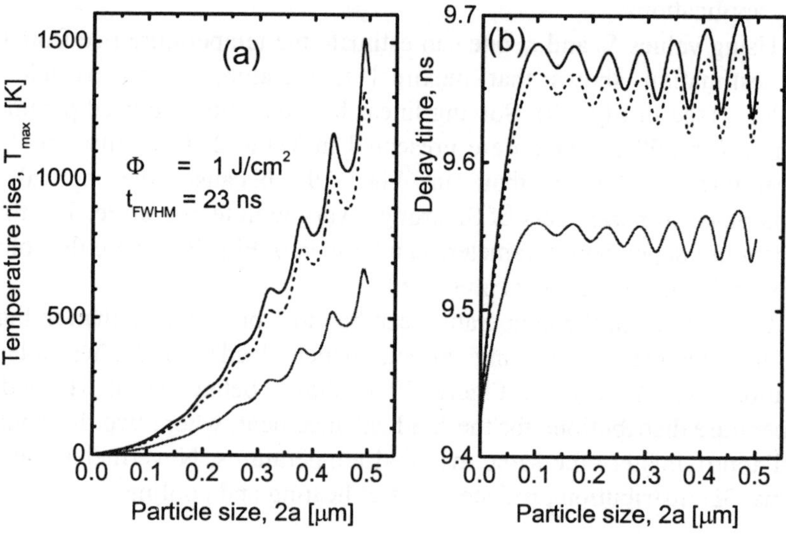

Fig. 19. (a) Maximal temperature, under the particle. (b) Delay time, when temperature reaches its maximal value (b). Field enhancement factor, S_0, and r_0 values are taken from Fig. 19. Two solid curves correspond to upper and lower limits in Fig. 19 b. The dash curves are calculated for $\kappa = 1.42$ W/cm K. Laser pulse width, t_{FWHM} = 23 ns and fluence $\Phi = 1$ J/cm^2.

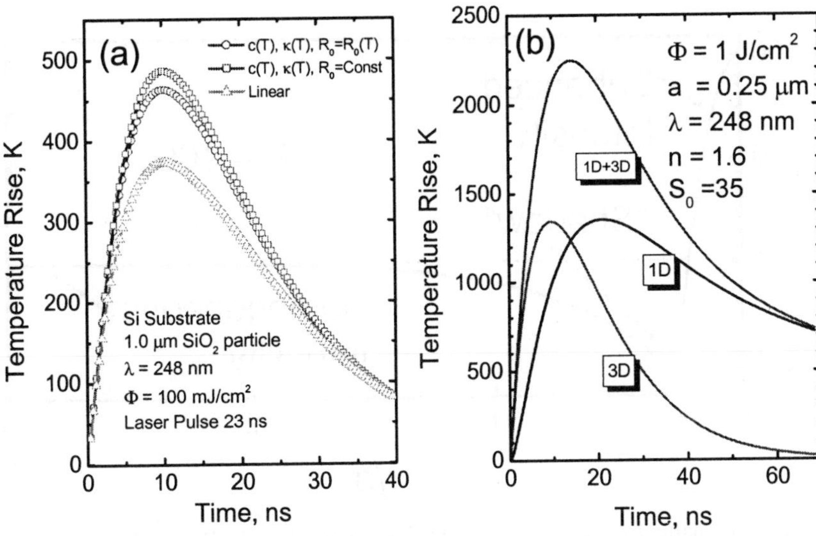

Fig. 20. (a) Temperature profile at the central point under the particle, calculated with non-linear heat equation by finite difference method (FDM) (Ozisik, 1994). Solution of linear heat equation yields the lowest maximal temperature. (b) Comparison of 3D (with field enhancement effect) and 1D solutions (without field enhancement) of linear heat equation.

In reality, the temperature under the particle is influenced by complex distribution of intensity; see e.g. Fig. 4 or Fig. 9. This distribution consists of 3 parts: 1) out of particle at $r > a$ it tends to homogeneity 1D intensity I_0; 2) in the region of enhanced radiation at $r < r_0$ it tends to enhanced field intensity $S_0 I_0$; 3) in the region of "shadow", $a < r < r_0$, intensity is rather small. One can neglect oscillations within the shadow region and the particle edge, and approximate the total intensity distribution by a sum of three Gaussian distributions (see in Fig. 21):

$$I(r,t) = I_0(t)\left[1 + S_0 e^{-r^2/r_0^2} - e^{-r^2/a^2}\right], \tag{63}$$

where $r_0 = a/\sqrt{S_0}$, this provide conservation of energy. We shall call this simplified distribution (63) as "1D + 3D heat model". An example of the calculation of the temperature rise with this model is shown in Fig. 20 b.

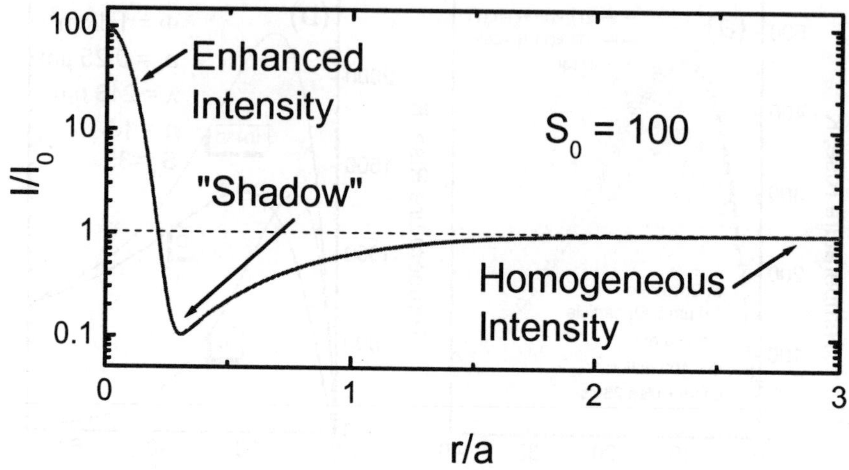

Fig. 21. Schematic for the intensity distribution on the substrate surface under the transparent particle. It can be approximated by the sum of three Gaussian distributions (63).

6. Dynamics of the particle, 3D effects

A particle on the surface performs its motion in the field of adhesion potential, which consists Van der Waals attraction and elastic forces. When the temperature is smaller than the melting temperature, then the main cleaning effect of laser action is related to material thermal expansion.

When the substrate expands, the position of the substrate surface, z_s, varies with time, $z_s = z_s(t)$, $z_s(0) = 0$. The particle displacement is a function of time t, say $z = z_f(t)$. Then the varying deformation parameter, $\delta(t)$ at time t can be expressed as

$$\delta(t) = z_s(t) + z_p(t) - z_f(t) + \delta_0 . \tag{64}$$

Here term $z_p(t) = \alpha_T^{(p)} a T_s(t)$ describes the effect of the particle heating due to thermal contact. We assume that the temperature of the particle is the same as the substrate surface temperature, $\alpha_T^{(p)}$ is the *linear* thermal expansion coefficient for the particle (we use additional superscript to distinguish the particle and substrate materials). If no external load acts on the particle, the initial deformation parameter δ_0 is expressed in DMT theory by

$$\delta_0 = \frac{1}{8}\left[\frac{9a\langle\hbar\omega\rangle^2}{2\pi^2 h^4 E^{*2}}\right]^{1/3}, \quad \frac{1}{E^*} = \left(\frac{1-\sigma_p^2}{E_p} + \frac{1-\sigma_s^2}{E_s}\right), \tag{65}$$

where $\sigma_{p,s}$ and $E_{p,s}$ are the Poisson coefficients and Young's modulus for the particle (index "*p*") and substrate (index "*s*"), $\langle\hbar\omega\rangle$ presents the Lifshitz constant.

The acceleration due to the elastic force can be expressed (Lu, 2000) by:

$$\frac{4}{3}\pi a^3 \rho_p \frac{d^2 z_f(t)}{dt^2} = \frac{4}{3}\sqrt{a}\,E^*\left[\delta(t)^{\frac{3}{2}} - \delta_0^{\frac{3}{2}}\right], \tag{66}$$

where ρ_p is the density of the particle and we neglect the energy losses due to plastic deformations and sound generation (Luk`yanchuk, 2001; Arnold, 2002). When the substrate expands, the position of the substrate surface, z_s, varies with time, $z_s = z_s(t)$. This expansion is the driving force for dry laser cleaning. The initial conditions for Eq. (66) are

$$\frac{dz_f}{dt}\Big|_{t=0} = \frac{dz_p}{dt}\Big|_{t=0}, \quad z_f\big|_{t=0} = 0. \tag{67}$$

The surface position, $z_s(t)$, is given by the z-component of the vector of displacement, $\mathbf{u} = \mathbf{u}(x, y, z, t)$ at $z = 0$. To find this displacement vector one should solve the equation of thermal elasticity (Landau & Lifshitz, *Theory of Elasticity*; Sokolnikoff 1956):

$$\rho_s \ddot{\mathbf{u}} = \frac{E_s}{2(1+\sigma_s)}\Delta\mathbf{u} + \frac{E_s}{2(1+\sigma_s)(1-2\sigma_s)}\,\mathbf{grad}\,div\,\mathbf{u} - \frac{\alpha_T^{(s)} E_s}{(1-2\sigma_s)}\,\mathbf{grad}\,T\,, \tag{68}$$

where $\alpha_T^{(s)}$ is the *linear* thermal expansion coefficient for substrate and $T = T(x, y, z, t)$ is the temperature distribution within the substrate.

The solution of the equation (68) for the stationary problem (laser beam with Gaussian intensity distribution) was presented in Welsh, 1988. This solution is valid for the calculation of the stresses and strains. Nevertheless displacement, e.g. u_z - component, which follows from this formal solution, contains logarithmic divergence, related to $1/r$ spatial distribution of the stationary thermal field at big distances (Arnold, 2002). Meanwhile, it is clear that the stationary limit is senseless, i.e. in the real physical situation the corresponding integrals should be cut off, and such calculations were presented recently (Luk`yanchuk, 2001). Nevertheless the cutting procedure is not strictly defined and it is more precise to use nonstationary solution (Luk`yanchuk, 2002 a, b). For different estimations we shall use parameters, listed in Table 3.

Table 3. Parameters used in the calculations (taken at T = 300 K). Optical parameters are given for $\lambda = 248$ nm (KrF excimer laser). Hamaker constant is given for the pair of identical materials, i.e. Si-Si, etc. Data for Hamaker constant (or Lifshitz constant) are taken from Bowing, 1988; Visser, 1972; Heim, 1999.

Parameter	Values for different materials			
	Si	Ge	NiP	SiO_2
Absorption Coefficient [cm^{-1}]	$1.7\ 10^6$	$1.6\ 10^6$	$5.7\ 10^5$	1
Absorptivity	0.39	0.35	0.71	0.95
Density [g/cm^3]	2.3	5.33	8.9	2.2
Thermal Conductivity [W/cm K]	1.42	0.73	0.14	0.0146
Heat Capacity [J/g K]	0.72	0.31	0.54	0.74
Melting Temperature [K]	1685	1210	1200	1873
Young Modulus [$dynes/cm^2$]	$1.3\ 10^{12}$	$8.2\ 10^{11}$	$2.0\ 10^{12}$	$7.3\ 10^{11}$
Poisson Ratio	0.28	0.3	0.31	0.18
Linear Thermal Expansion [K^{-1}]	$2.6\ 10^{-6}$	$6.0\ 10^{-6}$	$12.0\ 10^{-6}$	$0.54\ 10^{-6}$
Hamaker Constant, A, [J]	$2.5\ 10^{-19}$	$3.1\ 10^{-19}$	$1.2\ 10^{-19}$	$0.5\ 10^{-19}$
Lifshitz Constant, $\langle \hbar\omega \rangle$, [eV]	6.54	5.18	3.14	1.32

a) *1D model*.

This case is the simplest. We use solution $T = T(z, t)$ of 1D heat equation. From this solution one can estimate the surface displacement by

$$z_s(t) = 2(1+\sigma_s)\alpha_T^{(s)} \int_0^\infty T(z,t)dt \ . \tag{69}$$

The additional multiplier $2(1+\sigma_s)$ in (69) appears when one considers the "quasi-1D approach", where displacements and stresses are considered to be equal zero at infinity (at *x-y* plane). This term without a detailed explanation was written in Prokhorov, 1990; Vicanek, 1994. The detailed examination of the thermal elasticity problem, which precisely introduces this term, is given in Arnold, 2002. For the "true" 1D problem multiplier $2(1+\sigma_s)$ is absent. Experiment of Dobler, 1999, where corresponding displacement and accelerations were measured experimentally, was closer to "true" 1D problem rather than "quasi-1D approach".

Because of the energy conservation, one can write condition

$$A\int_0^t I(t_1)dt_1 = c\rho \int_0^\infty T(z,t)dz \ , \tag{70}$$

here c and ρ are heat capacity and density of the heated substrate respectively, A is absorptivity of the surface. If one considers a smooth pulse shape (61), then for the displacement holds the following formula:

$$z_s(t) = 2(1+\sigma_s)\frac{\alpha_T^{(s)} A\Phi}{c\rho}\left[1-\left(1+\frac{t}{\tau}\right)e^{-t/\tau}\right]. \tag{71}$$

For the rectangular laser pulse with the pulse duration τ_ℓ, a similar dependence is given by

$$z_s = 2(1+\sigma_s)\frac{\alpha_T^{(s)} A\Phi}{c\rho}\left[\frac{t}{\tau_\ell}\Theta_H\left(1-\frac{t}{\tau_\ell}\right)+\Theta_H\left(\frac{t}{\tau_\ell}-1\right)\right], \tag{72}$$

here $\Theta_H(x)$ is the unit step function (Heaviside function). Pay attention that the rectangular pulse produces infinite accelerations at $t=0$ and $t=\tau_\ell$.

The 1D model has natural restriction, related to absence of inward motion of the surface during cooling stage. To take into account this effect one should use the 3D thermo-elastic model. Effects related to temperature dependencies of parameters (solution of non-linear heat equation) can also play important role.

Although the heating process happens only within a short time interval, it determines both the kinetic energy and the elastic potential energy necessary to overcome the adhesion force.

Condition for the particle removal can be written from the energetic consideration Lu, 2000:

$$\frac{8}{15}E^*\sqrt{a\,\delta(t)^5}+\frac{4}{3}\pi\,a^3\frac{\rho v^2}{2}\geq\frac{1}{2}P_\ell\,\delta(t)+\frac{\langle\hbar\omega\rangle a}{8\pi h}. \tag{73}$$

Here v is understood as a relative velocity: $v=\dot{\delta}(t)$. Another condition is the so-called “force criterion”, which follows from (66) and (40) at quasi-static conditions:

$$-\frac{4}{3}\pi\,a^3\rho_p\ddot{z}_f=\frac{4}{3}\sqrt{a}\,E^*\left[\delta_0^{\frac{3}{2}}-\delta(t)^{\frac{3}{2}}\right]\geq\frac{\langle\hbar\omega\rangle a}{8\pi h^2}. \tag{74}$$

Pay attention that the force criterion needs necessary condition $\ddot{f}<0$, i.e. it works just on the stage of particle deceleration. The transition between the two (force and energy) criterions was discussed in Arnold, 2002.

An example of solution is shown in Fig. 22. The pulse shape with fluence 1 J/cm^2 and dynamics of the surface temperature rise $T_s=T_s(t)$ are shown in Fig. 22 a. The pulse duration is considered to be 23 ns, and the initial temperature is $T_\infty=300\,K$. Maximal surface temperature reaches 1561 K, which is close to Si melting temperature T_m = 1685 K. With higher fluences we exceed the melting temperature.

Different components of the deformation parameter (64) are shown in Fig. 22 b. The expansion, related to the thermal expansion of the particle is very small, $z_p \ll z_s$, mainly due to the small value of the particle’s thermal expansion coefficient, $\alpha_T^{(p)}$. Deformation parameter δ oscillates with amplitude ≈ 1.5 Å around the equilibrium value $\delta_0\cong 3$ Å. For the SiO_2 particle with a = 0.5 μm used in calculations, period of oscillations consists ≈ 6.4 ns. It is approximately 20 times longer than the period of the first mode of the particle sound vibration, $T_p = 4a / u_s$ (Landau & Lifshitz, "*Theory of Elasticity*"), where $u_s = 5.5 \cdot 10^5$ cm/s is sound velocity in quartz. We do not take into account these "high frequency" sound vibrations. We also neglect damping in "low frequency" oscillations. In principle, this damping occurs due to sound generation (Arnold, 2002) and due to plastic deformations in the particle-substrate contact area (Luk`yanchuk, 2001). Oscillation dynamics in dry laser cleaning was suggested in Luk`yanchuk, 2001, 2002 a, b; Arnold, 2002. This dynamics permits resonant enhancement.

Fig. 22 c presents characteristic velocities: $v_s = \dot{z}_s$, $v_p = \dot{z}_p$, $v_f = \dot{z}_f$ and $\dot{\delta}$. Although the condition $z_p \ll z_s$ is fulfilled for all interesting parameters of laser cleaning with 248 nm radiation and SiO_2 particles on the Si surface, velocities of expansion can be comparable, e.g. for 5 μm particles near the cleaning threshold, $v_{s\max} \approx 2 v_{p\max}$.

Fig. 22 d presents characteristic accelerations: $a_s = \dot{v}_s$, $a_p = \dot{v}_p$, and $a_f = \dot{v}_f$. Maximal accelerations due to substrate surface expansion can be smaller than maximal accelerations produced by particle heating, i.e. $a_{p\,max} \gg a_{s\,max}$. It holds for the big particles. The dash line presents the critical acceleration for the "force criterion":

$$a_c = -\frac{3}{24\pi^2} \frac{\langle \hbar\omega \rangle}{\rho_c a^2 h^2}. \tag{75}$$

With fluence 1 J/cm^2 $\ddot{\delta}$ is close to reach this critical value. The "energy criterion" (73) presents a line on the plane of parameters $\{\delta, \dot{\delta}\}$. When the phase trajectory of the system crosses the line, the condition for the particle removal is fulfilled (see in Fig. 23). The "energy criterion" yields higher threshold fluence than those, which follow from the "force criterion". For small particles with $a \leq 1$μm the removal occurs mainly due to big deformations δ.

With bigger particles kinetic energy starts to play an important role: phase trajectory touches the energy criterion curve at the point with non-zero value of velocity $\dot{\delta}$.

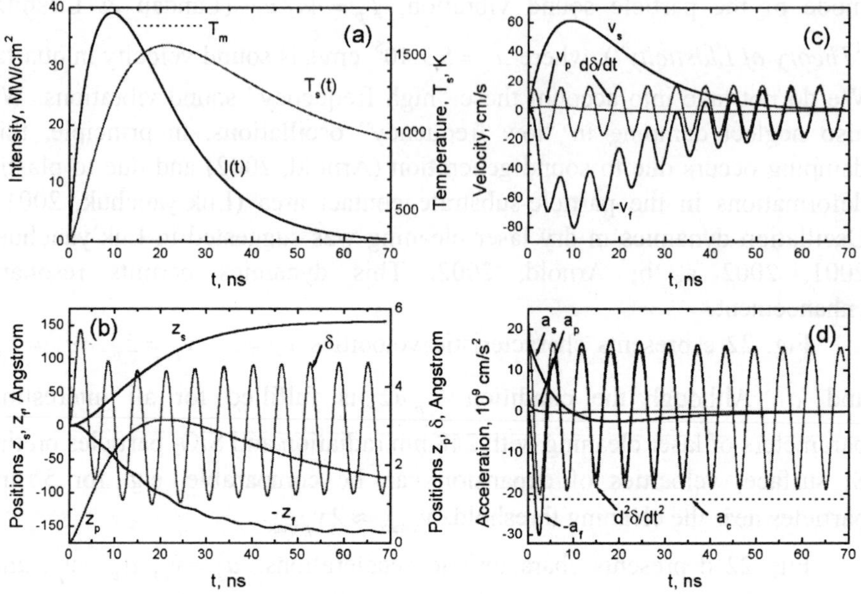

Fig. 22. Dynamics of dry laser cleaning for 1 μm SiO_2 particle on Si surface: a) laser pulse shape and the surface temperature for laser pulse length 23 ns and fluence 1 J/cm^2; b) different components of displacement and deformation in accordance with (64); c) characteristic velocities: $v_s = \dot{z}_s$, $v_p = \dot{z}_p$, $v_f = \dot{z}_f$; d) characteristic accelerations: $a_s = \dot{v}_s$, $a_p = \dot{v}_p$, and $a_f = \dot{v}_f$. Value a_c corresponds to critical acceleration, which follows from the "force criterion" (74).

Although 1D model explains qualitatively behavior of the threshold fluence versus particle size, it yields threshold fluences by the order of magnitude higher than the experimental (see in Fig. 24). It means that some important physics is missing with 1D model.

Another discrepancy is related to minimal size of the particle, which can be removed in dry laser cleaning. 1D model predicts the minimal size $2a \approx 1$ μm (a smaller size needs a surface temperature above the melting temperature). Meanwhile many groups reported about cleaning efficiency up to the size 0.2 μm in dry laser cleaning. Within the frame of 1D model there are no oscillations in threshold fluence, related to optical resonance.

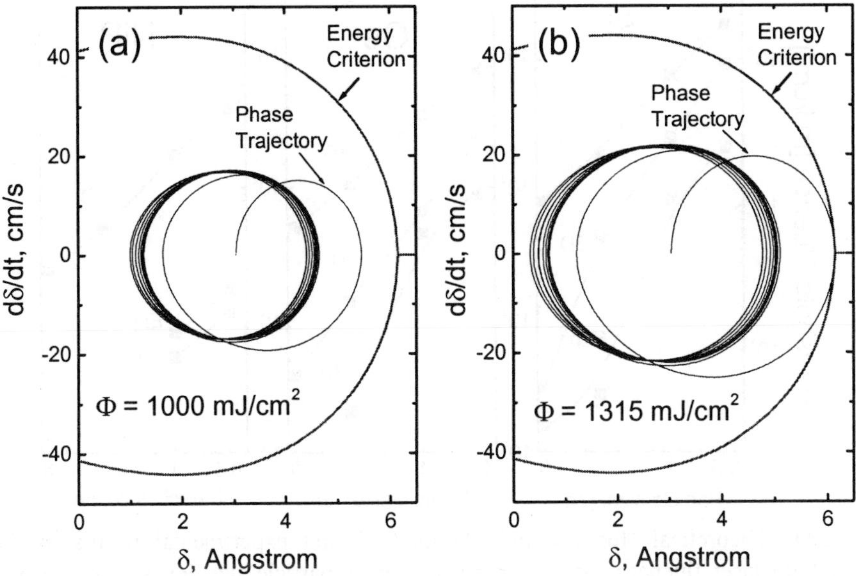

Fig. 23. Energy criterion curve, calculated from (73) and phase trajectory calculated from integration of (66), (67). Particle size is $1\mu m$. Pictures (a) and (b) correspond to different laser fluences.

b) *3D model*.

To find the surface position, $z_s(t)$, we have to use solution of 3D heat equation and solution of 3D thermal elasticity problem (see in Appendix A Chapter 2). We assume the Gaussian distribution (57) for the near-field light intensity, $I(r,t) = I(t)e^{-r^2/r_0^2}$. Then in the center of the laser beam we have:

$$z_s(t) = u_z = 2\,\alpha_T^{(s)}\,(1+o_s)\frac{A\,r_0^2\alpha\,\chi}{\kappa}$$
$$\int_0^t dt_1 I(t-t_1)\frac{2}{r_0^2+4\chi\,t_1}\int_0^\infty dz_1\left[1-\sqrt{\pi}\varsigma\, e^{\varsigma^2} erfc\varsigma\right]F(z_1,t_1), \tag{76}$$

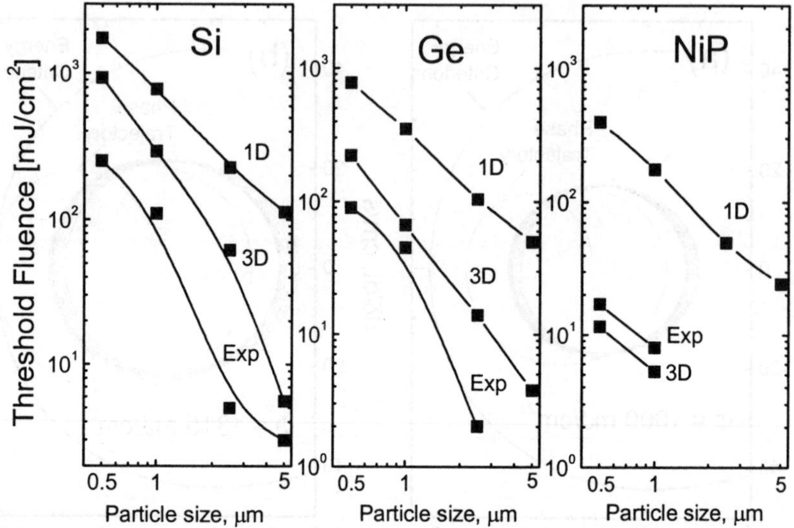

Fig 24. Theoretical (for 1D and 3D models) and experimental results of the threshold laser fluences for SiO_2 particles versus particle size for Si, Ge and NiP substrates. Excimer laser $\lambda = 248$ nm, pulse duration 23 ns.

where $\varsigma = z_1 / \sqrt{r_0^2 + 4\chi t_1}$, and F- function is given by formula (60). Solving the Newton equation (66) for the particle motion in elastic potential with driving force (76) one can see the main peculiarities of the particle dynamics with 3D model (see in Fig. 25).

Although 3D expansion with the same intensity yields smaller z_s values than 1D expansion, it is faster. As a result characteristic deformation velocities and accelerations can be comparable or even greater than in 1D case. The calculations shows that the threshold fluence for 3D model with a = 1 μm typically by the order of magnitude smaller than for 1D model (see in Fig. 24).

Further improvement in the position of theoretical curve can be done with 1D + 3D model. This model considers the distribution of laser intensity, given by (63). Surface expansion is a combination of results given by formulae (71) and (76).

With growth of the particle size the enhancement factor oscillates and possible maximum values becomes higher (see in Fig. 7). For example, with a = 1 μm, λ = 248 nm and n = 1.6 the nearest maximum yields S = 343.

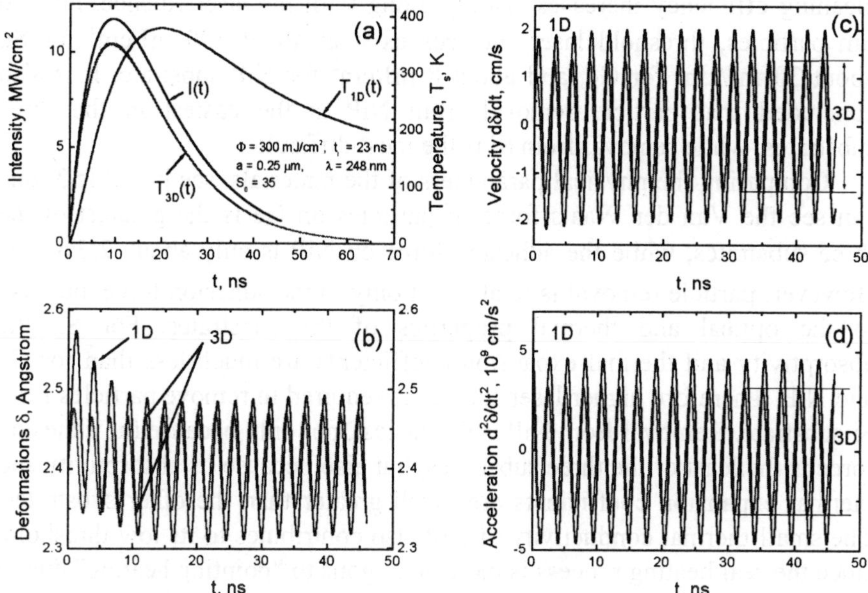

Fig. 25. Substrate expansion and particle movement for SiO_2 particle on the Ge substrate. Laser fluence is 19.2 mJ/cm^2. Enhanced near-field light intensity is modeled as Gaussian beam with a radius of 0.1 μm. a) Laser intensity, *I*(*t*) and corresponding 1D and 3D temperature dynamics. Combined 1D+3D case is shown as well. b) Displacements *z* for the particle (*p*) and substrate (*s*) for 1d and 3D heating. c) particle displacement *f*(*t*) for 1D and 3D heating and d) corresponding particle velocities.

It is clear that such high degree of the enhancement factor will be limited by the particle surface quality and it shapes deviation from the ideal sphere.

Silicon substrate is a particular material with a strong temperature dependences of optical and thermophysical parameters. It needs solution of nonlinear heat equation. One can see in Fig. 20a that the nonlinearities yield a higher surface temperature. The same effect produces combination of 1D and 3D model compared to 3D model (see in Fig. 20b). Both effects move the theoretical curve into the "correct direction" i.e. closer to experiment but we did not analyze the nonlinear problem in more detail.

The different cleaning efficiencies of 1.0 μm silica particle from Si, Ge and NiP surfaces were examined, see in Fig. 26. It is found that the laser cleaning efficiency increases sharply along with the laser fluence. For 1.0 μm particles, threshold laser fluences exist at about 100 mJ/cm² for Si, about 30 mJ/cm² for Ge, and about 8 mJ/cm² for NiP substrate. It is also concluded that particle removal from NiP is the easiest in the three substrates, while removal from Si is the most difficult.

Comparing the physical parameters of the materials, see in Table 3, one can see the Van der Waals force of particles on Ge is the greatest of the three substrates, while the adhesion force on NiP is only about $2/3$ of Ge. However, particle removal is related not only to the adhesion force, but also to the optical and thermal properties of the substrates. For Si, the absorptivity and thermal expansion coefficients are much less than for Ge and NiP. Therefore higher laser fluence is required to remove particles from Si substrate. Removal from NiP is the easiest, not only because the adhesion force is the least of the three substrates, but also because its absorptivity and thermal expansion coefficients are much greater than the other substrates. The small thermal conductivity of NiP also contributes to its low threshold, since the real heating process is more analogous to "pointing heating" due to the near-field effect. This experiment demonstrates that substrate thermal expansion certainly plays an important role in dry laser cleaning mechanism.

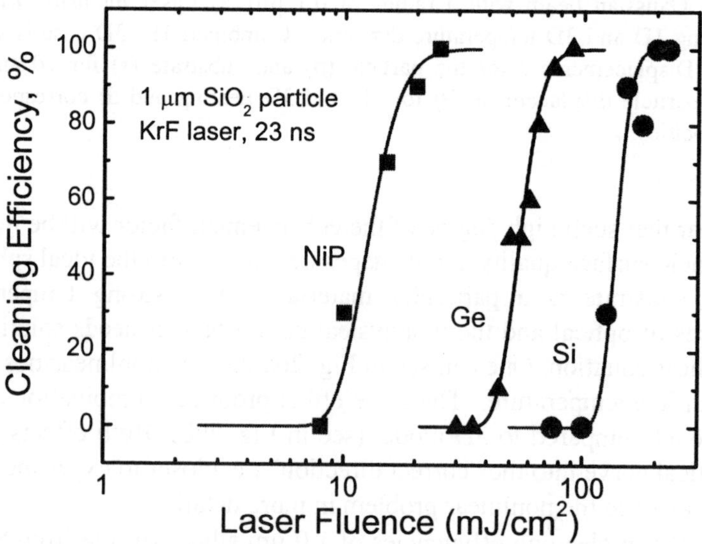

Fig. 26. Cleaning curves for NiP, Ge and Si substrates (from Zheng, 2001).

Nevertheless the 1D thermal expansion model consistently yields values, which are by one-two orders of magnitude higher than experimental data (see in Fig. 24). At the same time 3D model yields values much closer to experiment. For example, for NiP substrate 3D model practically coincides with experiment. Pay attention that 3D effects in heating and deformation play an important role of for both, transparent and absorbing particles.

7. Comparison with experimental results

In this section we will discuss experimental findings that allow a verification of the theoretically predicted intensity enhancement in the near field of particles resting on a surface. Besides their interest from a fundamental point of view, these experiments are of great importance for applications such as laser cleaning of particles on surfaces.

7.1. Local substrate ablation – a probe for optical near fields

Clearly, any conclusive evidence for the presence of intensity enhancement in the near field of particles at surfaces should rely on directly accessible experimental facts. Indirect indications such as the removal of particles at lower laser fluences than predicted by theory (Lu, 2000a) can be considered as a hint, however, they strongly rely on the correctness of the respective theory.

The most common approach for probing near fields at small particles is the application of Scanning Near Field Optical Microscopy (SNOM) (Durig, 1986). Although this is a well-established technique it inevitably disturbs the field distribution around the particle by the presence of the probe itself.

We therefore chose a new and different approach in order to probe undisturbed near field distributions as were found in laser cleaning applications (Leiderer, 2000; Lu 2000c, 2002; Münzer, 2001; Mosbacher, 2001; Münzer, 2002; Huang, 2002 a, b). In this approach we use the substrate, that is already present anyway, as the imaging medium, and consequently do not rely on the introduction of an additional probe into the field. Surface modifications due to structural transitions, defects formation, oxidation, melting or evaporation that are induced by the intensity enhancement at the enhanced laser intensity can be detected by optical, electron or scanning probe microscopy.

7.1.1 Morphology of near field-induced damage sites

Typical examples for such modification sites are shown in Fig. 27. In order to clearly relate any observed modification to the former particle sites, we have marked the initial positions of the colloidal spheres by evaporating a thin (10 nm) SiO_2 layer onto the sample after application of the colloidal particles and before the laser irradiation (Leiderer, 2000). After the application of the laser pulse the particles have been removed, and the sample was inspected in a Scanning Electron Microscope (SEM). As can be seen from Fig. 27 a), the particles size and their position could be determined by a contrast in the SEM pictures due to the different thickness of the oxide layers in and outside the particle site. By a comparison of the hole diameters on wafers with and without the additional oxide layer we have checked that this layer did not influence the optical properties of the particles.

The displayed image represents a shadowgraph of a cluster of six particles, which were removed. At the center of each particle a hole was created, just at the positions where the numerical calculations predict the highest laser intensity. Similar holes were found also for smaller particles (see Fig. 27 b). The laser used here had a FWHM of 150 fs and a wavelength of λ = 800 nm. By the choice of this short pulse length we minimized the heat diffusion in the silicon substrate and made sure that any ablation pattern reflected the actual intensity distribution at the particle.

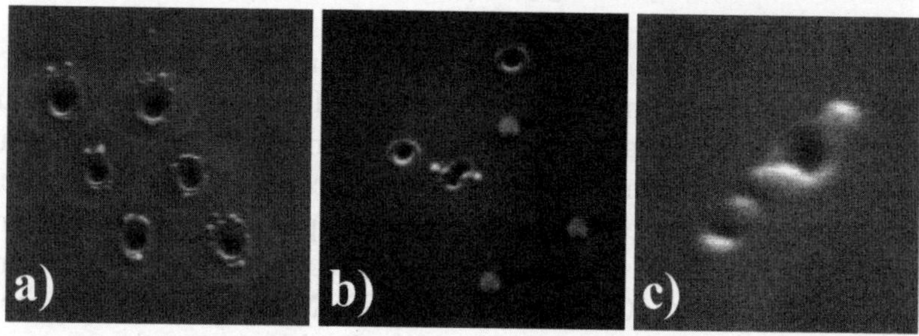

Fig. 27. Holes generated due to local field enhancement at laser-irradiated particles on a silicon wafer. In 27 a) and 27 b) the particles were PS spheres with a diameter of 1700 and 320 nm, respectively; in 27 c), irregular alumina particles with an average size of 400nm were used. The side lengths of the displayed squares are 6800 nm, 4970 nm, and 1300 nm for a), b) and c), respectively (from Leiderer, 2000).

As it turned out in all experiments where we removed particles by laser irradiation with ps and fs pulses, this removal was always accompanied by the appearance of such holes. Hence the particle removal – or "cleaning" process can be directly related to the ejection of material from the wafer, which then also leads to the observed crater formation. One might argue that realistic contaminants that have to be removed by laser cleaning applications are typically not of spherical shape. Also for irregular contaminants such as Al_2O_3 powder, however, field enhancement takes place – but in contrast to the spherical particles in asymmetric patterns – giving rise to the generation of holes also. An example of such a structure is shown in Fig. 27 c).

While in laser assisted particle removal the generation of holes due to local substrate ablation is an undesired, defect creating process, it may as well be utilized for nanolithographic surface structuring purposes (Leiderer, 2000; Lu, 2000c, 2002; Münzer, 2001; Münzer, 2002; Huang, 2002 a, b). By controlled application of a colloidal suspension, deposition of isolated PS spheres at any desired concentration onto the substrate is possible. Such isolated spheres can be used to create single holes. An example is shown in Fig. 28. In order to decrease the hole's size, the illumination wavelength was decreased to 400 nm and particles with a diameter of 370 nm have been illuminated. AFM imaging of the holes created this way reveal a hole diameter of about 100 nm – about one quarter of the applied laser

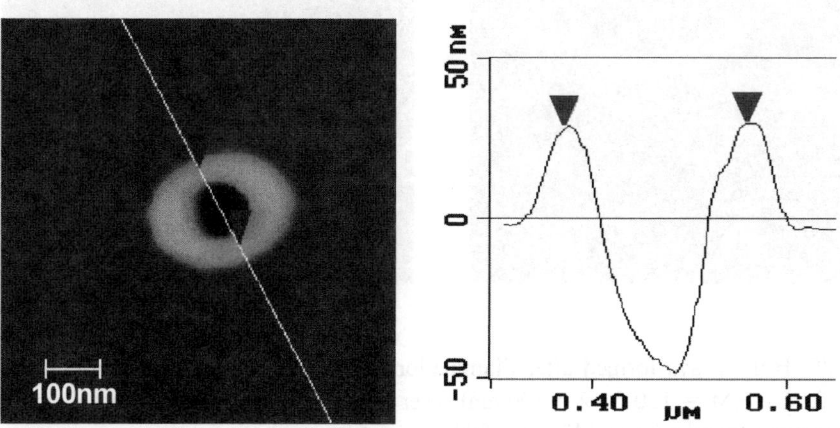

Fig. 28. AFM image of a singe hole created by irradiation of a 370 nm PS sphere with laser pulses from a frequency doubled Ti: Sapphire-laser ($\lambda = 400$ nm) (from Münzer, 2002).

wavelength or even one eight's part of the wavelength (Huang, 2002a).

An exciting possibility besides the fabrication of single holes is the exploitation of self-organization processes, e.g. the utilization of 2-D colloidal monolayer, for the structuring process (Burmeister, 1998). Illumination of an array of particles by an ultrashort laser pulse leads to an array of holes. This is shown in Fig. 29 where hexagonal monolayer of polystyrene spheres arranged on silicon and germanium substrates have been used to create hexagonal hole patterns. Being a parallel technique this method allows the structuring of large substrate areas and can result in a million holes and more for a single shot, limited only by the size of the laser spot. A similar technique was used to produce nanosize patterning in metallic films shown in Fig 30 (Huang, 2002 a, b). It was demonstrated that ns excimer laser produced a lattice of holes with a size below 50 nm (using 140 nm PS particles, see in Fig. 31).

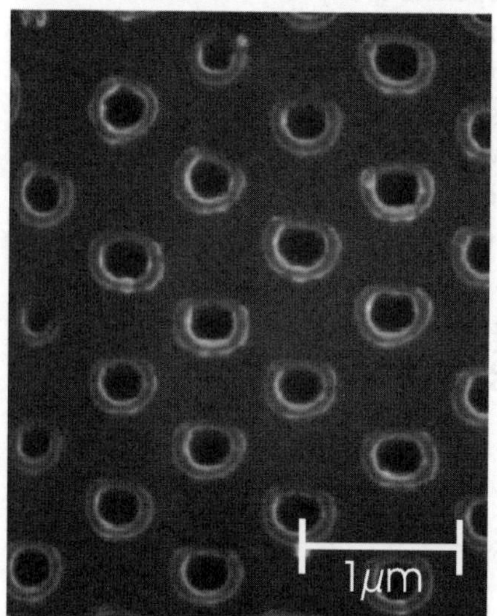

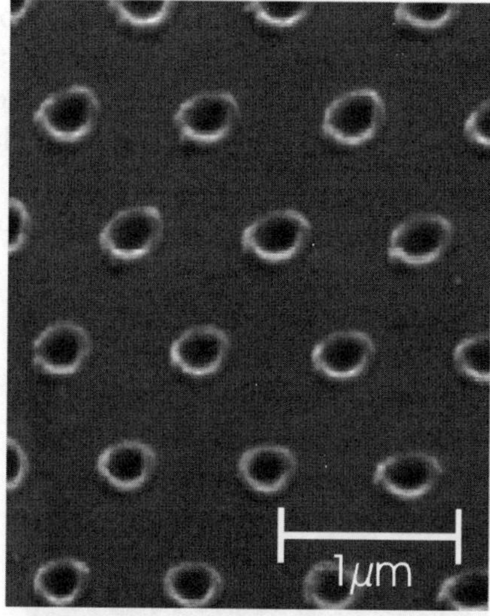

Fig. 29. Hole arrays formed after illumination of a hexagonal colloidal monolayer with fs (FWHM = 150 fs, λ = 800 nm) laser pulses. Left: silicon substrate, right: germanium substrate (from Münzer, 2002).

7.1.2. Parameters influencing field enhancement induced ablation

As already discussed above, the laser pulse length considerably influences the morphology of the field enhancement induced defect patterns in the substrate. Whereas laser pulses with FWHM in the ps and sub-ps range lead to well-pronounced ablation patterns, nanosecond laser pulses lead to a completely different defect shape. An SEM image of such a shallow, broadened melting pool is shown in Fig. 32.

Another parameter that might influence the field enhancement process considerably is the presence of the substrate itself. It was pointed out (Luk`yanchuk, 2000), that in numerical calculations the intensity is increased by a factor of 1.5 as a consequence of the presence of a silicon

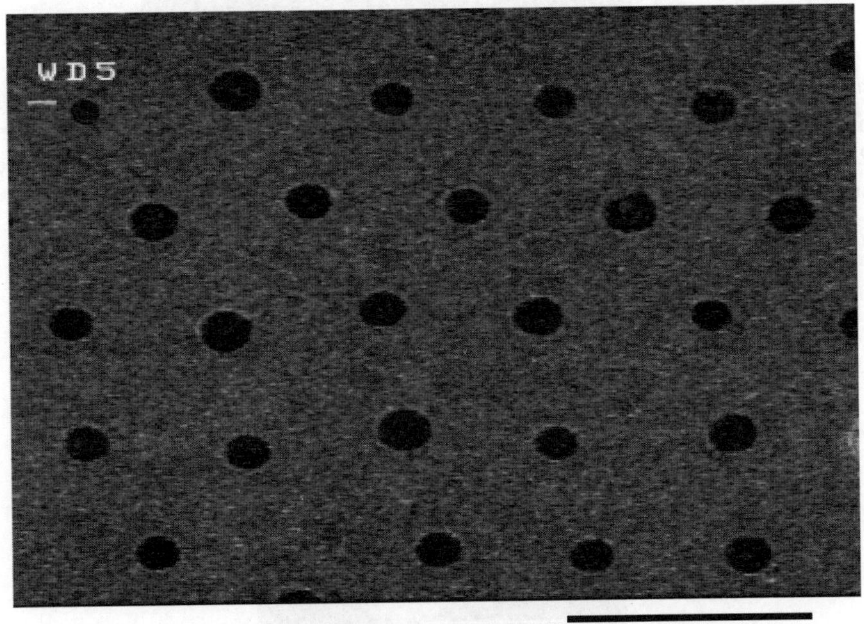

Fig. 30. SEM image of periodic pit arrays formed after illumination of isolated 1.0 µm SiO_2 particles on an Al film surface by a single laser pulse (FWHM = 23 nm, λ = 248 nm) with a laser fluence of 300 mJ/cm^2 (from Huang, 2002 a, b).

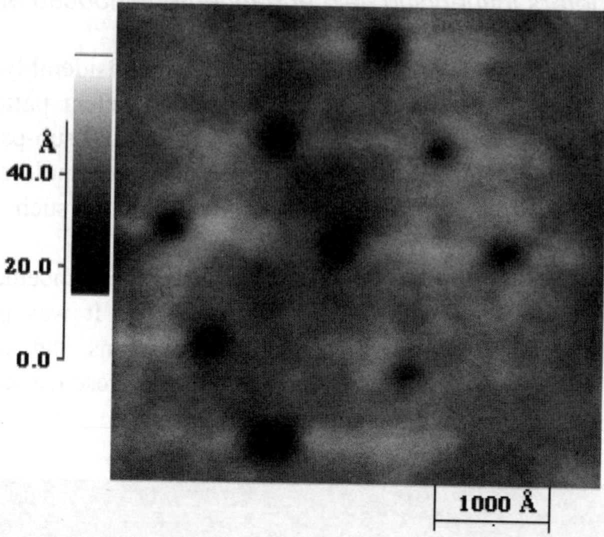

Fig. 31. SEM image of periodic pit arrays formed after illumination of isolated 140 nm polystyrene particles on an Al film surface by a single laser pulse (FWHM = 23 nm, λ = 248 nm) with a laser fluence of 300 mJ/cm^2 (from Huang, 2002). The smallest holes are of diameters ≈ 300 Å (from Huang, 2002 a).

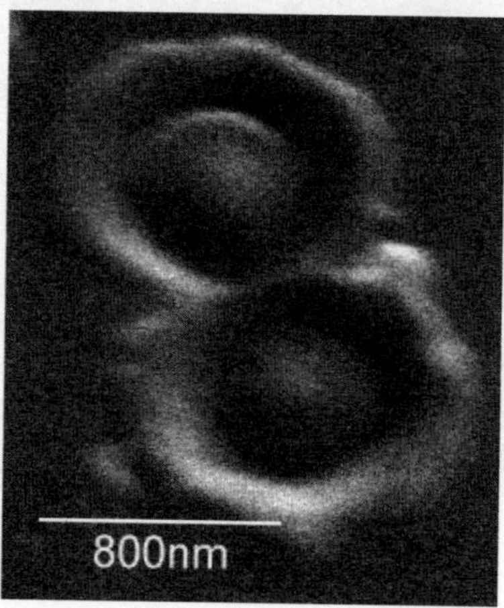

Fig. 32. Scanning electron microscope image of a typical melting site induced by field enhancement after ns laser illumination. In comparison to fs laser induced hole structures the morphology of the damage site is clearly smeared out due to heat diffusion in the silicon substrate (from Mosbacher, 2002).

substrate. Also the calculations based on the 3D model in the above sections strongly suggest, that the substrate should have a major influence on the process.

Motivated by these predictions we have conducted a set of experiments that allow the imaging of the actual field distribution at particles *on* surfaces and compared these experimental findings to calculations for free space (purely Mie-theory) and calculations including the substrate. The results of this comparison are shown in Fig. 12.

Neglecting the influence of the substrate we computed on the basis of the Mie theory the field of polystyrene particles in the substrate plane by an algorithm taken from Barber, 1990. The patterns resulting from these calculations are shown in Fig. 12a for a 320 nm (bottom) and an 800 nm (top) particle illuminated by a plane wave with a wavelength of 800 nm. They exhibit a double-peak structure with a distance between the maxima of about 300 nm for both particle sizes, whereas the absolute value of the intensity enhancement differs by a factor of 4.5.

In a next step we compared these calculations to the experimental results shown in Fig. 12b. As previously shown (Münzer, 2001), fs laser irradiation of particles on surfaces creates an ablation pattern underneath each illuminated particle. For constant laser fluences, all sites of ablation exhibit the same morphology. Underneath particles 320 nm in diameter, the ablation pattern shows a double hole structure, yet the distance between the maxima is obviously smaller than for the free space calculation. The pattern formed underneath an 800 nm particle does not show two peaks at all, but exhibits an elliptical shape.

Apparently, the Mie calculations differ substantially from the experimental findings. However, this theory holds only for an illuminated sphere in free space, thus neglecting the influence of the substrate. Consequently, these differences can be attributed to the influence of the silicon surface on the field distribution (see above in the Section 3).

In order to investigate this in more detail we performed semianalytical field calculations using the Multiple Multipole (MMP) technique (Hafner, 1990). Advantage was taken of the mirror symmetries of the problem and ring multipoles (were used to model the scattered fields both in the substrate and in vacuum). Optical constants were taken from Palik, 1998. The result of these calculations is shown in Fig. 12c. The calculated field distributions fit the experimental findings well, if for the particles 800 nm in diameter we compare them to the results obtained in the substrate (10 nm below the surface) and for the smaller particles with those above (10 nm above the surface). The reason for this needs some further investigation. Besides the effect of ablation in the central part of the hole one observes the formation of a rim around the ablated area. In the development of this rim structure the dynamics of the melt influenced by viscosity and surface tension may play a role.

7.2. Near-field effects in the laser cleaning process

The above-described experimental results clearly show the importance of near field effects for the field of laser cleaning. In the following will illustrate these consequences.

7.2.1. Experimental details

For dry laser cleaning we determined the removal efficiency and the cleaning thresholds in laser fluence for contaminants of different sizes and materials. We deposited individual spherical colloidal polystyrene[1] (PS) on industrial silicon wafers in a spin coating process described in Mosbacher, 2002b. Prior to particle application the wafers were cleaned in an ultrasonic bath in isopropanol (IPA). The samples prepared in this way were irradiated by a single Nd: YAG laser pulse (λ = 532 nm, FWHM = 8 ns) either in ambient conditions (relative humidity 30-40%) or in high vacuum (HV, 10^{-6} mbar). In ambient cleaning a flow of pressurized, filtered air was used to blow away the removed particles and to prevent their redeposition.

Under ambient conditions particle removal in the cleaned area (about 1 mm^2) was detected by a light scattering technique (Mosbacher, 2000). A 5 mW HeNe laser illuminated a spot with a diameter of 0.5 mm, which corresponds to several hundred particles monitored, its scattered light was detected by a photomultiplier. The monitored area was much smaller than the illuminated area, therefore in this case the laser fluence can be considered as almost homogeneous. In HV we determined the fluences necessary for particle removal by inspecting the illuminated spot with an optical microscope prior and after the laser pulse. We measured the threshold cleaning fluence relatively to the melting threshold of Si,

[1] IDC, Portland, Oregon, USA

monitoring the reflected light of the HeNe laser with ns time resolution. As the laser fluence for the onset of melting of silicon is well known, this can be used for a conversion of relative fluences into absolute numbers.

7.2.2. Variation of the size parameter

Within the frame of the Mie theory the extent of intensity enhancement in the near field of the particle critically depends on the size parameter $q = \pi d/\lambda$, where d denotes the particle diameter and λ the applied wavelength. When the secondary scattering of reflected radiation is taken into account (see Section 3) both parameters d and λ influenced independently. Experimentally there exist two approaches of varying q parameter in laser cleaning experiments, either by changing the particle size or by varying the wavelength. We will discuss both in the following.

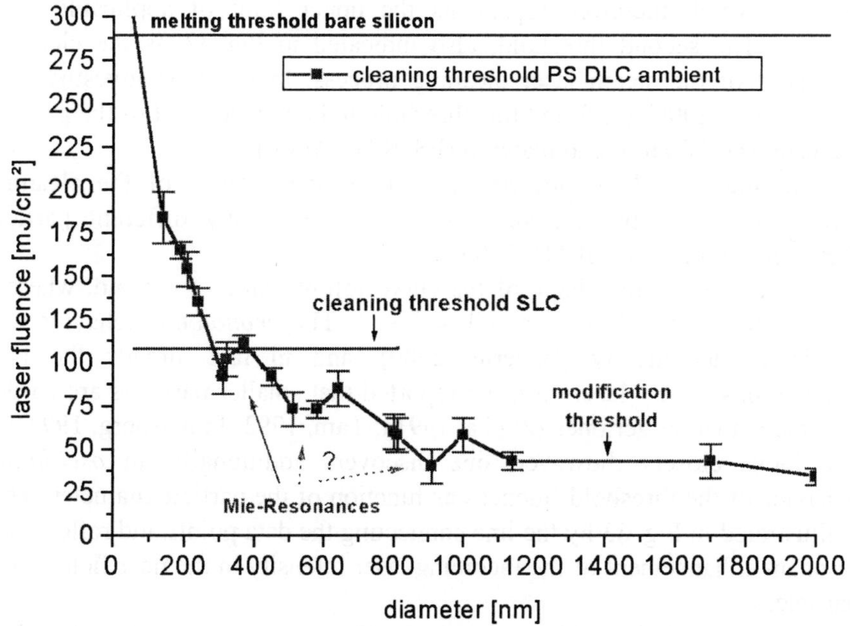

Fig. 33. Threshold in the applied laser fluence for particle removal in DLC in ambient air. Particles smaller in diameter than 110 nm could not be removed (from Mosbacher, 2002a).

7.2.2.1. Variation of the particle size

First we have studied particle removal in ambient conditions. This environment represents the conditions that may be found in a possible future application of the process.

In Fig. 33 the thresholds in applied laser fluence for the removal of PS particles are plotted as a function of the particle size. From this diagram one can obtain a lot of information on the process, especially on the role of field enhancement.

First we would like to bring to mind two very important thresholds in the laser cleaning process. One of them is the threshold for the onset of melting of the bare substrate. As dry laser cleaning is aimed for an industrial application, any change of the structure of the silicon wafer, i.e. the silicon substrate and a native oxide layer of specified thickness, as induced by melting has to be strictly avoided. From experiments (Kurz, 1983; Lowndes, 1983; Boneberg, 1993) this melting threshold is known to be about 280 mJ/cm^2, which therefore represents the upper limit of applicable laser fluence. The second threshold, also indicated in Fig. 32, is the cleaning threshold of the steam laser cleaning process. In previous investigations (Mosbacher, 2000) we found this threshold to be independent of the particle diameter (60-800 nm) and material (PS, SiO_2, Al_2O_3).

In order to obtain information on the dependency of the cleaning threshold on the particle size we investigated many different particle diameters in the range of 110-2000 nm.

At first sight the shape of the curve follows an r^{-k} – trend, where r denotes the particle radius and $1 < k < 2$. This *monotonic* behavior was predicted theoretically (Bäuerle, 2000), and in fact already the first publications on dry laser cleaning reported that smaller particles are harder to remove than larger ones (Zapka, 1991a; Tam, 1992; Engelsberg, 1993). If one looks closely, however, one discovers additionally an *oscillating* behavior of the threshold fluences as function of the particle diameter. This is illustrated in Fig. 33 by the line connecting the data points and reflects the resonant enhancement of the incoming laser intensity in the near field of the particles.

However, it should be pointed out, that the line connecting the data points is just a guide to the eye and does probably not describe the exact field enhancement efficiency as function of the particle diameter. The number of the discrete particle sizes used in our experiments is not sufficient to resolve this dependency. Nevertheless the resonances account for the deviation of the curve from a smooth, monotonic shape.

Besides decreasing the applied laser fluence necessary to remove particles, field enhancement underneath the particles is responsible for surface damage by local melting/ablation of the substrate, as shown in Section 7.1. Depending on the pulse length used for the cleaning laser, and hence the thermal diffusion length during this laser pulse, local ablation leads either to the formation of steep holes (ultrashort pulses <100 ps) or shallow melting/ablation sites (nanosecond pulses).

In dry laser cleaning using ultrashort pulses the removal of a particle is always accompanied by the formation of a hole, i.e. the hole formation threshold is identical with cleaning threshold. Consequently in this case the particles are removed by the momentum transfer of the ablated species rather than by thermal substrate expansion – local ablation acts as a cleaning mechanism (Mosbacher, 2001). This shows very clearly that for a damage free dry laser cleaning ultrashort pulses are not suitable.

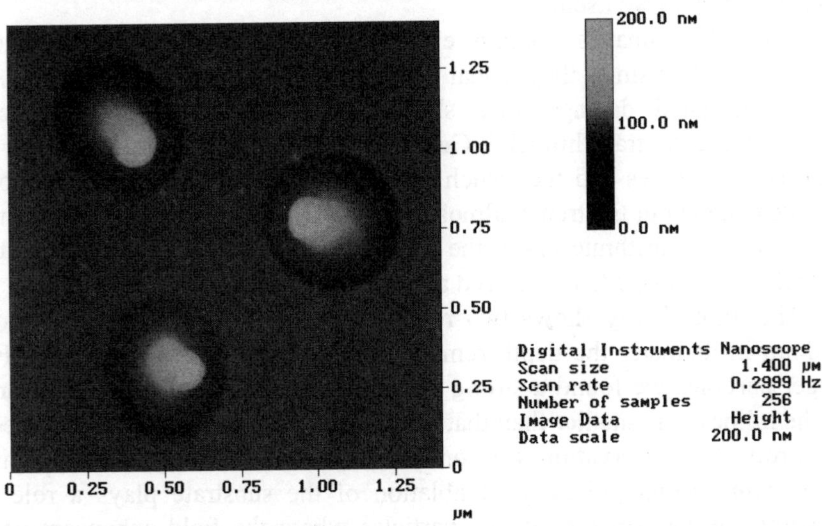

Fig. 34. Atomic force microscopy (AFM) image of a damage site after dry laser cleaning using ns pulses. The damage site was created by an agglomerate of three PS spheres 800 nm in diameter. Note the typical damage morphology: a shallow trench surrounds a central hillock (from Mosbacher, 2002 and 2002c).

Against this background also for the case of ns pulses we focused our interest on the determination of the *local melting/ablation* thresholds in addition to the cleaning thresholds. Instead of the melting threshold of the bare silicon surface this damage threshold represents the true upper limit for the applicable laser fluence and is by its nature particle dependent. For the determination we made use of the Gaussian spatial beam profile of our cleaning laser. Due to this profile a spatial variation in the cleaned area corresponds to a variation in the locally applied laser fluence. In a post process analysis we investigated the cleaned areas of our samples with an AFM. By this method we imaged the field enhancement induced damage sites (cf. Fig. 12 for an SEM image), a typical example can be seen in Fig. 34.

Imaging damage sites at different locations in the cleaned areas and especially at their borders, which correspond to the cleaning threshold fluence, we determined the damage threshold for each particle size. For all particles investigated the cleaning threshold was identical with the damage threshold. Damage free dry laser cleaning was *not* possible applying the laser parameters we used!

The AFM images contain even more information on the particle removal mechanism as they reveal quantitative topographic information. All the investigated damage sites showed the same features: a "trench" surrounded a central "hillock". Generally spoken the hillock was lower at high laser fluences and the trench deeper, for low laser fluences a hillock was detectable but the trench almost disappeared. In Fig. 35 we have plotted in a double logarithmic graph the mean trench depths for the investigated particles for damage that occurred at the cleaning/damage threshold.

This plot clearly shows two regimes: for particles smaller than about 250 nm in diameter the depth remains almost constant at about 1 nm. For larger particles we found a strong increase in the trench depth, the volume of the hillock was smaller than that of the trench – ablation had taken place.

From this observation we conclude, that even for dry laser cleaning using *nanosecond* pulses local ablation of the substrate plays a role as cleaning mechanism for "large" particles where the field enhancement is higher and thus provides fluences high enough for ablation. At smaller particles field enhancement probably causes local melting, but no ablation.

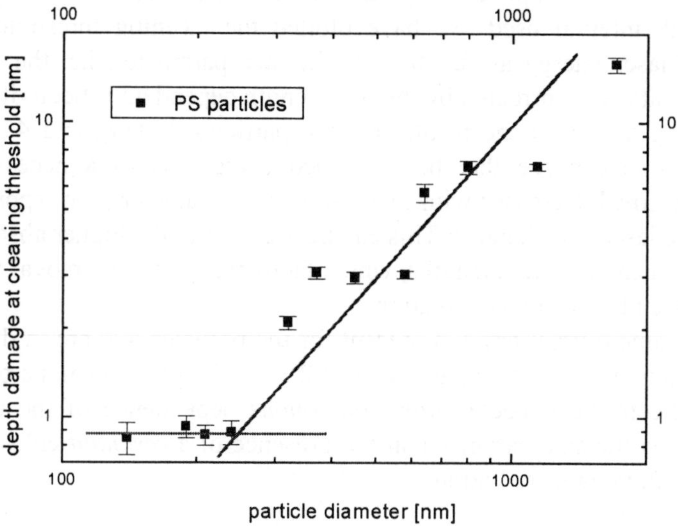

Fig. 35. Trench depth of the damage sites as a function of the particle size when the particles were removed in DLC applying the threshold cleaning fluence. For particles larger than 250 nm in diameter the depth increases strongly with the particle size, for smaller particles it remains almost constant (from Mosbacher, 2002 and 2002c)

7.2.2.2. Variation of the laser wavelength

A doubtless probing of the optical resonances in dry laser cleaning is not possible by a variation of particle sizes, as they are available only in certain, discrete diameters. On the contrary, it is possible to vary the laser wavelength continuously using an optical parametrical oscillator (OPO). For our dry laser cleaning experiments we chose a wavelength interval from 500 nm to 630 nm. It contains the wavelength of the frequency doubled Nd: YAG laser (532 nm) used in the previous experiments, and the refractive index of silicon – one parameter that influences the field distribution at the particle - does not vary too much in this wavelength range.

In Figure 36 we show the experimental results for the wavelength-dependent cleaning thresholds of particles with diameters of 370, 400 and 450 nm, respectively. The laser wavelength was varied in steps of 5-10 ± 0.2 nm, comparable to the standard deviation of the particle diameter size distribution. As the optical properties of the silicon substrate vary in the wavelength interval used, we have plotted the cleaning threshold in the *absorbed* laser energy as function of the size parameter, i.e. the incident laser intensity was corrected by the wavelength dependent reflectivity.

Taking a look at the results for the particles 370 nm and 450 nm in diameter, one can see that the absorbed energy per area necessary for particle removal decreases with increasing size parameter (corresponding to decreasing laser wavelength). This can be ascribed to the higher absorptivity of silicon at smaller wavelengths, which facilitates particle removal via field enhancement induced local ablation.

The same overall trend is present for the particles 400 nm in diameter, as well. However, at a size parameter of 2.3 – 2.4 a plateau can be observed which reflects the expected *non-monotonous* dependence of the cleaning threshold on the size parameter in the presence of a *resonant* enhancement of the incoming laser radiation.

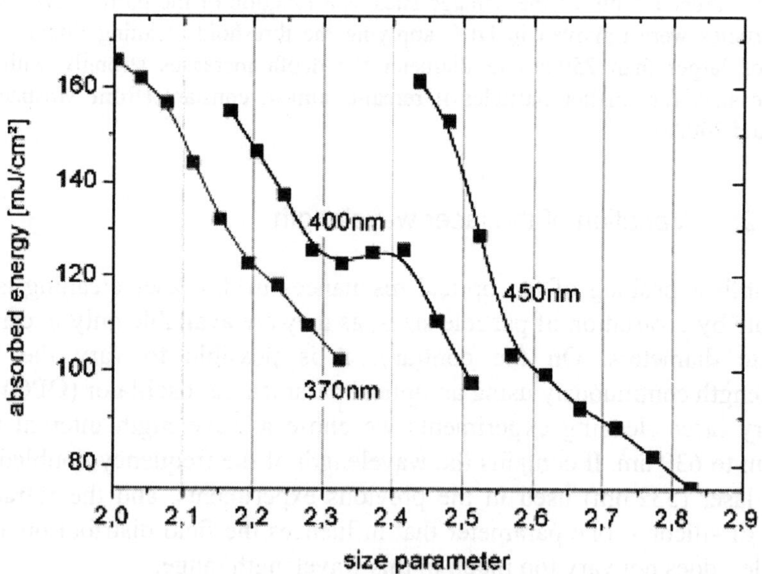

Fig. 36. Variation of the dry laser cleaning threshold as function of the size parameter when the laser wavelength is varied (from Mosbacher, 2002d).

7.2.3. Influence of the incident angle

Usually in laser cleaning studies the samples to be cleaned are irradiated at an incident angle perpendicular to the substrate surface. Consequently, the maximum intensity of the laser light due to field enhancement is located right underneath the particle in the substrate surface plane. Especially for particles larger in size than the wavelength, where the field enhancement pattern is quite similar to a "point-focus" known from geometrical optics, the location of this maximum intensity is crucial for the cleaning performance as it leads to the local surface ablation.

A deviation from the perpendicular incidence results in a displacement of the maximum field intensity away from the contacting point. Such a displacement should result in a drop in the cleaning efficiency, and indeed this could be confirmed experimentally (Zheng, 2001). Silicon samples with particle contamination (SiO_2, diameter 2.5μm) on the surface have been irradiated by a KrF excimer laser (FWHM ≈ 23 ns, λ=248 nm) at incident angles ranging from 0°-30°. The incident laser fluence of 43 mJ/cm² was chosen well above the cleaning threshold fluence of about 5 mJ/cm², which should result in a complete particle removal. However, as the experimental results displayed in Figure 37 show, this is not the case for incident angles different from 0°. On the contrary, the efficiency decreases with increasing incident angel, and for angles above 15° no particle removal was detected.

Besides the cleaning efficiency we plot the uniform laser intensity as function of the incident angle. This intensity is proportional to $A(\theta)\cos\theta$, where θ is the incident angle, and A is the absorptivity. The variation of incident angle from 0° to 15° causes the effective laser fluence to drop less than 3.5% (see in Fig. 37) – negligible in comparison with the efficiency drop.

From the calculation based on the Mie theory (Zheng, 2001) follows that the near-field light intensity in agreement with the experimental results declines from 100% to zero, when the tilting angle increases from 0° to 15°.

7.2.4. Influence of the surface roughness

Another nontrivial consequence of near field focusing can be found from the dependence of the cleaning efficiency on the surface roughness. The roughness by itself leads to a decrease in the Hamaker force because of a smaller contact region. The theoretical prediction shows that due to this the

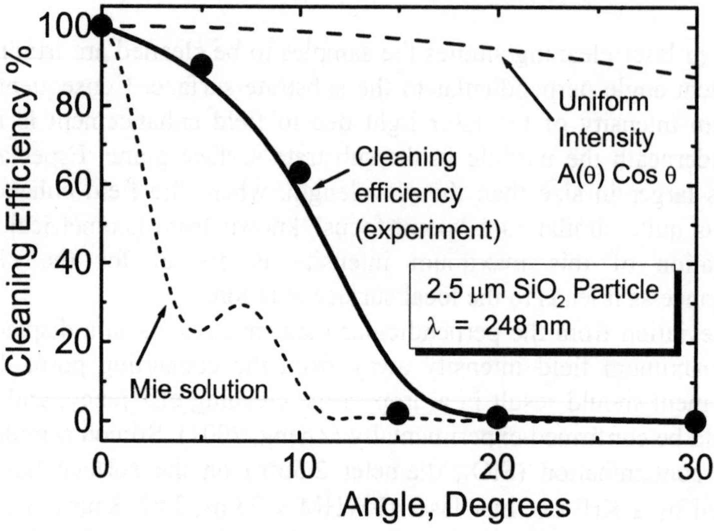

Fig. 37. Cleaning efficiency versus incident angle (from Zheng, 2001). The uniform light intensity and the Mie solution are nornalized to 100 % at normal incidence.

particle adhesion force drops with the presence of very small roughness (Whitehouse, 1970; Schaefer, 1995; Soltani, 1995). It was confirmed experimentally in static experiments with AFM (Schaefer, 1995). At the same time the roughness leads to a "lifting" of the particle above the surface, which results in a fast decrease in the near-field light intensity. Roughness amplitude of 50 nm reduces the laser intensity by a factor of two (see in Fig. 10). Thus, in spite of the reduction of the Hamaker force, one should expect a decrease of cleaning efficiency due to fast decrease of laser intensity with particle lifting.

The cleaning curves for different samples (with different roughness amplitude) are shown in Fig. 38. The SiO_2 particle size was 1.0 μm. One can clearly see decrease in the cleaning efficiency versus surface roughness.

8. Conclusion

The theory of dry laser cleaning, based on 1D thermal expansion of the substrate was the dominant model for the last 10 years. This theory predicts qualitatively correct results; nevertheless it yields the quantitative

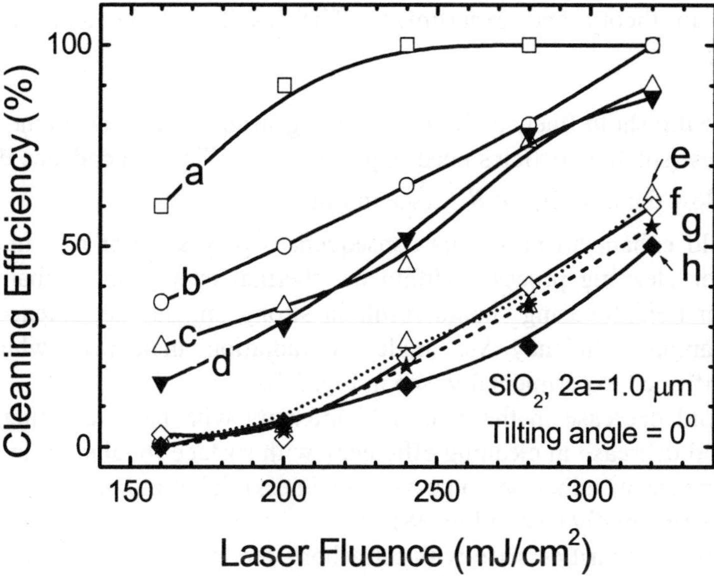

Fig. 38. Cleaning curves for samples with different roughness. The surface roughness increases with the alphabatic order from $\delta = 0$ (a) to $\delta \approx 50$ nm (h). From Zheng, 2001.

discrepancy with experiment by one-two orders of the magnitude. One can see from Figs. 24, that the 1D theory cannot explain experimental results on dry laser cleaning for both small and big particles. It means that some important physics is missing with 1D model. The results of recent papers (Arnold, 2002) do not confirm the 1D thermal expansion mechanism either.

The main idea of our latest examinations was that instead of a 1D model one should use a 3D model, based on the ideas of near field focusing and optical resonance effect, which are very sensitive to wavelength, particle size, incident angle and surface morphology.

Our examinations show, that for transparent spherical particles with a size comparable or even smaller than the laser wavelength, the optical resonance and substrate reflection produce under the particle an "enhanced" near-field light intensity, which may exceed the incident laser intensity by one or two orders of magnitude.

The present work was devoted to the examination of the mentioned 3D effects in theory and experiments. The results can be summarized as follows:

1) The threshold fluence for laser cleaning strongly depends on the particle sizes (smaller particles need higher fluence). This dependence does not follow the law $\Phi_{th} \propto R^{-2}$ (see in Fig. 33).
2) Field enhancement and its consequences play a major role in the dry laser cleaning process. Within the thermal expansion mechanism the near-field focusing should result in strong angular dependence in the cleaning efficiency vs. angle of radiation incidence, which was confirmed experimentally (see in Fig. 37).
3) Rapid decrease in the near field intensity with particle lifting should yield decrease in cleaning efficiency with surface roughness (in spite of decrease in adhesion force vs. roughness). This effect was confirmed experimentally (see in Fig. 38).
4) New peculiarities arise with ultrashort, femtosecond laser pulses. First the enhanced intensity leads to a local ablation of the substrate, which acts as a cleaning mechanism and results in particle ejection via momentum transfer (Mosbacher, 2002a). Second this enhanced intensity is the main source of substrate damage in the dry laser cleaning process.
5) The theory (Luk`yanchuk, 2000) predicts a rapid variation in the cleaning efficiency with a change of wavelength (for fixed size of particles) due to optical resonance effect. To check this prediction experiments with a tunable laser have been carried out, which revealed evidence of optical resonances (Mosbacher, 2002d).

Acknowledgement

We wish to thank Prof. S. I. Anisimov and Dr. N. Arnold for discussions. B. L. is thankful to Russian Basic Research Foundation (grants 01-02-16136, 01-02-16189). The research at University of Konstanz is supported by the European Union (grant n° ERBFMRXCT98 0188), the Konstanz Center for Modern Optics and Wacker Siltronic, Germany.

References

Abrikosov A. A., Gorkov L. P., Dzyaloshinski I. E., *Methods of Quantum Field Theory in Statistical Physics*, (Prentice-Hall, Englewood Cliffs, New Jersey, 1965)

Arnold N., Schrems G., Mühlberger T., Bertsch M., Mosbacher M., Leiderer P., Bäuerle D., *Proc. SPIE,* vol. **4426**, pp. 340-346 (2002); see also Chapter 2 in this book

Assendel'ft E. Y., Beklemyshev V. I., Makhonin I. I., et al., *Sov. Tech. Phys. Lett.* **14**, 1006 (1988 a)

Assendel'ft E. Y., Beklemyshev V. I., Makhonin I. I., et al., *Sov. Tech. Phys. Lett.* **14**, 1494 (1988 b)

Barber P. W., Hill S. C., *Light Scattering by Particles: Computational Methods*, (World Scientific, Singapore 1990)

Bäuerle D., *Laser Processing and Chemistry*, 3[d] ed., (Springer, Berlin, 2000)

Beklemyshev V. I., Makarov V. V., Makhonin I. I., Petrov Yu. N., Prokhorov A. M., Pustovoy V. I., JETP Letters, vol. **46**, 347 (1987)

Bobbert P. A., Vlieger J., *Physica A*, vol. **137**, 209 (1986 a)

Bobbert P. A., Vlieger J., Greef R., *Physica A*, vol. **137**, 243 (1986 b)

Bohren C. E., Huffman D. R., *Absorption and Scattering of Light by Small Particles*, (John Willey & Sons, 1983)

Boneberg J., *Metallische Dünnfilmschmelzen von Halbleiterschichten nach ns-Laser-Annealing*, Hartung Gorre, Konstanz (1993)

Born M., Wolf E., *Principles of Optics*, 7-th Edition, (Cambridge University Press, 1999)

Bowing R. A., In (Mittal, 1988), pp. 129

Boykov D. A., Kolomenskii A. A., Maznev A. A., Maishev Yu. V., Panfilov Yu. V., Ravich A. M., *USSR Patent* #4872125/25 (August, 1991)

Burmeister F., Schäfle C., Keilhofer B., Bechingr C., Boneberg J., Leiderer P., *Adv. Mater.* **10**, 495 (1998).

Derjaguin B. V., *Kolloid Z.*, **69**, 155 (1934)

Derjaguin B. V., Muller V. M., Toporov Yu. P., *Journal of Colloid and Interface Science*, **53**, 314 (1975); **73**, 293 (1980)

Dobler V., Oltra R., Boquillon J. P., Mosbacher M., Boneberg J., Leiderer P., *Appl. Phys. A*, **69**, 335 (1999)

Donovan R. P., Locke B. R., Ensor D. S., In (Mittal, 1988), pp. 43-56

Durig U., Pohl D.-W., Rohner F., *J. Appl. Phys.*, **59**, 3318 (1986)

Dzyaloshinski I. E., Pitaevsky L. P., *JETP*, **9**, 1282 (1959)

Dzyaloshinski I. E., Lifshitz E. M., Pitaevsky L. P., *JETP*, **10**, 161 (1960)

Dzyaloshinski I. E., Lifshitz E. M., Pitaevsky L. P., *JETP*, **10**, 165 (1960)

Engelsberg A. C., *Proc. MRS*, vol. **315**, 255 (1993)

Fields M. H., Popp J., Chang R. K., In "*Progress in Optics*", vol. **41**, Ed. by E. Wolf (Elsevier, 2000)

Grigoriev I. S., Meilikhov E. Z. (Eds.), *Handbook of Physical Quantities*, (CRC Press, Boca Raton, 1997)

Hafner C., *The Generalized Multipole Technique for Computational Electromagnetism* (Artech House, Boston, 1990)

Halfpenny D. R., Kane D., *J. Appl. Phys*., vol. **86**, 6641 (1999)

Hamaker H. C., *Physica*, **4** (10), 1058 (1937)

Heim L. O., Blum J., Preuss M., Butt H. J., *Phys. Rev. Lett*., **83**, 3328 (1999)

Huang S. M., Hong M. H., Luk'yanchuk B. S., Lu Y. F., *Laser assisted nanofabrications on metal surfaces with optical near field effects*, Proc. SPIE, vol. **4760** (2002a)

Huang S. M., Hong M. H., Luk`yanchuk B. S., Zheng Y. W., Song W. D., Lu Y.F., Chong T. C., *Pulsed laser-assisted surface structuring with optical near-field enhanced effects,* J. Appl. Phys., (2002b)

Imen K., Lee S. J., Allen S. D., *Appl. Phys. Lett*., **58**, 203 (1991a)

Imen K., Lee S. J., Allen S. D., *U. S. Patent* Serial No. 4, 987, 286 (January 1991b)

Israelachvili Ya., *Intermolecular and Surface Forces*, 2^{d} ed., (Academic Press, London, 1991)

Johnson K. L., Kendall K., Roberts A. D., *Proc. Roy. Soc*., **A 324**, 301 (1971)

Johnson K. L., In: "*Theoretical and Applied Mechanics*", Ed. by W. T. Koiter, p.133 (North-Holland 1976)

Karlov N. V., Kirichenko N. A., Luk'yanchuk B. S., *Laser Thermochemistry. Fundamentals and Applications,* (Cambridge International Science Publishing, Cambridge, 2000)

Kelley J. D., Hovis F. E., *Microelectronic Engineering* **20**, 159 (1993)

Kerker M., *The scattering of Light*, (Academic Press, New York & London, 1969)

Kerker M., *Selected Papers on Light Scattering*, Proc. SPIE, vol. **951** (Part One), (1989), see Section 4 "*Optical Resonances*"

Klimchitskaya G. L., Mohideen U., Mostepanenko V. M., *Phys. Rev. A*, vol. **61**, 062107 (2000)

Kolomenskii A. A., Maznev A. A., *Sov. Tech. Phys*. Lett., **17**, 62 (1991)

Kurz H., Lompré L. A., Liu J. M., *J. de Physique* **10**, C5 (1983)

Landau L. D., Lifshitz E. M, *Course of Theoretical Physics* (Pergamon Press, 1994, 1998, 1999):
Vol. 1, *Mechanics*, 3rd Edition
Vol. 2, *The Classical Theory of Field*, 4th Edition
Vol. 3, *Quantum Mechanics*, 3rd Edition
Vol. 4, *Quantum Electrodynamics*, 2nd Edition
Vol. 5, *Statistical Physics*, Part 1, 3rd Edition
Vol. 6, *Fluid Mechanics*, 2nd Edition
Vol. 7, *Theory of Elasticity*, 3rd Edition
Vol. 8, *Electrodynamics of Continuous Media*, 2nd Edition
Vol. 9, *Statistical Physics*, Part 2 (see Lifshitz E. M., Pitaevsky L. P.)
Vol. 10, *Physical Kinetics*

Lee L. H., *Fundamentals in Adhesion*, (Plenum Press, New York, 1990)

Leiderer P., Boneberg J., Dobler V., et al, *Proc. SPIE*, vol. **4065**, 249 (2000)

Leontovich M. A., Rytov S. M., *JETP*, **23**, 246 (1952)

Lifshitz E. M., *JETP,* **29**, p. 94 (1955)

Lifshitz E. M., Pitaevsky L. P., *Statistical Physics*, Part 2, §§ 80-82 (Pergamon Press, 1980)

Lowndes D. H., Wood R. F., Westbrook D., *Appl. Phys. Lett.* **43**, 258 (1983)

Lu Y.F., Song W.D., Ang B.W., Chan D.S.H., Low T.S., Appl. Phys. A **65**, 9 (1997)

Lu Y. F., Zheng Y. W., Song W. D., *J. Appl. Phys*. **87**, 1534 (2000a)

Lu Y. F., Song W. D., Luk`yanchuk B. S., Hong M. H., Zheng Y. W., In: "*Laser Solid Interactions for Materials Processing*", Editors: D. Kumar, D. P. Norton, C. B. Lee, K. Ebihara, X. Xi, 2000 MRS Spring Meeting, San Francisco, CA, Vol. **617**, J1.4 (2000b)

Lu Y. F., Zhang L., Song W. D., Zheng Y. W., Luk'yanchuk B. S., *JETP Letters*, vol. **72**, 457 (2000c)

Lu Y. F., Zhang L., Song W. D., Zheng Y. W., Luk'yanchuk B. S., *Proc. SPIE*, vol. **4426**, 143 (2002)

Luk'yanchuk B. S., Zheng Y. W., Lu Y. F., *Proc. SPIE*, vol. **4065**, 576 (2000)

Luk'yanchuk B. S., Zheng Y. W., Lu Y. F., *Proc. SPIE,* vol. **4423**, 115 (2001)

Luk'yanchuk B. S., Zheng Y. W., Lu Y. F., *RIKEN Review*, No. **43**, 37 (2002a)

Luk'yanchuk B. S., Zheng Y. W., Lu Y. F., *Proc. SPIE,* vol. **4426**, 284 (2002b)

Luk'yanchuk B. S., Huang S. M., Hong M. H., *3D effects in dry laser cleaning*, Proc. SPIE, vol. **4760** (2002c)

Madelung O. (Ed.), *Semiconductors-Basic Data*, 2nd Edition, (Springer, Singapore, 1996)

Maugis D., *Journal of Colloid and Interface Science*, **150**, 243 (1992)

Maugis D., Gauthier-Manuel B., *In:* (Rimai D. S., 1995), p. 49

Mishenko M. I., Hovenier J. W., Travis L. D. (Eds.), *Light Scattering by Nonspherical Particles*, (Academic Press, New York, 2000)

Mittal K. L. (Ed.), *Particles on Surfaces: Detection, Adhesion and Removal*, vol. **1** (Plenum Press, New York, NY, 1988)

Mizes H., Loh K. G., Ott M. L., Miller R. J. D., In: "*Particles on Surfaces: Detection, Adhesion and Removal*", Ed. by K. L. Mittal (Marcel Dekker, New York, 1995)

Moody J. E., Hendel R. H., *J. Appl. Phys*., vol. **53**, 4364 (1982)

Morse P. M., Feshbach H., *Methods of Theoretical Physics*, vol. 2 (McGraw-Hill, 1953)

Mosbacher M., Dobler V., Boneberg J., Leiderer P., *Appl. Phys. A* **70**, 669 (2000).

Mosbacher M., Münzer H. -J., Zimmermann J., Solis J., Boneberg J., Leiderer P., *Appl. Phys.* **A 72**, 41 (2001)

Mosbacher M., Bertsch M., Münzer H.-J., Dobler V., Runge B.-U., Bäuerle D., Boneberg J., Leiderer P.: *Proc. SPIE*, vol. **4426**, 308 (2002)

Mosbacher M., Münzer H.-J., Bertsch M., Dobler V., Chaoui N., Siegel J., Oltra R., Bäuerle D., Boneberg J., Leiderer P., In: *„Particles on Surfaces 7: Detection, Adhesion and Removal"*, Ed. by K. L. Mittal, VSP Publishing (2002b, in press).

Mosbacher M., Dobler V., Bertsch M., Münzer H.-J., Boneberg J., Leiderer P., In: *„Particles on Surfaces 8: Detection, Adhesion and Removal"*, Ed. by K. L. Mittal, VSP Publishing (2002c, in press).

Mosbacher M., Dubbers O., Münzer H.-J., Bertsch M., Leiderer P., submitted (2002d)

Muller V. M., Yushchenko V. S., Derjaguin B. V., *Journal of Colloid and Interface Science*, **77**, 91 (1980); **92**, 92 (1983)

Münzer H.-J., Mosbacher M., Bertsch M., Zimmermann J., Leiderer P., Boneberg J., *Journ. Microscopy*, vol. **202**, 129 (2001)

Münzer H.-J., Mosbacher M., Bertsch M., Dubbers O., Burmeister F., Pack A., Wannemacher R., Runge B.-U., Bäuerle D., Boneberg J., Leiderer P., *Proc. SPIE*, vol. **4426**, 180 (2002)

Ordal M. A., Bell R. J., Alexander R. W., Jr., Long L. L., Querry M. R., *Appl. Opts*., **24** (24), 4493(1985)

Osborne-Lee I. W., In (Mittal, 1988), pp. 77

Ozisik M. N., Czisik M. N., *Finite difference methods in heat transfer*, (CRC Press, 1994)

Palik E. D. (Ed.): *Handbook of Optical Constants of Solids*, (Academic Press, Washington, 1985)

Palik E. D. (Ed.): *Handbook of Optical Constants of Solids*, (Academic Press, Boston, 1998)

Prokhorov A. M., Konov V. I., Ursu I., Mihailescu I., *Laser Heating of Metals* (Adam Hilger, Bristol, 1990)

Qiu T. Q., Longtin J. P., Tien C. L., *Journ. of Heat Transfer*, **117**, 340 (1995)

Rimai D. S., DeMejo L. P., Mittal K. L. (Eds.), *Fundamentals of adhesion and interfaces* (Utrecht, VSP Netherlands, 1995)

Rose M. E., *Elementary Theory of Angular Momentum* (Willey, New York, 1957), pp. 48-57

Rytov S. M., *Theory of Electromagnetic Fluctuations and Thermal Radiation*. Publ. of USSR Academy of Sciences, Moscow 1953 (English Translation, AFCRL TR 59-162)

Shaefer D. M., Carpenter M., Gady B., Reifenberger R., Demejo L. P., Rimai D. S., *In:* (Rimai D. S., 1995), p. 35; J. Adhesion Sci. Technol. **9**, 1049 (1995)

Shiles E., Sasaki T., Inokuti M., Smith D. Y., *Phys. Rev. B*, vol. **22**, 1612 (1980)

Sik J., Hora J., Humliche J., *J. Appl. Phys*., vol. **84**, 6291 (1998)

Sokolnikoff I. S., *Mathematical Theory of Elasticity*, 2nd Ed., (McGraw-Hill, 1956)

Soltani M., Ahmadi G., Bayer R. G., Gaynes M. A., J. *Adhesion Sci. Technol.* **9**, 453 (1995)

Stratton J. A., *Electromagnetic Theory*, (McGraw-Hill, New York & London, 1941)

Tam A. C., Leung W. P., Zapka W., Ziemlich W., *J. Appl. Phys.* **71**, 2217 (1992)

Vicanek M., Rosch A., Piron F., Simon G., *Appl. Phys. A*, **59**, 407 (1994)

Visser J., *Advances in Colloid and Interface Science*, **3**, 331 (1972)

Van de Hulst H. C., *Light Scattering by Small Particles*, (Dower Publ., New York, 1981)

Welsh L. P., Tuchman J. A., Herman I. P., *J. Appl. Phys.*, **64**, 6274 (1988)

Whitehouse D. J., Archard J. F., Proc. Roy. Soc. **A 310**, 97 (1970).

Wojcik G. L., Vaughan D. K., Galbraith L. K., *Proc. SPIE*, vol. **777**, 21 (1987)

Wolfram S., *Mathematica*, 4th Ed., (Wolfram Media/Cambridge University Press, 1999)

Ye H., Sun C. Q., Huang H., Hing P., *Appl. Phys. Lett.*, Vol. **78**, No. 13, 1826 (2001)

Zapka W., Asch K., Meissner K., *European Patent* EP 0297506 A2, 4 January 1989

Zapka W., Ziemlich W., Tam A. C., *Appl. Phys. Lett.*, **58**, 2217 (1991a)

Zapka W., Tam A. C., Ziemlich W., *Microelectronics Eng.*, **13**, 547 (1991b)

Zheng J., In *7th Annual Review of Progress in Applied Computational Electromagnetics (ACES)*, Conference Proceedings, Monterrey, pp. 170-173 (1990)

Zheng Y. W., Lu Y. F., Mai Z. H., Song W. D., *Jpn. J. Appl. Phys.*, vol. **39**, Part 1, 5894 (2000)

Zheng Y. W., Luk'yanchuk B. S., Lu Y. F., Song W. D., Mai Z. H., *J. Appl. Phys.*, vol. **90**, 2135 (2001)

Part 3. Steam Laser Cleaning

Chapter 4

PULSED LASER CLEANING OF PARTICLES FROM SURFACES & OPTICAL MATERIALS

D. M. Kane, A. J. Fernandes, D. R. Halfpenny

The main focus of laser cleaning research at Macquarie University, Sydney, Australia, has been developing techniques for cleaning applications in the optics, optoelectronics and photonics industries. Dry/raised humidity cleaning techniques are favoured for small scale optical, integrated-optic and photonic devices. All of wet, damp and dry laser-cleaning methods are applicable to larger scale optics. All of our studies to date have measured the laser cleaning result effected by a single laser pulse. Excellent removal efficiencies of alumina particles (0.3 μm –tens of microns) from glass surfaces have been obtained. Also we have used "dry" particles rather than particles in suspension. Use of dry particles leads to significant agglomeration, thus, our studies evaluate the impact of particle agglomeration on laser cleaning results. In the sections that follow we firstly present some tables summarizing the results from experimental laser cleaning studies, from the research literature published at the time of writing. Then, secondly, we review our own laser cleaning research from the perspectives of development and critical evaluation of the experimental and analysis methods, contrasting laser cleaning results achieved with different UV pulsed laser systems, and accurately measuring laser cleaning threshold fluence.

Keywords: Laser cleaning, surface cleaning, particle removal, industrial applications of lasers, optical materials

PACS: 81.65.Cf, 81.05.t, 42.62Cf

1. Introduction

The program of laser cleaning research at Macquarie University has, as its main focus, developing techniques for cleaning applications in the optics, optoelectronics and photonics industries. Wet chemical, ultrasonic and plasma cleaning are all utilised, currently, in optical contexts, but there is a need for alternate, non-contact, cleaning techniques that can be integrated readily with multiple-station production processes in the above mentioned industries, particularly for small scale components. Dry/raised humidity cleaning techniques are favoured for such small scale optical, integrated-optic and photonic devices as it is undesirable to introduce solvents to the small area surfaces, which, subsequently, may prove difficult to dry or remove without leaving residues which degrade the optical performance. All of wet, damp and dry laser cleaning methods are applicable to larger scale optics.

Our studies reported to date have looked at the removal of micron and sub-micron sized particles from various glass surfaces using three different UV nanosecond pulsed laser systems (Halfpenny (1999, 2000), Kane (2001), Fernandes (2001)). All of these studies have been completed in a humid environment and are thus categorized as raised humidity or “damp” laser cleaning. All of the studies have measured the laser cleaning result effected by a single laser pulse. This is a point of contrast with much of the laser cleaning literature where multiple pulses are more usually applied. Excellent removal efficiencies of alumina particles (0.3μm –tens of microns) from glass surfaces using (i) high beam quality UV-copper vapour lasers (Halfpenny (1999, 2000), (ii) a XeCl excimer laser and (iii) a KrF excimer laser (Kane (2001), Fernandes (2001)) have been obtained. Contrasting the results from these studies has given useful insights into the dry/damp laser cleaning process. Another point of contrast with the studies of other groups around the world is the method used to prepare the particles on the surface. We have used “dry” particles rather than particles in suspension. Use of dry particles leads to significant agglomeration. As agglomerates are expected in real application situations we have chosen to continue studying laser cleaning where both agglomerates and single particles are present in order to evaluate the impact of particle agglomeration on the results. We are currently extending these laser cleaning studies to include removal of particles from coated glass surfaces (most optical components have reflectance altering coatings of various materials) and, removal of hydrocarbon contaminants on small-scale optics.

In addition to the parametric studies of laser cleaning of particles from glass we have made related discoveries that include: developing a method for semi-permanently dehydroxylating glass surfaces rendering them more hydrophobic than native glass (Halfpenny (1996), Halfpenny (2000)); and an energy efficient method of “machining” glass using assistance from medium densities of particles on the surface (Kane (2000)). In the latter, the volume of material removed is increased ~1000 fold compared to laser ablation by a single laser pulse when using the particle-assisted-laser-material removal (PALMR).

A primary focus of our research has been the development of experimental methods and systems for measuring cleaning efficiency by detailed image analysis of the particles on the same area of the surface before and after processing. We have used in-line optical microscopy and image processing of digital images using commercial software for this. Many groups have now developed and/or adopted this methodology but at the beginning of our own researches in this area, published reports tended to include low resolution microscopic images of a sample which had a laser cleaned and an untreated area side by side. We have critically appraised the use of optical microscopy for quantifying laser cleaning effectiveness, particularly in the context where agglomerated particles introduce depth of focus concerns. Also, the uncertainty in cleaning efficiency caused by image processing steps such as thresholding of the images, to define the boundaries between particles and surface, when there is a distribution of particle sizes present, has been quantitatively evaluated.

Other experimental protocols developed are those for accurately measuring the laser cleaning threshold fluence (Halfpenny (1999), Fernandes (2001)). This has demonstrated that threshold fluence is independent of particle density. In our studies it is the laser cleaning threshold fluence for a single pulse from the UV laser. This is not necessarily the same as the multiple pulse laser cleaning threshold measured in multiple-pulse studies. However, the method we have developed, and its application in our studies, gives clear evidence that many laser cleaning threshold fluences quoted in the literature are overestimated as they are often the fluence which gives clearly visible removal of most of the particles rather than the fluence at which removal of any particles is initiated.

From our perspective there are two overarching research questions in laser cleaning we are seeking to contribute to answering.

i) Given a specific material surface, with micron and sub-micron particles of a specific material on that surface, what laser/light source parameters are required to remove all the particles from the surface without causing surface damage or unwanted modification?

ii) Can knowledge in this field be developed to the point where reasonably reliable predictions can be made, without recourse to an endless series of specific-material characterization studies?

From results of laser cleaning research overall (we review experimental studies in the next section) we already know that laser cleaning outcomes depend on: the substrate material; the particle material; the substrate "preparation"; the particle "preparation"; the particle size; the degree of agglomeration of the particles; the environment in which the laser cleaning is carried out, the wavelength of the laser used, the angle of incidence of the laser beam, the fluence of the laser pulses, the duration of the laser pulse, the pulse shape; the pulse sequencing, and possibly the spatial coherence of the laser, and focusing of the laser beam. Thus, it is not at all clear at this time whether an affirmative answer to the questions posed can be realized. It is clear that new methods of laser cleaning are still being developed and the interpretation of results from experiments and theory is evolving rapidly, as is expected in a relatively young research field. Thus, a chapter such as we present here, represents an opportunity to contrast and re-assess some our earlier research in the context of the rapidly growing field of knowledge.

In the sections that follow we firstly present some tables summarizing the results from experimental laser cleaning studies, from the research literature published at the time of writing. We apologize in advance for any we may not have uncovered. We have been maintaining and updating these tables over a period of some years and present them here as a useful resource for contrasting studies that have been completed and the variations in methods and parameters reported in those studies. Then, secondly, we review our own laser cleaning research from the perspectives of development and critical evaluation of the experimental and analysis methods, contrasting laser cleaning results achieved with different UV pulsed laser systems, and accurately measuring laser cleaning threshold fluence. Space does not allow us to review the reinterpretation of the discovery we made of a systematic laser induced optical surface damage mediated by the presence of a high density of particles on the glass surface (Kane 2000a). Further experiments on this subject have shown it to involve cracking and explosive removal of a "lump" of material rather than being an ablative process as at first speculated. Nor do we review UV pulsed laser dehydroxylating of glass surfaces as a means of increasing the hydrophobicity of the surface (Halfpenny 2000), an application which also has significance in biological and industrial settings. However, we note for the reader that our laser cleaning studies have generated these related discoveries that have become important programs of research.

2. Review Tables of experimental pulsed laser cleaning of contaminants (mostly particles) from surfaces

There is a rapidly growing research literature reporting experimental studies of removal of particles and contaminants from various material surfaces by laser cleaning. Many different laser systems have been used and many different experimental protocols have been adopted. We have summarized the literature in table form, grouped by surface material. Laser cleaning of silicon surfaces for the semiconductor industry has been the most investigated material system. A table for silicon is presented for this reason. A second table for glass and related optical material surfaces is given because of its relevance to Macquarie University research projects. A third table including other surface materials (germanium, metals, polymers, etc) is included for completeness.

2.1. Overview of the information in the Tables

The tables include particle/contaminant material, size and deposition technique; laser cleaning technique used; laser used, pulse length, the fluence used, pulse sequencing and threshold fluence for cleaning (though it is often not specified how this has been measured); technique for analysing particle removal; the percentage of particles removed or a comment on how laser cleaning effectiveness was determined, and the reference for the source of the data. Where there are blanks in the table it indicates that the information was not given in the original source. Where there is no value for the percentage of particles removed nor a comment, it indicates that the evidence of laser cleaning was given in image form.

In order to present a large amount of information in the main tables, in abbreviated form, there are a series of keys to define abbreviations used. These are also useful in themselves as they summarize the broad range of experimental systems and techniques that have been applied in laser cleaning studies.

Table 2.1.1 is the key for the method of depositing the particles (and occasionally contaminants) on the surface. The majority of studies have used spin coating of a suspension of the particles in a liquid. The liquids used have included water, acetone, ethanol, methanol and isopropyl alcohol. In other experiments drops of a suspension are air jet or pressurized spray dried. All of particle generators, dusting, dipping, electrostatic attraction and laser-assisted deposition have been used. There has been little attention given to date of how the adhesion of the particles to the surface is effected by the method used to deposit the particles on the surface. There are subsequent ramifications for using the results of the experiments as models of real-life cleaning situations where the particles have arrived "naturally"

Table 2.1.1. Key for Abbreviations on particle/contaminant deposition technique used in laser cleaning summary Tables 2.2 – 2.4.

Key	**Deposition Technique**
SCCS	Spin-coating of colloidal solution/suspension
EA	Electrostatic attraction
Nat	Natural/pollution
PMS	Particle Measuring System Particle generator
DT	Dip/tap
Du	Dusted on
LAD	Laser assisted deposition
Dr	Surface drawn on
Glue	Organic glue
Res	Residue after etching
Pack	During packaging process
Fab	During fabrication / exposure to air
Lay	Layer
WS	Water suspension
Asus	Acetone suspension
Esus	Alcohol/ Ethanol suspension
Msus	Methanol suspension
Isus	Isopropyl alcohol suspension
Sol	Solution
AJ	Air jet drying
LiqC	Liquified CCl_2F_2 spray can drying

on the surface. A small number of studies have dealt with such real-life situations of natural or pollutant particles or contaminants; or residues from packing and fabrication processes.

All the studies fall into four main types of laser cleaning. These include wet, steam, dry and dry in a humid environment. All these techniques are described in detail in other chapters in this book. The keys in Table 2.1.2 for the type of laser cleaning are all additional information on the main technique that has been applied. They indicate whether wet laser cleaning has been mediated by a liquid layer or vapour. A range of different liquids and vapors has been used. They indicate whether dry laser cleaning has been assisted by background humidity, which in some studies has been carefully controlled. The laser cleaning technique is sometimes enhanced by surface acoustic waves, shock waves, laser beam focusing or by combining with other cleaning techniques like ultrasonic. Manipulation of the electrochemical potential of the surface can also lead to lower adhesion of the particles and hence easier removal. Thus, a range of possibilities for consideration in all future studies results from collecting and considering the techniques used and thought about to date.

Table 2.1.2. Key for abbreviations on additional information of method of laser cleaning in summary Tables 2.2 – 2.4.

Key	**Type of Laser Cleaning**
LFE	Liquid film enhanced
SuD	Surface displacement investigated
SAW	Nonlocal laser cleaning using laser-generated SAWs
LAS	Laser-assisted shock wave
U	Ultrasonics
AF	Acetone film
EF	Ethanol film
AlF	Water/alcohol 300nm layer
IF	Isopropanol film
WA/WAV	Water/alcohol, Water/alcohol vapour
WV/ WVC	Water vapour, Water vapour condensed layer
RH/HH	Raised humidity, High humidity
RH40	Relative humidity 40%
RH100	Relative humidity 100%
LB	Surface scanned with line beam
Lyte	Sample immersed in electrolyte to fix electrochemical potential of the surface

Table 2.1.3 gives the abbreviations for comments used to describe the effectiveness of laser cleaning, particularly in earlier studies where the results were often evaluated by visual inspection of micrographs. Terms like optimal were adopted to indicate the laser parameters that gave the "best" laser cleaning outcomes seen in the studies. Comments on whether the laser cleaning was effective, ineffective, incomplete, improved relative to some other conditions (changing laser beam angle of incidence, increasing number of pulses used, for example) were given in the reports. The standard practices now adopted lead to quantitative results so that improvements in outcomes can be measured in a substantive way.

Table 2.1.4 lists all the pulsed lasers that have been applied in laser cleaning studies and their wavelengths. It should be noted that continuous wave lasers have been used successfully to remove hydrocarbon contamination (eg. fingerprint removal from a surface) but pulsed lasers are required for particle removal as demonstrated by the chapters addressing the theory of laser cleaning in this book. The long wavelength TEA CO_2 lasers have been used for wet laser cleaning (with two exceptions) because of the high absorption of this wavelength in water. All the lasers with wavelengths between 193 nm and 1.06 μm, listed in Table 2.1.4, have been used with some success for particle removal as detailed in the summary tables, which follow. But generally better laser cleaning, in terms of lower threshold fluence for cleaning and larger fraction of particles removed from the irradiated area, are obtained with shorter wavelengths. This generalisation does not hold when the particle is spherical and of a size of the order of the wavelength when Mie scattering effects lead to focusing of the light in the forward direction, and hence increased effective fluence at the surface (Luk'yanchuk (2000)). Multiple scattering between the particle and surface also becomes important. Most of the experimental studies to date have not investigated this regime but Curran (2000) has observed a cleaning efficiency that varies with wavelength using a wavelength tunable laser.

Table 2.1.5 gives the abbreviations for, and lists the techniques that have been used for analyzing particle removal from a surface in laser cleaning studies. The most important techniques are light scattering, optical microscopy including dark field microscopy, and scanning electron microscopy (SEM). Cameras, which have been used in conjunction with the techniques including CCD, video and polaroid. Frame grabbers have been used to digitize the images for subsequent analysis. This has ranged from measuring the particle density before and after, counting particles before

Table 2.1.3. Key for abbreviations on effectiveness of laser cleaning in summary Tables 2.2 – 2.4.

Key	Comment on Laser Cleaning Effectiveness
Opt	Optimal
NE/ IR	Not effective, Incomplete removal
Eff	Efficient
NI	Normal Incidence
LA	Increased at glancing beam incidence (large) angle
BGI	Beam at glancing angle cleans area larger than beam spot
NC	No improvement with changing beam incidence angle
IE	Improved efficiency
NAGSW	Not as good as shorter wavelength
LT	Particles >0.5μm removed with some efficiency
IN	Increases with number of pulses
OPI	One pulse insufficient
Acc	Accelerations of 10^{10}m/s^2 remove 0.05μm particles

Table 2.1.4. Key for lasers used in laser cleaning studies and their wavelengths as used in summary Tables 2.2 – 2.4.

Key	Laser	Wavelength
	Excimer	Unspecified
	ArF	193 nm
	KrF	248 nm
UV CVL	Frequency doubled copper vapour laser	255.3 nm
Nd:YAG^4	Frequency Quadrupled Nd:YAG	266 nm
	XeCl	308 nm
	Nitrogen	337 nm
Nd:YLF^3	Frequency tripled Nd:YLF	349 nm
Nd:YAG^3	Frequency tripled Nd:YAG	355 nm
Nd:YLF^2	Frequency doubled Nd:YLF	523 nm
Nd:YAG^2	Frequency doubled Nd:YAG	532 nm
	Dye laser	583 nm
Nd:YAG^O	Nd:YAG/OPO	638 nm
Nd:YLF	Nd:YLF	1.047 μm
Nd:YAG	Nd:YAG	1.06 μm
$CO_2^{9.6}$	TEA CO_2	9.6 μm
$CO_2^{10.6}$	TEA CO_2	10.6 μm

and after, analyzing area coverage by particles before and after, or expressing the area covered by particles rationed to the laser beam area. A number of other techniques have monitored the surface directly including surface directly including surface photovoltage measurement, acoustic wave monitoring, electron probe microanalysis, Auger spectroscopy or atomic force microscopy. Particle counting systems have been used to count the particles ejected from the surface. Little thought has so far been given to which technique(s) are best. Optical microscopy and associated image processing is the technique most easily incorporated for in-line monitoring but the systems so far developed have been limited to resolving particles greater than 0.3 μm across in size. SEM can image nanometer-sized particles but the sample must be removed for analysis making it impossible to monitor the same area before and after laser cleaning. It is difficult to recover information on particle size distributions from light scattering and as an indirect measurement it is less preferred.

2.2. Silicon - wafers, hydrophilic and membrane masks

Table 2.2 is a summary of laser cleaning studies that have been completed on silicon, including hydrophilic silicon and silicon stencil membrane masks. Probably the most significant results from the silicon laser cleaning studies come from contrasting the work of Vereecke (1999, 2000), undertaken in semiconductor fabrication cleanroom conditions, with many of the other studies, undertaken in standard research laboratory conditions. In the latter case hydrocarbon and moisture have the opportunity to condense on to the silicon wafers, from the air, leading to an uncontrolled variable. It is clear from the results that this additional moisture and hydrocarbon contamination can assist in removing particles from the silicon surface. Where moisture is absent (Vereecke (1999)) the cleaning efficiency is much reduced, but can be improved by deliberately raising the relative humidity of the environment. Thus, it is made clear that the cleanliness of the surface before the particles are added is an important factor in determining the subsequent results of laser cleaning.

Particles of many different materials have been investigated for removal from silicon wafers, by laser cleaning. Spherical particles, such as silica spheres and polystyrene spheres, represent good test particles for comparison with theory. Theory has been developed for "ideal" spherical

Table 2.1.5. Abbreviations for descriptor of analysis technique used in laser cleaning studies listed in tables 2.2 – 2.4.

Key	***Analysis Technique***
LS	Light scattering
OM	Optical microscope
SEM	Scanning electron microscopy
SR	Si photodiode monitors specular reflection of HeNe beam from substrate
PMS	Particle Measuring Systems Inc, SAS 3600 particle counting system
CANS	Censor ANS100 Light scattering particle counter
SPV	Surface photovoltage monitored
HI	Heterodyne interferometer (surface displacement)
IP	Before & after images, image analysis software
PD	Comparing particle densities before & after
AWM	Acoustic wave monitoring
VB	Video-based, PC-driven system counts particles > ~75μm diameter
OSA	Automated optical surface analyzer (limit of resolution 1μm)
AES	Surface analysed by Auger Electron Spectroscopy
EPM	Surface analysed before & after by electron probe microanalysis
AFM	Atomic Force Microscopy
PC	Particle counter
ECM	Emission current monitoring
DFM	Dark field microscopy
AR	Efficiency = clean area / laser spot area
Pol	Polaroid camera
CCD	CCD camera

particles (Luk'yanchuk (2000)). Alumina is common as it is readily available in a range of particle sizes due to its application as a polishing compound. SiO_2 and Si_3N_4 were the particle materials chosen by Vereecke (1999) as appropriate models for contaminants actually occurring in semiconductor fabrication processes. Also, only particle sizes of less than 1 μm are interesting in the silicon context as larger sized particles are readily removed by the standard wet chemical cleaning etc., processes already in place in the industry. Excellent cleaning efficiencies (>95%) for 0.1 μm particles of several materials (PSL, CML, SiO_2) from hydrophilic silicon have been measured using a KrF laser by Meunier (1999) and Wu (1999a, 1999b, 2000).

Table 2.2. Laser cleaning of silicon.

Contaminant	Particle Size	Deposition	Type of LC	Threshold Fluence	Fluence Used	% Removed
Silicon wafers						
Particles	1 - 9μm	-	Wet, WVC	-	2J/cm^2	>90
Fe_2O_3 - front surface	0.1-2μm	PMS	Dry	-	350mJ/cm^2	LT
Fe_2O_3 - front surface	0.3-2μm	PMS	Steam, WVC	-	200mJ/cm^2	Opt
Fe_2O_3 - back surface	0.3-2μm	PMS	Steam, WVC	-	200mJ/cm^2	Opt
Au	-	-	-	-	-	-
-	-	-	Dry, SuD	-	-	-
Al_2O_3, SiC, BC, CeO_2	>0.1μm	EA or SCCS	Wet, WVC	650mJ/cm^2	2.9-3.2J/cm^2	Opt
Al_2O_3	mixed	ASus	Wet, WVC	-	30J/cm^2	Opt
Tungsten	1-2μm	Du, LiqC	Dry,	-	2.1 J/cm^2	>95
Al_2O_3	1-10μm	Du	Dry, SAW	-	10^{-3}J/cm^2 using SAW	Acc
Al_2O_3	1-10μm	Du	Dry, SAW	-	10^{-3}J/cm^2 using SAW	Acc
Al_2O_3	9.5μm	ASus	Wet, WV	2.14J/cm^2	-	-
Al_2O_3	5μm	Asus	Wet, WV	2.23J/cm^2	-	-
PS (Polystyrene Spheres)	1μm	ASus	Wet, WV	2.22J/cm^2	2.43J/cm^2	92
Al_2O_3	9.5μm	Asus	Wet, WV	1.45J/cm^2	-	-
Al_2O_3	5μm	Asus	Wet, WV	1.49J/cm^2	-	-
PS	1μm	Asus	Wet, WV	1.41J/cm^2	-	-
Silica spheres	800nm	WS	Steam, WAV	110mJ/cm^2	180mJ/cm^2	>90
PS	800nm	WS	Steam, WAV	110mJ/cm^2	-	-
Particles	0.1-10μm	-	Dry	-	3.34 J/cm^2 (energy flux)	-
Quartz	mixed	MSus,AJ	Dry	135mJ/cm^2	300mJ/cm^2	~60
Silica spheres	0.5μm	WS, AJ	Dry	225mJ/cm^2	325mJ/cm^2	~75
Silica spheres	1μm	WS, AJ	Dry	100mJ/cm^2	325mJ/cm^2	~95
Silica spheres	2.5 μm	WS, AJ	Dry	<5mJ/cm^2	25mJ/cm^2	100
Silica spheres	5μm	WS, AJ	Dry	<5mJ/cm^2	25mJ/cm^2	100
PS	800nm	SCCS & ISus	Steam, AlF	50mJ/cm^2, UT	100mJ/cm^2	>90
PS	800nm	SCCS & ISus	Steam, AlF	50mJ/cm^2, UT	75mJ/cm^2	>90
PS	800nm	SCCS & ISus	Steam, AlF	50mJ/cm^2, UT	75mJ/cm^2	>90
PS	800nm	SCCS & ISus	Steam, AlF	20mJ/cm^2 ,UT	75mJ/cm^2	>90

Table 2.2.

Laser Used	No. of pulses	Pulse Length	Analysis Technique	Reference
$CO_2^{10.6}$	single	200ns	SR	Allen 1997
KrF	-	22ns	PMS	Beaudoin 1998
KrF front irradiation	2 laser cleanings scanning cycles	22ns	PMS	Beaudoin 1998
KrF front irradiation	2 laser cleanings scanning cycles	22ns	SPV	Beaudoin 1998
$Nd{:}YAG^{O}$	-	6ns	SEM	Curran 2000
$Nd{:}YAG^{2}$	-	-	HI	Dobler 1999
$CO_2^{10.6}$	multiple	0.25μs	PD, PMS	Heroux 1996
$CO_2^{10.6}$	5	1μs	OM	Imen 1991
Nd:YAG	100	24ns	OM,Pol	Kelley 1993
Nitrogen	50	10ns	OSA	Kolomenskii 1998
Nd:YAG	50	10ns	OSA	Kolomenskii 1998
$CO_{2,}^{9.6}$	single	-	OM	Lee 1993
$CO_2^{9.6}$	single	-	OM	Lee 1993
$CO_2^{9.6}$	single	-	OM	Lee 1993
$CO_2^{10.6}$	single	-	OM	Lee 1993
$CO_2^{10.6}$	single	-	OM	Lee 1993
$CO_2^{10.6}$	single	-	OM	Lee 1993
$Nd{:}YAG^{2}$	single	7ns	LS	Leiderer 1998
$Nd{:}YAG^{2}$	single	7ns	LS	Leiderer 1998
KrF	-	-	-	Lizotte 1996
KrF	100	23ns	OM	Lu 1997a,d, 1998b
KrF	200	23ns	SEM	Lu 2000c, Zheng 2001a
KrF	200	23ns	SEM	Lu 2000c, Zheng 2001a
KrF	200	23ns	SEM	Lu 2000b,c; Zheng 2001a
KrF	200	23ns	SEM	Lu 2000b, 2000c; Zheng 2001a
$Nd{:}YAG^{2}$	20	8ns	OM	Mosbacher 1999, Oltra 2000
$Nd{:}YAG^{2}$	20	2.5ns	OM	As above
Dye laser	20	2.5ns	OM	As above
Dye laser	20	30ps	OM	As above

Table 2.2. (cont.)

Contaminant	Particle Size	Deposition	Type of LC	Threshold Fluence	Fluence Used	% Removed
PS, SiO_2, Al_2O_3	(60/235/300 /500/800nm) (500/800nm) (300nm)	SCCS & ISus	Steam, WA	110 mJ/cm^2, UT: Universal Threshold	170mJ/cm^2	>90
Al_2O_3 & Epoxy film	1 μm 300nm film	-	Steam, WA & Dry	-	110mJ/cm^2	Opt
Al_2O_3	0.1 μm	ESus, WS	Steam, WAV	-	120 mJ/cm^2	Opt
Al_2O_3 & Epoxy film	1μm 1μm film	-	Steam & Dry & U	-	110mJ/cm^2	Opt
SiO_2	1μm	WS, SCCS	Dry, RH40	-	300mJ/cm^2	84+/-8
Si_3N_4	0.3 μm	WS, SCCS	Dry, RH40	-	300mJ/cm^2	33+/-4
SiO_2	0.3 μm	WS, SCCS	Dry, RH40	-	300mJ/cm^2	12+/-7
SiO_2	1 μm	WS, SCCS	Dry, RH100		300mJ/cm^2	72+/-10
Si_3N_4	0.3 μm	WS, SCCS	Dry, RH100		300mJ/cm^2	78
SiO_2	0.3 μm	WS, SCCS	Dry, RH100		300mJ/cm^2	88+/-6
Al_2O_3	0.1 μm	ESus	Steam, WAV	-	120mJ/cm^2	Opt
Al_2O_3	0.3 - 1.5 μm	ESus	Dry	-	300mJ/cm^2	Opt
Copper	1 μm	SCCS	Dry	-	180 mJ/cm^2	100
Tungsten	1 μm	SCCS	Dry	-	280 mJ/cm^2	NE
Tungsten	1μm	SCCS	Dry, LAS	-	10^{12}W/cm^2 focal point	100
Silica spheres	0.5μm	WS, AJ	Dry	<5 mJ/cm^2	~10 mJ/cm^2	~100
Silica spheres	0.5μm	WS, AJ	Dry	<5 mJ/cm^2	~15 mJ/cm^2	~100
Alumina	1μm	-	Steam, AF	80 mJ/cm^2	-	-
Alumina	1μm	-	Steam, EF	90 mJ/cm^2	-	-
Si_3N_4	0.15 – 0.30μm	WS, SCCS	Dry	-	-	LA
SiO_2	0.3μm	WS, SCCS	Dry	-	-	NC
Alumina	0.1-0.3μm	ESus, WS	Dry, LB	-	200-600 mJ/cm^2	100

Table 2.2. (cont.)

Laser Used	No. of pulses	Pulse Length	Analysis Technique	Reference
Nd:YAG[2]	20	7ns	OM, LS, SEM	Mosbacher 2000
KrF	multiple (steam more effective)	16ns	OM	Park 1994
KrF	20	16ns	SEM	Tam 1992, Zapka 1993 a, b
KrF	multiple	-	OM	Tam 1998
KrF	1	30ns	LS, CANS	Vereecke 1999
KrF	12	30ns	LS, CANS	Vereecke 1999
KrF	1	30ns	LS, CANS	Vereecke 1999
KrF	1	30ns	LS, CANS	Vereecke 1999
KrF	12	30ns	LS, CANS	Vereecke 1999
KrF	1	30ns	LS, CANS	Vereecke 1999
KrF	20	15ns	SEM	Zapka 1993 a,b
KrF	4	15ns	SEM	Zapka 1993 a, b
Nd:YAG[4]	10	10ns	OM	Lee 2001
Nd:YAG[4]	10	10ns	OM	Lee 2001
Nd:YAG[4]	3 pulses (3 shock waves)	10ns	OM	Lee 2001
Nd:YAG[3]	-	7 ns	SEM	Zheng 2001 a
Nd:YAG[2]	-	7 ns	SEM	Zheng 2001 a
Excimer	-	-	-	Lu 2001
Excimer	-	-	-	Lu 2001
KrF	-	30ns	LS, CANS	Vereecke 2000
KrF	-	30ns	LS, CANS	Vereecke 2000
KrF	50	-	SEM	Kumar 1998

Table 2.2. (cont.)

Contaminant	Particle Size	Particle Deposition	Type of LC	Threshold Fluence	Fluence Used	% Removed
Hydrophilic Silicon:						
Al_2O_3	0.1 µm	PMS	Steam, WV	143mJ/cm^2	154mJ/cm^2	~90
SiO_2	0.1 µm	PMS	Steam, WV	143 mJ/cm^2	180 mJ/cm^2	~95
PSL (polystyrene latex)	0.1 µm	PMS	Dry	76 mJ/cm^2	320 mJ/cm^2	~100
CML (carboxylate-modified latex)	0.1 µm	PMS	Dry	-	353 mJ/cm^2	~99
Al_2O_3	0.2 µm	PMS	Dry	-	314 mJ/cm^2	~40
SiO_2	0.1-0.2 µm	PMS	Dry	-	326 mJ/cm^2	~15
Silicon Membrane Stencil Masks						
Al_2O_3	>0.35 µm	ESus	Dry	-	350 mJ/cm^2	Opt
Silica spheres	1200 nm	Sol	LFE	260 mJ/cm^2	-	-
Silica spheres	500 nm	Sol	LFE	260 mJ/cm^2	260 mJ/cm^2	99.7
Silica spheres	250 nm	Sol	LFE	<400 mJ/cm^2	-	-
Alumina	0.2-2 µm	Sol	LFE	.28-.35 J/cm^2	350 mJ/cm^2	99
Alumina	0.2-2 µm	Sol	LFE	.26-.32 J/cm^2	-	-
Alumina	0.2-2 µm	Sol	LFE	-	690 mJ/cm^2	Eff

2.3. Glass and related optical surfaces

There have been fewer experimental studies of removing particles from glass and related optical material surfaces. Table 2.3 summarizes the studies to date. Dry laser cleaning is the preferred method for optical materials, especially for optical components of small physical scale. Materials that have been studied have included microscope slides and cover slips (borosilicate glass), fused silica, quartz, aluminum coated mirrors, lithium niobate (an important material in integrated optic devices) and calcium fluoride. Laser cleaning of optical materials is most definitely an area in

Table 2.2. (cont.)

Laser Used	No. of pulses	Pulse Length	Analysis Technique	Reference
KrF	4 cleaning scanning cycles	22ns	PMS	Meunier 1999, Wu 2000
KrF	5 cleaning scanning cycles	22ns	PMS	Meunier 1999, Wu 2000, 1999a
KrF	2 cleaning scanning cycles	22ns	PMS	Meunier 1999, Wu 2000, 1999a ,b
KrF	2 cleaning scanning cycles	22ns	PMS	Meunier 1999, Wu 2000
KrF	4 cleaning scanning cycles	22ns	PMS	Meunier 1999, Wu 1999a,b
KrF	4 cleaning scanning cycles	22ns	PMS	Meunier 1999, Wu 1999b
KrF	4	16ns	SEM	Zapka 1991
XeCl	-	-	DFM	Zapka 2000
XeCl	1	-	DFM	Zapka 2000
XeCl	-	-	DFM	Zapka 2000
XeCl	44	-	DFM	Zapka 2000
KrF	-	-	DFM	Zapka 2000
KrF	1	-	DFM	Zapka 2000

which laser cleaning has real promise to be an industrially significant process, in photonic component manufacture, in particular. The manufacturing processes are beginning to develop along similar lines to semiconductor fabrication – clean room production lines and remote control cluster tools. Currently, much manufacture is still labor intensive, involving precision alignment of small optical components by skilled technicians. The laser cleaning processes can be developed in line with the increasing sophistication and automation of the manufacturing processes.

In the studies to date the emphasis has been on the removal of micron sized and larger particles, which lead to scattering and reduced throughput in bulk optical systems. Thus, in current optics and photonics applications smaller particles have not been as much of a concern as they are in semiconductor fabrication where the particle dimension of 0.1 microns is determined by the track width on the chips. These small particles will become an increasing focus due to their potential impact in planar waveguide optics, for example.

Optical materials contrast with silicon in being transparent throughout the visible and near infrared regions of the electromagnetic spectrum. However, all optical materials become absorbing in the UV or VUV and hence shorter laser wavelengths are preferable for dry laser cleaning. Some excellent cleaning efficiencies are listed in Table 2.3. The experimental studies of laser cleaning of glass undertaken at Macquarie University are discussed in more detail in section 3.

2.4. Other material surfaces

Table 2.4 summarises the experimental laser cleaning studies of particles and contaminants from other materials surfaces such as the semiconductor germanium, metals, polymers, magnetic head sliders and disks etc. This shows that the complete range of materials that have been investigated to date is very broad and the results are not always sufficiently good to be useful in practical applications.

3. Experimental studies of laser cleaning particles from glass at Macquarie University

3.1. Single pulse laser cleaning studies

3.1.1. "Dip and Tap" sample preparation

Alumina particles of various sizes (0.1, 0.3, 1 and 3 μm used in UV copper vapour laser cleaning studies (Halfpenny 1999), and mostly 1 μm in XeCl and KrF laser cleaning studies (Kane 2001, Fernandes 2001) from Baikowski (irregularly shaped, agglomerate free) have been used. The samples were prepared using a "dip and tap" method where the slide was either dipped in a large volume of the particles or a generous coating of particles was spooned onto the slide. Sharp tapping and fast flow of dry air was then used to remove particles, which were loosely bound to the slide surface. This method leads to medium densities of particles (10-40 % coverage by area) on solvent cleaned slides, and lower densities on ultrasonically cleaned slides (0.1 –7 %, average 1.8 %). Samples have also been prepared using laser assisted deposition. This leads to significant differences in the nature of the agglomerates on the surface but does not affect the laser cleaning threshold fluence when these changes in agglomeration are taken into account. Thus, only the "dip and tap" method has been used in the studies presented here.

Two different glasses were used as substrates in the experiments, glass microscope slides and fused silica (SiO_2). While silica is the basic constituent of both of these glasses their optical properties were vastly different. They were chosen as examples of glasses commonly found in the field of optics (as well as in a myriad of other fields) and in order to provide contrasting information on the laser cleaning effect. The dimensions of the glass microscope slides were 25.4 by 76.2 mm^2 and 1mm thick. The exact composition of the glass used to make the microscope slides was unobtainable from the manufacturers, and manufacturers of similar slides claimed proprietary information on the composition of their glass as well. However, they appear to be soda-lime silica glass. The optical properties were measured using a CARY 500 spectrometer. The reflectance was 4.40 %, the transmittance 0.56 % and the absorption 94.74 %, all at 255 nm.

The pure silica glass used in these experiments was Corning 7940 silica, a high purity synthetic amorphous silicon dioxide manufactured by flame hydrolysis. It has a very low thermal expansion coefficient and excellent optical qualities with exceptional transmission in the ultraviolet. The silica samples used were ground using opaline compound to a flatness of $\lambda/10$, with a surface quality of 10/5. Typical optical properties measured for the pure silica samples at 255 nm were reflectance 5.56 %, transmittance 89.73 % and absorption 4.71 %. It can be seen from these values that the optical properties of the pure silica samples are notably different from those of the microscope slide at the UV-CVL wavelength.

3.1.2. Dry/Damp laser cleaning

Three lasers have been used and contrasted in our laser cleaning studies. The characteristics of these lasers are summarized in Table 3.1.1.

Table 2.3. Laser cleaning of glass and related optical materials.

Substrate	Contaminant	Contaminant Size	Deposition Technique	Type of LC	Threshold Fluence mJ/cm^2	Fluence Used	% Removed
Microscope Slides Fused Silica	Al_2O_3	0.3-3μm & agglomerates	dip/tap	Dry, HH	62-93	Up to 0.5 J/cm^2	100
Microscope Slides	Al_2O_3	1μm & agglomerates	dip/tap	Dry, HH	330-380	Up to 10 J/cm^2	95
Microscope Slides	Al_2O_3	1 μm & agglomerates	Laser assisted	Dry HH	320-350	Up to 10 J/cm^2	100
Microscope Slides	Al_2O_3	1 μm & agglomerates	dip/tap	Dry, HH	200-400	Up to 1.2 J/cm^2	97
Microscope Cover Slide	Finger prints	N/A	-	Dry	-	450 mJ/cm^2	100
Microscope Cover Slide	Finger prints	N/A	-	Dry	-	70 mJ/cm^2	IR
Microscope Cover Slide	Finger prints	N/A	-	Dry	-	420 mJ/cm^2	IR
Quartz	Paint, fingerprints, particles & haze	-	-	Dry	-	2-20J/cm^2 average flux	Optimal
Quartz	Finger prints	N/A	-	Dry	-	400 mJ/cm^2	IR
Quartz	Finger prints	N/A	-	Dry	-	400 mJ/cm^2	100
Quartz	Al	<3μm	MSus, AJ	Dry	40	150mJ/cm^2	100
Quartz	Al	<3μm	MSus, AJ	Dry	30	120mJ/cm^2	100
Quartz	Al	<3μm	MSus, AJ	Dry	10	110mJ/cm^2	100
Quartz	Al	<3μm	MSus, AJ	Dry	50	100mJ/cm^2	24
Quartz	Al	<3μm	MSus, AJ	Dry	50	100mJ/cm^2	100
Quartz	Cu	<20μm	Esus, AJ	Dry	80	400mJ/cm^2	~65
Quartz	Cu	<20μm	Esus, AJ	Dry	40	200mJ/cm^2	optimal
Quartz	Cu	<20μm	Esus, AJ	Dry	20	180mJ/cm^2	optimal
Quartz	Al	<20μm	Esus, AJ	Dry	40	150mJ/cm^2	optimal
Quartz	Al	<20μm	Esus, AJ	Dry	30	120mJ/cm^2	optimal
Quartz	Al	<20μm	Esus, AJ	Dry	10	110mJ/cm^2	optimal

Table 2.3.

Laser Used	No. of pulses	Pulse Length	Analysis Technique	Reference
UV CVL	Single	35ns	OM, CCD, IP	Halfpenny 1999 Halfpenny, 2000
XeCl	single	8 ns	OM, CCD, IP	Kane 2001
XeCl	single	8 ns	OM, CCD, IP	Kane 2001
KrF	single	12 ns	OM, CCD, IP	Fernandes 2001
KrF	2	20ns	EPM	Lu 1994c
KrF	18000	20ns	EPM	Lu 1994c
$CO_2^{10.6}$	50	100ns	EPM	Lu 1994c
KrF	multiple	34ns - duty cycle	-	Engelsberg 1995
KrF front irradiation	1200	20ns	EPM	Lu 1994c
KrF back irradiation	600	20ns	EPM	Lu 1994c
Nd:YAG	500	7ns	OM, IP	Lu 1997c
Nd:YAG^2	500	7ns	OM, IP	Lu 1997c
Nd:YAG^4	500	7ns	OM, IP	Lu 1997c
KrF front irradiation	100	23ns	OM, IP	Lu 1997b,c, 1998a
KrF back irradiation	100	23ns	OM, IP	Lu 1997b,c, 1998a
Nd:YAG	500	7ns	OM, IP	Lu1998b,d, 1999a, 2001
Nd:YAG^2	500	7ns	OM, IP	Lu1998b,d, 1999a, 2001
Nd:YAG^4	500	7ns	OM, IP	Lu1998b,d, 1999a, 2001
Nd:YAG	500	7ns	OM, IP	Lu 1998b, 1998d
Nd:YAG^2	500	7ns	OM, IP	Lu 1998b, 1998d
Nd:YAG^4	500	7ns	OM, IP	Lu 1998b, 1998d

Table 2.3. (cont.)

Substrate	Contaminant	Contaminant Size	Deposition Technique	Type of LC	Threshold Fluence mJ/cm^2	Fluence Used	% Removed
Al mirror coating	Dust (mainly quartz sand)	several to 100s μms	Natural	Dry	50	160+/-30 mJ/cm^2 optimal (range 50-1000mJ/cm^2)	optimal
$LiNbO_3$	Tungsten	down to 1-2 μm	Du, LiqC	Dry	-	650 mJ/cm^2	>95
$LiNbO_3$	Tungsten	down to 1-2 μm	Du, LiqC	Dry	-	0.16J/cm^2	<50
$LiNbO_3$	Epoxy particles	1-2μm	Du, LiqC	Dry	-	650 mJ/cm^2	<5
CaF_2	Tungsten	1-2μm	Du, LiqC	Dry	-	2.1 J/cm^2	>95

Table 2.4. Laser cleaning other materials.

Substrate	Contaminant	Contaminant Size	Deposition Technique	Type of LC	Threshold Fluence	Fluence Used	% removed
Ge	Silica spheres	1μm	WS, AJ	Dry	30 mJ/cm^2	~90 mJ/cm^2	~100
Ge	Silica spheres	0.5μm	WS,AJ	Dry	~1 mJ/cm^2	~8 mJ/cm^2	~100
Ge	Silica spheres	0.5μm	WS, AJ	Dry	~2 mJ/cm^2	~10 mJ/cm^2	~100
Ge	Silica spheres	0.5μm	WS, AJ	Dry	~15 mJ/cm^2	~40 mJ/cm^2	~100
Cast Iron & Stainless Steel	-	-	-	Dry	-	1-1.5 J/cm^2	Opt
Stainless steel	Magic marker	N/A	Drawn	Dry	-	500 mJ/cm^2	Opt
Stainless steel	Magic marker	N/A	Drawn	Dry	-	830mJ/cm^2	Opt
Stainless steel	Iron oxide	-	-	Dry	0.5J/cm^2	-	-
Stainless steel	Cr oxide	-	-	Dry	0.8J/cm^2	-	-
Stainless steel	Al oxide	-	-	Dry	1.0J/cm^2	-	-
gold/SiO_2	Al_2O_3, SiC, BC, CeO_2	>0.1μm	EA, SCCS	Wet, WL	650mJ/cm^2	3.2-6.2J/cm^2	Opt

Table 2.3. (cont.)

Laser Used	No. of pulses	Pulse Length	Analysis Technique	Reference
KrF (w/o damage cf 193/308 /351nm excimer & Nd:YAG[3])	5 (optimal in range 1-1000)	30ns (optimal from 30ns, 50-200ns)	Light scattering with HeNe spectrometer	Mann 1996
Nd:YAG	100	20ns	OM, IP, Pol	Kelley 1993
Nd:YAG	100	20ns	OM, IP, Pol	Kelley 1993
Nd:YAG	600	20ns	OM, IP, Pol	Kelley 1993
Nd:YAG	100	24ns	OM, IP, Pol	Kelley 1993

Table 2.4.

Laser Used	No. of pulses	Pulse Length	Analysis Technique	Reference
KrF	200	23ns	SEM	Zheng 2001a
Nd:YAG[3]	-	7 ns	SEM	Zheng 2001a
Nd:YAG[2]	-	7 ns	SEM	Zheng 2001a
Nd:YAG	-	7 ns	SEM	Zheng 2001a
Nd:YAG	-	10-20ns	SEM	Boquillon, 1994
KrF	20	20ns	OM,SEM, AES	Lu 1994a
KrF	20	20ns	OM,SEM, AES	Lu 1994e
Nd:YAG & [2]	single	14.5ns	-	Psyllaki 2000
Nd:YAG & [2]	single	14.5ns	-	Psyllaki 2000
Nd:YAG & [2]	single	14.5ns	-	Psyllaki 2000
CO_2[10.6]	multiple	0.25ms	PMS, PD	Heroux 1996

Table 2.4. (cont.)

Substrate	Contaminant	Contaminant Size	Deposition Technique	Type of LC	Threshold Fluence	Fluence Used	% removed
Al	Al_2O_3, SiC, BC, CeO_2	>0.1μm	EA,SCCS	Wet, WL	650mJ/cm^2	>6.2J/cm^2	Opt
Al	Finger print	N/A	-	Dry	-	375mJ/cm^2	100
anodized Al	Oxide layer	20μm	-	Dry	-	-	OPI
anodized Al	Oxide layer	20μm	-	Dry	-	-	OPI
black anodized Al	Oxide layer	20μm	-	Dry	-	~1.3J/cm^2	Opt
Al coated S	Volcanic dust	-	Nat	Dry	-	380mJ/cm^2	90
Cu	Magic Marker	N/A	Drawn	Dry	-	500mJ/cm^2	Opt
Cu	Contaminants	-	Nat	Dry	-	0.8J/cm^2	Opt
Cu	Oil & grease layer	N/A	Layer	Dry	-	460 mJ/cm^2	100
Cu	Oil & grease layer	N/A	Layer	Dry	-	70 mJ/cm^2	Opt
Cu	Magic Marker	N/A	Drawn	Dry	-	420mJ/cm^2	Opt
Cu	Surface contaminants / magic marker	-	-	Dry	-	-	OPI
Cu	Cu	-	EtSus	Dry	0.15 J/cm^2	~0.8 J/cm^2	~80, NI
Cu	Cu	-	EtSus	Dry	0.01 J/cm^2	0.15 J/cm^2	130, BGI
Fe	Iron oxide	1020nm layer	-	Dry	-	0.5J/cm^2	Good
Fe	Iron oxide	1020nm layer	-	Dry	-	0.5J/cm^2	NAGSW
Fe	Iron oxide	1020nm layer	-	Wet, Lyte	-	0.5J/cm^2	Improved cf Dry
Polyimide	PS	800nm	SCCS	Dry (air)	~5mJ/cm^2	17.5mJ/cm^2	69+/-5
Polyimide	PS	320nm	SCCS	Dry (air)	~12mJ/cm^2	17.5mJ/cm^2	31+/-7
Polyimide	PS	110nm	SCCS	Dry (air)	~14mJ/cm^2	17.5mJ/cm^2	10+/-10
Polyimide	SiO_2	800nm	SCCS	Dry (air)	~7mJ/cm^2	17.5mJ/cm^2	95+/-5
Polyimide	SiO_2	400nm	SCCS	Dry (air)	~14mJ/cm^2	17.5mJ/cm^2	95+/-5
PMMA	PS	1700nm	SCCS	Dry (air)	~35mJ/cm^2	160mJ/cm^2	>80
PMMA	PS	800nm	SCCS	Dry (air)	~60mJ/cm^2	160mJ/cm^2	>80
PMMA	PS	320nm	SCCS	Dry (air)	~110mJ/cm^2	160mJ/cm^2	>80
PMMA	SiO_2	800nm	SCCS	Dry (air)	-	160mJ/cm^2	0
PMMA	SiO_2	400nm	SCCS	Dry (air)	-	160mJ/cm^2	0
Polyimide	SiO_2	400nm	SCCS	Dry (air)	~11mJ/cm^2	17.5mJ/cm^2	92+/-5
Polyimide	SiO_2	400nm	SCCS	Dry (air)	-	17.5mJ/cm^2	95+/-5
Polyimide	SiO_2	400nm	SCCS	Dry (air)	-	12.5mJ/cm^2	80+/-5
Polyimide	SiO_2	400nm	SCCS	Dry (air)	-	10mJ/cm^2	40+/-7

Table 2.4. (cont.)

Laser Used	No. of pulses	Pulse Length	Analysis Technique	Reference
CO_2[10.6]	multiple	0.25ms	PMS, PD	Heroux 1996
KrF	1	20ns	OM, SEM, AES	Lu 1994e
Nd:YAG	-	10ns	OM & SEM	Meja 1999b
Nd:YAG[2]	-	10ns	OM & SEM	Meja 1999b
KrF	4-10	25ns	OM & SEM	Meja 1999b
XeCl	8	-	VB	Kimura 1994
KrF	20	20ns	OM, SEM, AES	Lu 1994a
KrF	10	23ns	AWM	Lu 1997a
KrF	5	20ns	OM, SEM, AES	Lu 1994e
KrF	600	20ns	OM, SEM, AES	Lu 1994e
KrF	20	20ns	OM, SEM, AES	Lu 1994e
CO_2[10.6]	-	100ns	OM, SEM, AES	Lu 1994e
Nd:YAG[2]	10	10ns	AR	Lee 2000
Nd:YAG[2]	10	10ns	AR	Lee 2000
Nd:YAG[2]	1	14.5ns FWHM	OM & SEM	Meja 1999a
Nd:YAG	1	14.5ns FWHM	OM & SEM	Meja 1999a
Nd:YAG	1	14.5ns FWHM	OM & SEM	Meja 1999a
KrF	single	31ns	CCD, IP	Fourrier 2000
KrF	single	31ns	CCD, IP	Fourrier 2000
KrF	single	31ns	CCD, IP	Fourrier 2000
KrF	single	31ns	CCD, IP	Fourrier 2000
KrF	single	31ns	CCD, IP	Fourrier 2000
KrF	single	31ns	CCD, IP	Fourrier 2000
KrF	single	31ns	CCD, IP	Fourrier 2000
KrF	single	31ns	CCD, IP	Fourrier 2000
KrF	single	31ns	CCD, IP	Fourrier 2000
KrF	single	31ns	CCD, IP	Fourrier 2000
ArF	single	>31ns	CCD, IP	Fourrier 2000
ArF	20	>31ns	CCD, IP	Fourrier 2000
ArF	20	>31ns	CCD, IP	Fourrier 2000
ArF	20	>31ns	CCD, IP	Fourrier 2000

Table 2.4. (cont.)

Substrate	Contaminant	Contaminant Size	Deposition Technique	Type of LC	Threshold Fluence	Fluence Used	% removed
PMMA	SiO_2	400nm	SCCS	Dry (air)	18.5mJ/cm^2	25.5mJ/cm^2	~35
PMMA	SiO_2	400nm	SCCS	Dry (air)	-	25.5mJ/cm^2	~80
Polyimide	SiO_2	400nm	SCCS	Dry, RH	~10mJ/cm^2 (30% lower cf normal conditions)	-	-
Polyimide	SiO_2	800nm	SCCS	Dry RH	~5mJ/cm^2 (30% lower cf normal conditions)	-	-
PMMA	PS	320nm	SCCS	Dry, RH	~110mJ/cm^2 (no change cf normal conditions)	-	-
PMMA	SiO_2	400nm	SCCS	Dry, RH	no cleaning	n/a	n/a
IC mould	Organic (grease/ wax/resin)	-	Pack	Dry	100mJ/cm^2	400mJ/cm^2	optimal
Magnetic Head Slider	Epoxy resin	N/A	Glue	Dry	-	450mJ/cm^2	optimal
Magnetic Head Slider	Metal particles	micron sized	-	Dry	-	330 mJ/cm^2	optimal
Magnetic Head Slider	Metal particles	micron sized	-	Dry	-	187.5 mJ/cm^2	optimal
Magnetic Head	Epoxy resin	several micrometres thick	Glue	Dry	-	60 mJ/cm^2	100
Magnetic Head	Sn	-	ASus, AJ	Dry	25mJ/cm^2	100 mJ/cm^2	100
Magnetic Head	Al	-	ASus, AJ	Dry	25mJ/cm^2	100 mJ/cm^2	90
slider	Al	-	-	Dry	-	100 mJ/cm^2	IN
Slider	Sn	-	-	Dry	-	50 mJ/cm^2	IN
Slider	-	-	-	Steam	-	130 mJ/cm^2	IN
magnetic disk	SiC	-	Sol	Dry	-	200 (or 250) mJ/cm^2	100

Table 2.4. (cont.)

Laser Used	No. of pulses	Pulse Length	Analysis Technique	Reference
ArF	Single	>31ns	CCD, IP	Fourrier 2000
ArF	20	>31ns	CCD, IP	Fourrier 2000
KrF	Single	31ns	CCD, IP	Fourrier 2000
KrF	Single	31ns	CCD, IP	Fourrier 2000
KrF	Single	31ns	CCD, IP	Fourrier 2000
KrF	Single	31ns	CCD, IP	Fourrier 2000
KrF	10	23ns	OM, IP	Lu 2000a
KrF	25	20ns	OM, SEM, AES	Lu 1994a
KrF	30	20ns	OM, SEM, AES	Lu 1994a
KrF	150	20ns	OM, SEM, AES	Lu 1994a
KrF	2500	20ns	OM, SEM, AES	Lu 1994b
KrF	100	23ns	SEM	Lu 1996
KrF	100	23ns	SEM	Lu 1996
Excimer	Varies	-	-	Lu 2001
Excimer	Varies	-	-	Lu 2001
Excimer	Varies	-	OM	Lu 2001
KrF	50 (or 5)	23ns	-	Song 1998

Table 2.4. (cont.)

Substrate	Contaminant	Contaminant Size	Deposition Technique	Type of LC	Threshold Fluence	Fluence Used	% removed
NiP	Al_2O_3	1μm	-	Wet, IF	30mJ/cm^2	70mJ/cm^2	Optimal
NiP	Quartz	5μm	-	Dry	16mJ/cm^2	140mJ/cm^2	~65
NiP	Silica spheres	1μm	WS, AJ	Dry	8 mJ/cm^2	~20 mJ/cm^2	~100
TiN coating	Polymer M-CxFyOz (M=Ti,Al,Cu) & M-CFx	-	Res	Dry	-	250mJ/cm^2	Optimal
TiN coating	Polymer M-CxFyOz (M=Ti,Al,Cu) & M-CFx	-	Res	Dry	-	100mJ/cm^2	Optimal
Limestone sculpture	Black crusts	0.05mm	Nat	Wet, WL	-	400mJ/cm^2	Optimal
Limestone sculpture	Black crusts	0.05mm	Nat	Dry	-	400mJ/cm^2	Optimal
field emitter arrays	Oxide layer / surface contaminant	-	Fab	Dry	-	85mJ/cm^2	NE
field emitter arrays	Oxide layer / Surface contaminant	-	Fab	Dry	-	85mJ/cm^2	NE
field emitter arrays	Oxide layer / Surface contaminant	-	Fab	Dry	-	35mJ/cm^2	Removed

Table 3.1.1. Characteristics of the lasers used in the cleaning studies.

Laser	λ (nm)	Pulse length (ns)	Laser Beam Dimension at Cleaning Site (μm)	Coherence
UV-CVL	255	35	70-100 (ϕ)	High
XeCl	308	8	(200x800)–(400x1200)	Low
KrF	248	12	4000 x 7000	Low

Table 2.4. (cont.)

Laser Used	No. of pulses	Pulse Length	Analysis Technique	Reference
KrF	-	23ns	OM, IP	Lu 1998d,1999a, 1998c,2000d, 2000e,2001
KrF	100	23ns	OM, IP	Lu 1999b, 2000d,2000e, 1998d, 2001
KrF	200	23ns	SEM	Zheng 2001a
KrF at 45°	60	23ns	SEM/AFM	Lu 1998a
KrF at 45°	1200	23ns	SEM/AFM	Lu 1998a
Nd:YAG	10	6ns	LS, AWM	Cooper 1995
Nd:YAG	30	6ns	LS, AWM	Cooper 1995
Nd:YLF		15ns	ECM	Yavas 1998
Nd:YLF [2]		15ns	ECM	Yavas 1998
Nd:YLF [3]		15ns	ECM	Yavas 1998

The experimental layout is shown schematically, with a UV-CVL as the laser, in figure 1. The sample was placed at 45° to the incoming laser beam. The same basic layout has been used in all studies except the angle of incidence on the sample has been both 45° and normal in studies using the XeCl and KrF laser. A constant dry air flow was maintained across the surface of the sample to remove particles ejected from the surface by laser cleaning.

Data was collected by imaging an area of the sample surface before and after irradiation, using a customized microscope. A total magnification of 100x was used. Global Lab Image software package was used for image analysis. A rectangular region of interest (ROI, 530 μm x 350 μm) was defined that encapsulated the cleaned area and a gray scale threshold was

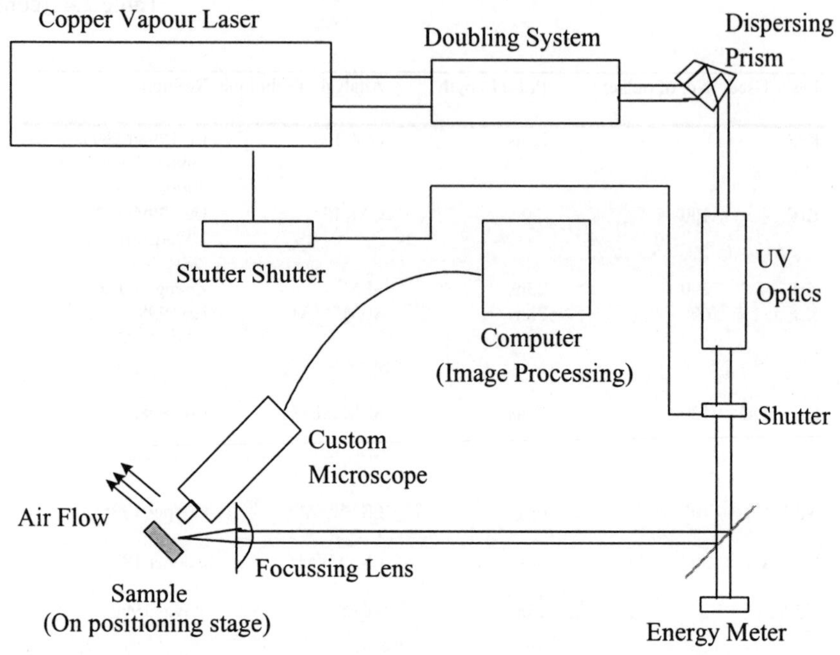

Fig. 1. Schematic of the experimental layout.

manually set to account for all contaminants on the surface. Only particle and agglomerate sizes above 0.3 μm were unambiguously resolvable in the digitised images. Two measures of cleaning efficiency were used. Firstly, the fractional area covered by particles and agglomerates in the ROI before and after laser cleaning, p_b and p_a respectively, were measured. Cleaning efficiency was then taken as $1\text{-}(p_a / p_b)$, expressed as a percentage. Secondly, the number of particles in the ROI, before and after laser cleaning, were counted, N_b and N_a, respectively. The cleaning efficiency is then defined as $1\text{-}(N_a / N_b)$ expressed as a percentage. The result can also be presented as histograms of particle numbers before and after laser cleaning.

These three studies together allow a number of laser characteristics and their effect on laser cleaning to be contrasted. These include: using lasers of similar wavelength but different coherence properties; using lasers of similar wavelength but significantly differing area of treatment and using lasers of similar coherence but differing wavelength. Also, laser cleaning of microscope slides and fused silica have been contrasted using the UV-CVL.

3.1.3. Before and after images – microscope slides and fused silica

Fig. 2 shows examples of laser cleaning alumina particles from microscope slides using the UV-CVL. These are contrasted with laser cleaning alumina particles from fused silica in Fig. 3.

It is apparent from Figs. 2 and 3 that there is very little difference in the laser cleaning results achieved on microscope slides and fused silica samples. More careful image analysis establishes that the threshold fluence for cleaning was very similar in both cases, though there was more scatter in the cleaning efficiency as a function of the single laser pulse fluence for fused silica compared to microscope slides. For the fused silica sample the area cleaned relative to the beam size (about 100 microns in diameter) was somewhat reduced and there was more evidence of particle and agglomerate movement. In both cases it must be remembered that the glass sample has been prepared by standard solvent wipe techniques that represent normal practice for cleaning optics.

This prepares a surface which still has some hydrocarbon contamination from the air and thus, the adhesion of the alumina particles to the surface is likely modified because of this. For a laser cleaning mechanism involving the thermal expansion of the glass and or the particles there would be a significant difference predicted for the laser cleaning results of these two glasses on the basis of their differing physical properties. Failure to observe this indicates that the laser cleaning has been enhanced by the condensation of moisture and hydrocarbons from the air in the normal laboratory environment (humidity is greater than 50%). Further investigation of the effect of humidity and hydrocarbon contamination on the surface are currently being carried out.

Further images of laser cleaning achieved with a XeCl excimer laser pulse and a KrF excimer laser pulse on microscope slide samples, with alumina particles, appear in figures 4 & 5. These demonstrate that excellent laser cleaning results are achieved with both lasers. The threshold fluence for cleaning is higher for both these lasers compared to the UV-CVL as is discussed in section 3.3.

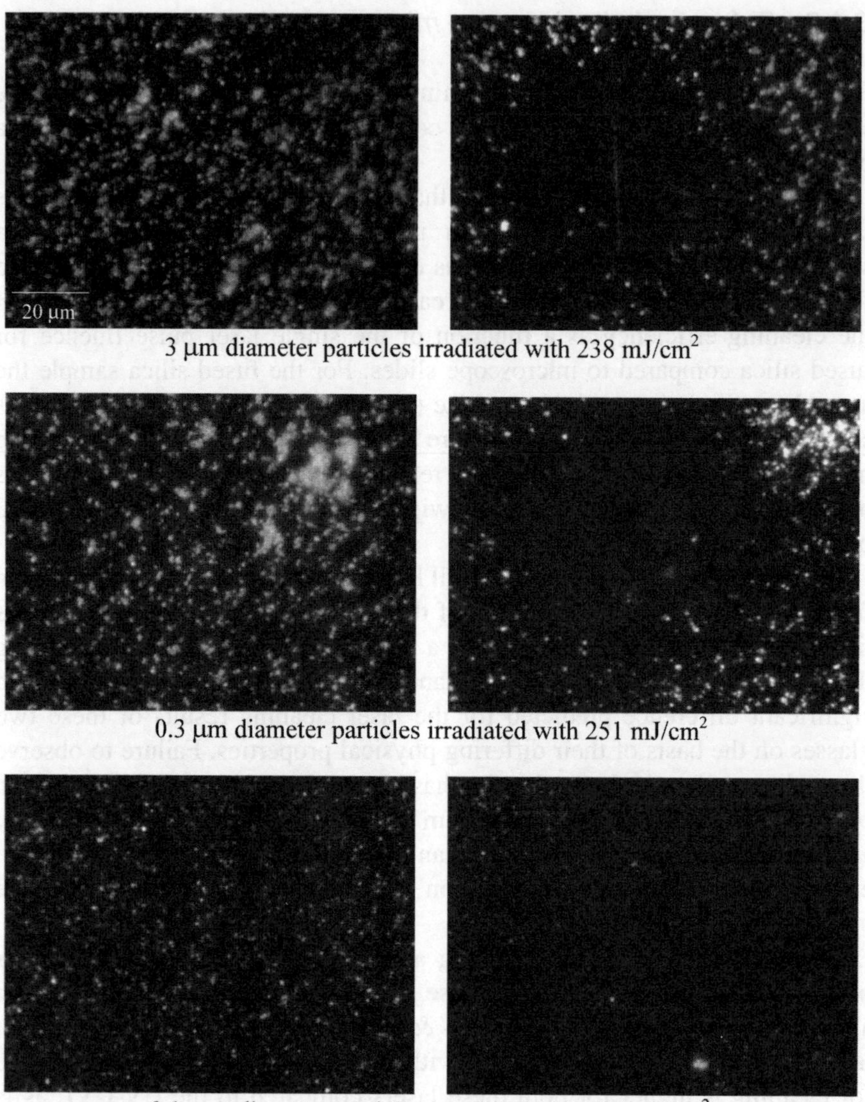

3 μm diameter particles irradiated with 238 mJ/cm^2

0.3 μm diameter particles irradiated with 251 mJ/cm^2

0.1 μm diameter particles irradiated with 399 mJ/cm^2

Fig. 2. Optical micrographs showing before (left) and after (right) images of a glass microscope slide that has been contaminated with Al_2O_3 particles of specified size and irradiated with a single 35 ns laser pulse from a UV-CVL. The laser beam was ~80 microns in diameter. The central region has been cleaned.

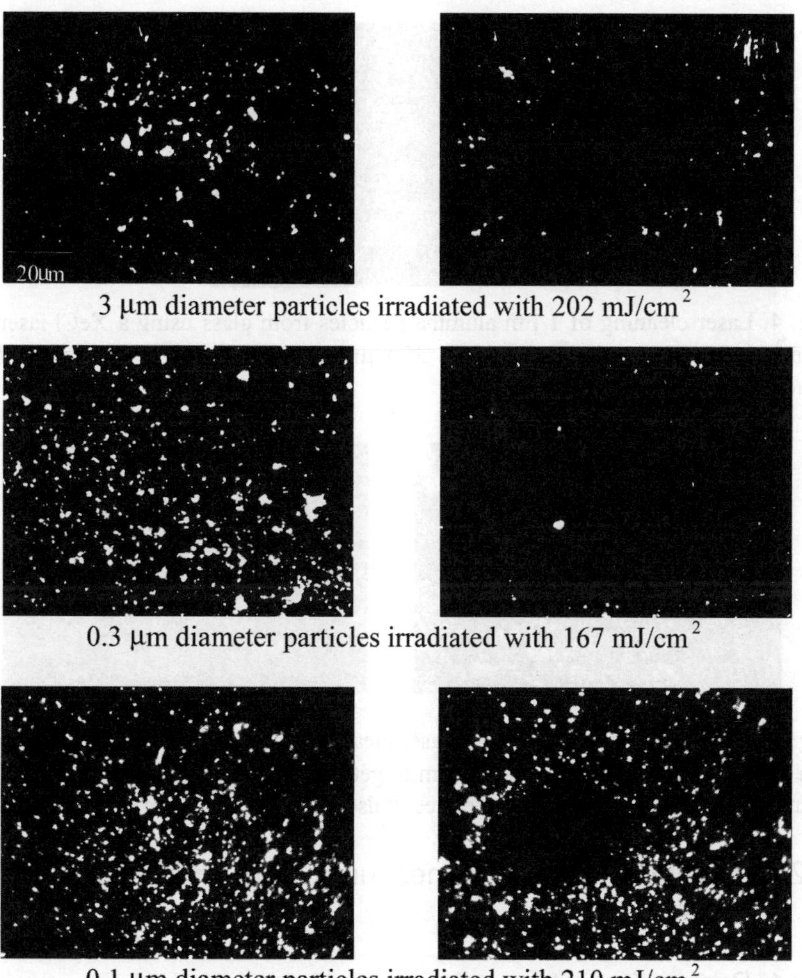

Fig. 3. Optical micrographs showing before (left) and after (right) images of a fused silica sample that has been contaminated with Al_2O_3 particles of specified size and irradiated with a single 35 ns laser pulse from a UV-CVL. The laser beam was ~80 microns in diameter. The central region has been cleaned.

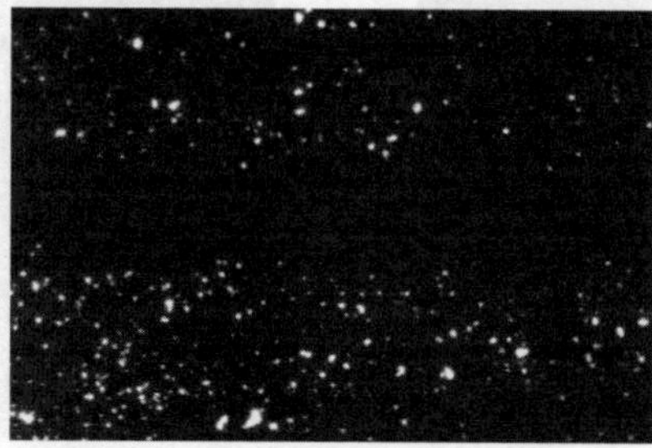

Fig. 4. Laser cleaning of 1 μm alumina particles from glass using a XeCl laser (~1 J/cm^2). Area depicted is 530 μm x 350 μm. Central band of sample has been irradiated and particles removed.

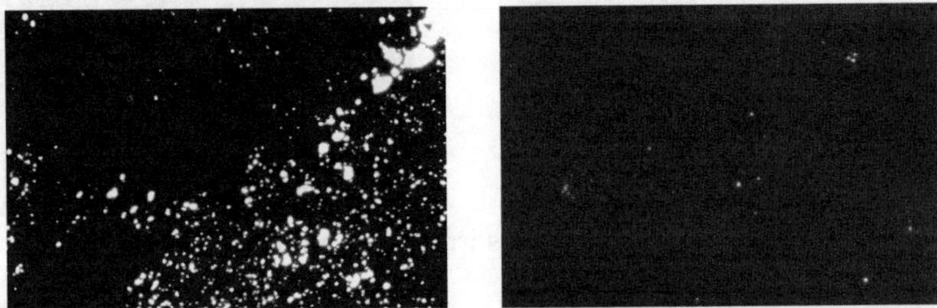

Fig. 5. Before (left) and after (right) laser cleaning images of a region 530 μm x 350 μm at the centre of the KrF laser beam target site on a microscope slide with 1 μm alumina particles and their agglomerates. Pulse fluence 850 mJ/cm^2.

3.2. Cleaning efficiency measured from optical microscopy images

3.2.1. Particle de-agglomeration and removal

Image analysis of the before and after optical micrographs has been used to determine the cleaning efficiency. Both the area covered by particles and agglomerates, and the number of particles and agglomerates has been used. The comparison of these two different measures has been used to gain insight into the impact of agglomerates. The agglomerates may show strong self adhesion and removal as a single particle, or may be broken up, possibly even leading to increasing rather than decreasing particle numbers which can still correspond to reduced area coverage. In the former case the cleaning efficiency by the two measures agree closely. In the latter case the cleaning efficiency from area covered by particles tends to be significantly higher than that from particle numbers. It must also be noted that measuring cleaning efficiency from area covered gives a positive weighting to larger particles and this must be critically evaluated to determine whether the smallest sized particles have been removed at all.

Fig. 6 shows an example of UV-CVL laser cleaning of alumina particles from a microscope slide. Five areas around that which is irradiated (area A) are also examined to quantify the effect of agglomerate break-up, and particle & agglomerate movement. Each of the areas is just large enough to encapsulate the beam size. The complete "square" region of interest was analysed. This means that the apparent laser cleaning effect in area A is less than would be obtained by analysing a circular region of interest encapsulating just the "cleaned" area. The before and after particle and agglomerate distributions, given as number of particles of different sizes appear in Fig. 7. In region A there is clear evidence of both reduction in particle number and reduction in agglomerate size. In all the other regions B-F there is clear evidence of reduction in agglomerate size, but no definitive evidence that particle removal has occurred. This is true even in region F which is a full beam width from the irradiated site. Similar results are seen in other sets of images irradiated at different fluences. The break up of agglomerates outside of the irradiated area is present for all nominal particle sizes used and becomes more prominent as the laser pulse fluence is increased.

There are several mechanisms that could be responsible for this disturbance of particles in areas B-F. If the substrate absorbs some laser light, as is the case here, then it will undergo a rise in temperature and an associated thermal expansion. While the extent of the temperature rise is contentious, it can be argued that a thermal expansion of the substrate may produce enough force to disturb the particles present on it, even if it does not remove them. The surface displacement caused by the thermal expansion may propagate outwards from the irradiated site and be responsible for the particle disturbances seen at a distance from the irradiated region. The particles and agglomerates also absorb some of the laser irradiation, which may be preferentially absorbed by any moisture that has capillary condensed between the particles of the agglomerate. If this moisture is vaporized by the absorbed laser radiation this would reduce, and possibly overcome the adhesive forces between the particles in the

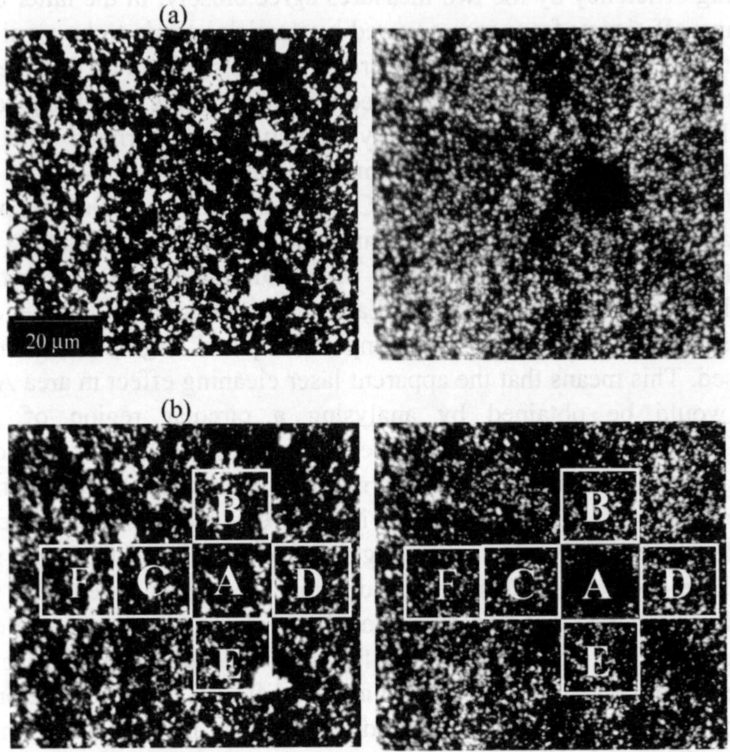

Fig. 6. (a) before and after images of 3 μm diameter particles irradiated with a 213 mJ/cm^2 UV-CVL pulse. (b) shows the same image with regions defined for the particle categorisation study.

agglomerate. Another possible mechanism for the particle disturbance is the generation of a shockwave during irradiation. Optical damage was sometimes seen during laser cleaning, which suggests such mechanical and acoustic waves may be present. Further studies are underway to investigate this further. It appears the small diameter of the laser beam may be an important factor in generating a shockwave in the substrate at modest fluences.

Regardless of the mechanism responsible, the disturbance of particles at the surface has implications for the practice of laser cleaning. Smaller particles have been shown to be harder to remove from surfaces, when the surface is chemically clean and the same would likely be true for smaller

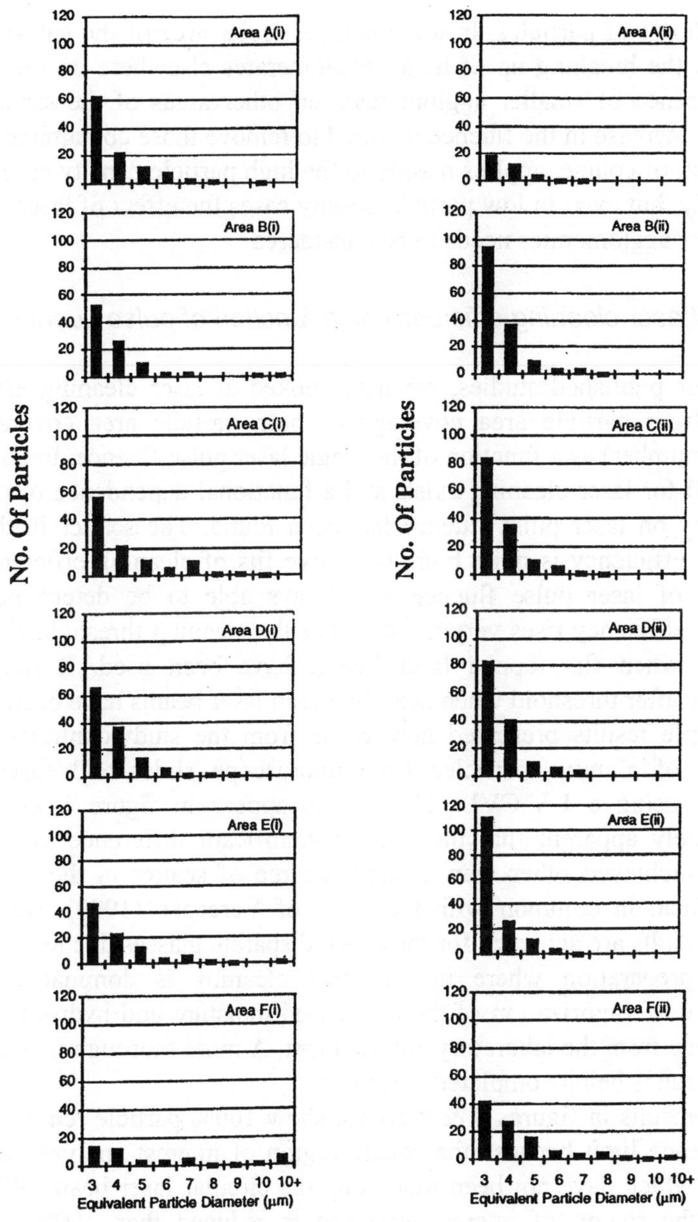

Fig. 7. Particle and agglomerate distribution histograms for areas in figure 6, both (i) before and (ii) after irradiation.

agglomerates of particles. Irradiating a particular area of the substrate may result in the breaking up of larger agglomerates elsewhere on the surface. The presence of smaller agglomerates on other areas of the surface may mean an increase in the fluence required to remove those contaminants. This argument, of course, applies mainly to the high particle density case seen in this study, but even in low particle density cases the effect of laser cleaning on nearby agglomerates needs to be considered.

3.2.2. Laser cleaning efficiency as a function of pulse fluence

In all our published studies, we have looked at laser cleaning efficiency (either from particle area coverage or both particle area coverage and particle number) as a function of the single laser pulse fluence. In all cases a threshold for laser cleaning exists and a functional dependence of cleaning efficiency on laser pulse fluence has been found. The scatter in the laser cleaning efficiency is quite high, but curve fits of cleaning efficiency as a function of laser pulse fluence are always able to be determined. The cleaning efficiency rises very rapidly after the cleaning threshold fluence is reached, when flat topped laser beams have been used. It rises more gradually after threshold when near-Gaussian laser beams have been used.

Sample results presented here come from the study-contrasting laser cleaning of alumina particles from microscope slides and fused silica samples, using a UV-CVL. The results appear in figure 8 & 9. It is immediately apparent that there is no significant difference in the laser cleaning achieved other than a larger degree of scatter in the results for fused silica. In common with the work of Vereecke (1999), we believe similar results are achieved for these two disparate glasses due to a common surface preparation where in the laser cleaning is dominated by the assistance of vaporization of the condensed moisture and hydrocarbons on the surface from the laboratory environment. A more thorough investigation of this issue is being completed currently.

The results in figures 8 & 9 do not show 100% particle removal in the high fluence limit because the square region of interest analyzed includes corners, which have not been treated by the circular laser beam. When the size of the region of interest analyzed is reduced then 100% cleaning efficiency has been obtained as shown in figure 10 for alumina particles on microscope slides cleaned using a UV-CVL.

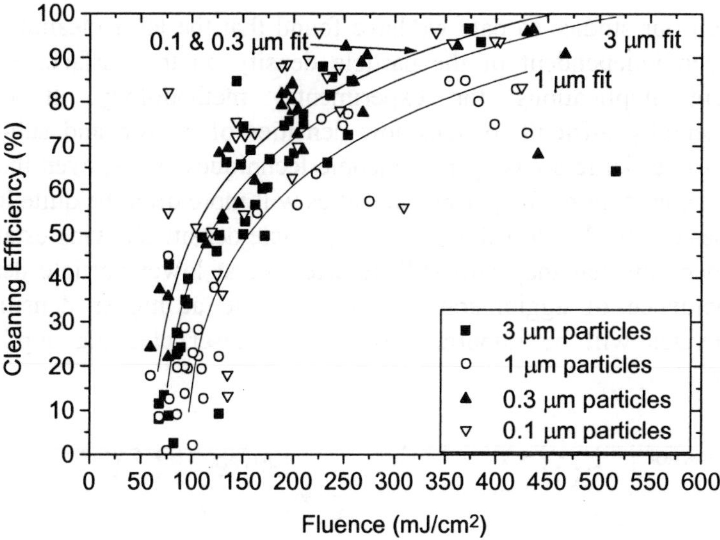

Fig. 8. Cleaning efficiencies (UV-CVL) for the four nominal alumina particle sizes from microscope slides.

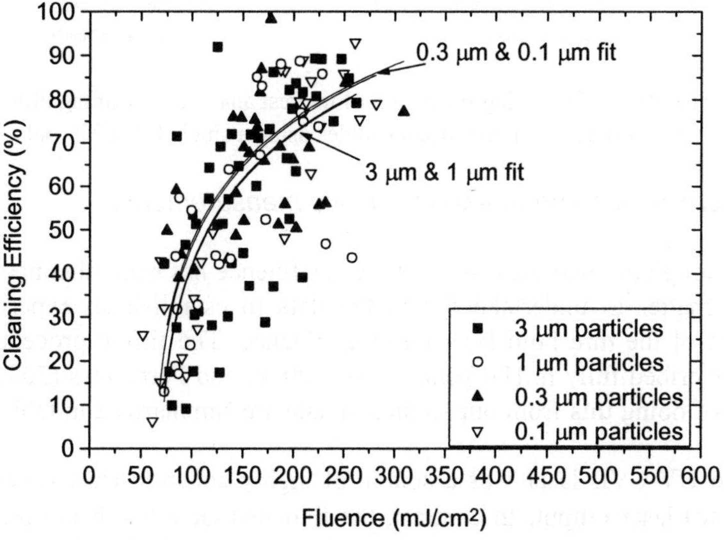

Fig. 9. Cleaning efficiencies for the four nominal alumina particle sizes from fused silica.

In all our studies to date we have found that the laser cleaning results have been independent of the particle density on the surface. This has important implications for experimental methodology. It can be experimentally difficult to track low densities of micron and sub-micron particles over large areas by microscopic techniques. Thus, even though in practical applications the particle densities will more often be quite low, the processes can be developed with higher particle densities having established experimentally that they will still be effective at lower particle densities. The incidence of agglomeration does increase at higher densities but agglomerates will be important to practical laser cleaning applications anyway.

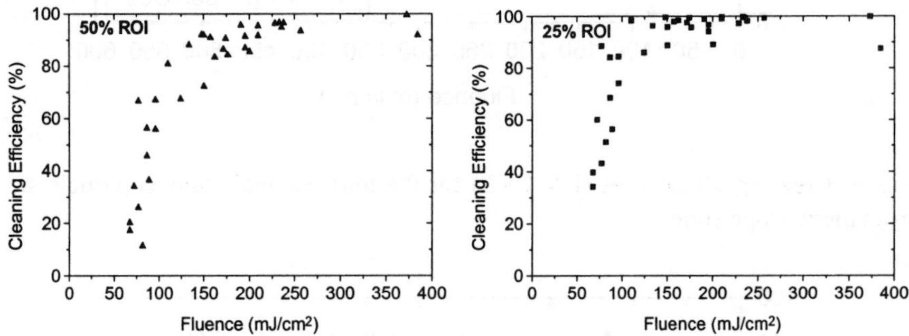

Fig. 10. The effect of reducing the region of interest analysed. Cleaning efficiency of 3 μm alumina particles from microscope slides using a single UV-CVL pulse.

3.2.3. Laser cleaning threshold fluence measurements

Once a graph of cleaning efficiency versus fluence has been obtained, it is a simple matter to undertake fits to the data to establish an experimental measure of the threshold laser cleaning fluence. The fitting procedure has been described fully in Halfpenny (1999, 2000) and Fernandes (2001). The results of doing this from our studies to date are summarized in tables 3.1 – 3.3.

The UV-CVL laser was a source of highly coherent (low divergence, <50 μrad) laser output. In contrast, the flattened Gaussian beam profile of most commercial excimer lasers (near Gaussian profile for short axis, flat topped, sloping at edges for long axis of the beam) has a beam divergence of

Table 3.1. Threshold laser cleaning fluence for removing different sized alumina particles from microscope slides using a single pulse from a UV-CVL.

Individual Particle Diameter (μm)	Threshold Fluence Range (mJ/cm^2)	Errors due to initial value sensitivity
0.1	62-67	+ 17
0.3	62-68	~ 0
1	91-93	+ 24
3	72-79	+ 5

Table 3.2. Threshold laser cleaning fluence for removing different sized alumina particles from fused silica using a single pulse from a UV-CVL.

Individual Particle Diameter (μm)	Threshold Fluence Range (mJ/cm^2)	Errors due to initial value sensitivity
0.1	70-72	+12
0.3	64-70	~ 0
1	71-78	~ 0
3	74-77	+16

Table 3.3. Threshold laser cleaning fluence for removing 1 μm alumina particles from microscope slides.

Laser	Particle Fractional Coverage	Threshold Fluence (mJ/cm^2)	Maximum Cleaning Efficiency
UV -CVL	10-40%	90-115	100%
XeCl	0.1-40%	330-380	95%
KrF	0.8-14%	200-400	97%

10 - 40 times this. The lower threshold for laser cleaning for the 255 nm UV-CVL compared to the 248 nm KrF laser was most likely because higher coherence can be an advantage in the laser cleaning process. The UV-CVL and KrF pulsed laser output is measured to be totally absorbed in the glass microscope slide so there is no differential absorption to explain the difference in the threshold fluence for laser cleaning.

When comparing the laser cleaning results for the XeCl and KrF excimer lasers the wavelength difference dominates. The measured absorption of the pulsed XeCl laser output in the glass slides was 93%, which was close to the 100% absorption of the KrF output. However, the fact that there was some transmission indicates that the output was being absorbed through the 1millimetre thickness of the slide whereas the KrF pulse may well be absorbed with an optical penetration depth, which was a fraction of the slide thickness. The uncertainty in the threshold fluence in the laser cleaning studies using the larger area KrF beam makes it difficult to make direct comparisons, but the threshold fluence using the longer wavelength was between 0.8 and 1.65 times that at the shorter wavelength. This would suggest the conclusion that the 60 nm difference in wavelength and the small difference in total absorption have little impact on the laser cleaning threshold fluence. The differences observed when comparing the XeCl laser results with the UV-CVL results would then also be interpreted as being largely due to the difference in coherence of the two lasers rather than the difference in wavelength. However, it is not possible to draw firm conclusions from the available experimental evidence at this time. Further investigation of the reasons for spatial variation in laser cleaning efficiency when using larger area beams needs to be completed.

4. Concluding remarks

This chapter has reviewed the overall field of experimental laser cleaning studies and described in more detail the laser cleaning of particles and agglomerates from glass and optical materials that are the focus of research at Macquarie University. The excellent results that can be achieved removing sub micron and micron sized alumina particles from glass indicate that this process has very real promise for utilisation in the optics, photonics and optoelectronics industries. We will continue this program of laser cleaning and related programs on surface modification and material removal for many years to come in order to bring this promise to reality. Currently we are engaged in projects to study removal of hydrocarbon contamination from optical materials, evaluating the effect of glass surface cleanliness before particles are introduced on the laser cleaning results, further developing a process for laser dehydroxylation of fused silica as a means of creating a hydrophobic glass surface, evaluating spatial variation of laser cleaning achieved with larger area beams and elucidating the mechanism of particle assisted laser material removal.

Acknowledgments

We would like to thank Professor Boris Luk'yanchuk for inviting us to submit this chapter to this book and also for the key role he has played in bringing the laser cleaning community together. We acknowledge the financial support our research has had from the Australian Research Council and Macquarie University. We also gratefully acknowledge representatives of Photonics Companies in Australia for defining some of their manufacturing problems so that we may include this perspective in our academically oriented research.

References

Allen S.D., Miller A.S., Lee S.J., *Laser assisted particle removal 'dry' cleaning of critical surfaces*, Materials Science & Engineering, **B49**, pp. 85-88 (1997)

Beaudoin F., Meunier M., Simard-Normandin M., Landheer D., *Excimer laser cleaning of silicon wafer backside metallic particles*, J. Vac. Sci. Technol., **A16** No. 3, pp. 1976-1979 (1998)

Boquillon J. P., Oltra R., *Principle of surface cleaning by laser impact,* European Conference on Lasers and Electro-Optics ECLEO, OSA Conference Digest (1994)

Cooper M. I., Emmony D. C., Larson J., *Characterization of laser cleaning of limestone*, Optics & Laser Technology, **27**(1), pp. 69-73 (1995)

Curran C., Lee J. M., Watkins K. G., *Laser removal of particles using tunable wavelength*, 2000 Conference on Lasers and Electro-Optics Europe, (IEEE Piscataway NJ, USA), IEEE Catalog Number 00TH8505 (2000)

Dobler V., Oltra R., Boquillon J. P., Mosbacher M., Boneberg J., Leiderer P., *Surface acceleration during dry laser cleaning of silicon*, Appl. Phys. A **69** (Suppl), pp. S335-S357 (1999)

Engelsberg A. C., *Laser-assisted cleaning proves promising*, Precision Cleaning Magazine (May 1995)

Fernandes A. J., Kane D. M., *Dry Laser cleaning threshold fluence – how can it be measured accurately?,* Proc. SPIE, vol. **4426** (2002)

Fourrier T., Schrems G., Mühlberger T., Heitz J., Arnold N., Bauerle D., Mosbacher M., Boneberg J., Leiderer P., *Laser cleaning of polymer surfaces*, Appl. Phys. **A 72**, pp. 1- 6 (2000)

Halfpenny D. R., Kane D.M., Lamb R. N., Gong B., *Creation of adhesion resistant silica surfaces with ultraviolet laser cleaning*, 1996 Conference on Optoelectronic and Microelectronic Materials and Devices (IEEE Piscataway NJ, USA), IEEE Catalog Number 96TH8197 (1996)

Halfpenny D. R, Kane D. M., *A quantitative analysis of single pulse ultraviolet dry laser cleaning*, J. Appl. Phys., **86**, pp. 6641-6646 (1999)

Halfpenny D. R., *Ultraviolet laser cleaning of glass,* PhD Thesis, (Macquarie University, 2000a)

Halfpenny D. R., Kane D. M., Lamb R. N., Gong B., *Surface modification of silica with ultraviolet laser radiation*, Appl. Phys. A. **71**, pp. 147-151 (2000b)

Heroux J. B., Boughaba S., Ressejac I., Sacher E., Meunier M., *CO_2 laser-assisted removal of submicron particles from solid surfaces*, J. Appl. Phys., **79** (6) pp. 2857-2862 (1996)

Imen K., Lee S. J., Allen S. D., *Laser-assisted micron scale particle removal*, Appl. Phys. Lett., **58** (2), pp. 203-205 (1991)

Kane D. M., Halfpenny D. R., *Reduced Threshold UV-Laser Ablation of Glass Substrates with Surface Particle Coverage and Associated Systematic Surface Laser Damage*, J. Appl. Phys. **87**, pp. 4548-4552 (2000)

Kane D. M., Fernandes A. J., *Laser Cleaning of Particles from Surfaces – Issues Relating to Sample Preparation*, Proc. SPIE, vol. **4426** (2002)

Kelley J. D., Hovis F. E., *A thermal detachment mechanism for particle removal from surfaces by pulsed laser irradiation*, Microelectronic Engineering, **20**, pp. 159-170 (1993)

Kimura W. D., Kim G. H., Balick B., *UV laser cleaning of astronomical mirror samples*, 1994 Conference on Lasers and Electro-Optics OSA, 1994.

Kolomenskii A. A., Schuessler H. A., Mikhalevich V. G., Maznev A. A., *Interaction of laser-generated surface acoustic pulses with fine particles: Surface cleaning & adhesion studies*, J. Appl. Phys., **84**, pp. 2404-2410 (1998)

Kumar D., Ata A., Mahajan U., Singh R., *Role of line-beam on the removal of particulate contaminations from solid surfaces by pulsed laser*, J. Electronic Materials, **27**, pp. 1104-1106 (1998)

Lee S. J., Imen K., Allen S. D., *Laser-assisted particle removal from silicon surfaces*, Microelectronic Engineering, **20**, pp. 145-157 (1993)

Lee J. M., Watkins K. G., Steen W. M., *Angular laser cleaning for effective removal of particles from a solid surface,* Appl. Phys. A **71**, pp. 671-674. 2000)

Lee J. M., Watkins K. G., *Removal of small particles on silicon wafer by laser-induced airborne plasma shock waves,* J. Appl. Phys., **89**, pp. 6496-6500 (2001)

Leiderer P., Boneberg J., Mosbacher M., Schilling A., Yavas O., *Laser cleaning of silicon surfaces*, Proc. SPIE, vol. **3274**, pp. 68-78 (1998)

Lizotte T. E, O'Keefe T. R., *Chemical free cleaning using excimer lasers,* Proc. SPIE, vol. **2703**, pp. 279-287 (1996)

Lu Y. F., Aoyagi Y., Takai M., Namba S., *Laser surface cleaning in air: mechanisms & applications*, Jpn. J. Appl. Phys., **33**, pp. 7138-7143 (1994a)

Lu Y. F, Aoyagi Y., *Laser induced dry cleaning in air - a new surface cleaning technology in lieu of CFC solvents*, Jpn. J. Appl. Phys., **33**: pp. L430-L433 (1994b)

Lu Y. F., Komuro S., Aoyagi Y., *Laser-induced removal of fingerprints from glass and quartz surfaces*, Jpn. J. Appl. Phys., **33**, pp. 4691-4696 (1994c)

Lu Y. F., Takai M., Komuro S., Shiokawa T., Aoyagi Y., *Surface cleaning of metals by pulsed-laser irradiation in air*, Appl. Phys. A **59**, pp. 281-288 (1994d)

Lu Y. F., Song W. D., Hong M. H., Teo B. S., Chong T. C., Low T. S., *Laser removal of particles from magnetic head sliders*, J. Appl. Phys., **80**, pp. 499-504 (1996)

Lu Y. F., Lee Y. P., Hong M. H., Low T. S., *Acoustic wave monitoring of cleaning and ablation during excimer laser interaction with copper surfaces*, Appl. Surf. Sci., **119**, pp. 137-146 (1997a)

Lu Y. F., Song W. D., Ang B. W., Hong M. H., Chan D. S. H., Low T. S., *A theoretical model for laser removal of particles from solid surfaces*, Appl. Phys. A **65**, pp. 9-13 (1997b)

Lu Y. F., Song W. D., Hong M. H., Chan D. S. H., Low T. S., *Laser cleaning of microparticles - theoretical prediction of threshold laser fluence*, Proc. SPIE, vol. **3097**, pp. 352-357 (1997c)

Lu Y. F., Song W. D., Ye K., Lee Y., Chan D., Low T., *A cleaning model for removal of particles due to laser induced thermal expansion of substrate surface*, Jpn. J. Appl. Phys., **36**, pp. L1304-L1306 (1997d)

Lu Y. F., Lee Y. P., Zhou M. S., *Laser cleaning of etch-induced polymers from via holes*, J. Appl. Phys., **83**, pp. 1677-1684 (1998a)

Lu Y. F., Song W. D., Tee C. K., Chan D. S. H., Low T. S., *Wavelength effects in the laser cleaning process*, Jpn. J. Appl. Phys., **37**, pp. 840-844. (1998b)

Lu Y. F., Zhang L., Song W. D., Chan D. S. H., *A theoretical model for laser cleaning of microparticles in a thin liquid layer*, Jpn. J. Appl. Phys., **37**, pp. L1330-L1332 (1998c)

Lu Y. F., Song W. D., Zhang Y., Low T. S., *Theoretical model and experimental study for dry and steam laser cleaning,* Proc. SPIE, vol. **3550**, pp. 7-18 (1998d)

Lu Y. F, Song W. D., Low T. S., *Laser cleaning of micro-particles from a solid surface - theory & applications*, Materials Chemistry & Physics, **54**, pp. 181-185 (1998e)

Lu Y. F., Song W. D., Zhang Y., Zheng Y. W., *Laser surface cleaning of electronic materials*, Proc. SPIE, vol. **3618**, pp. 278-289 (1999a)

Lu Y. F., Zheng Y. W., Song W. D., *An energy approach to the modelling of particle removal by pulsed laser irradiation*, Appl. Phys. A **68**, pp. 569-572 (1999b)

Lu Y. F, Song W. D, Hong M. H., Ren Z. M., Chen Q., Chong T. C., *Laser cleaning of IC mould and its real-time monitoring*, Jpn. J. Appl. Phys., **39**, pp. 4811-4813 (2000a)

Lu Y. F, Zheng Y. W., Song W. D., *Characterization of ejected particles during laser cleaning*, J. Appl. Phys., **87**, pp. 549-552 (2000b)

Lu Y. F., Zheng Y. W., Song W. D., *Laser induced removal of spherical particles from silicon wafers*, J. Appl. Phys., **87**, pp. 1534-1539 (2000c)

Lu Y. F, Song W. D., Hong M. H., Zheng Y. W., *Laser surface cleaning and real-time monitoring,* Proc. SPIE, vol. **4070**, pp. 331-337 (2000d)

Lu Y. F., Song W. D., Hong M. H., Ren Z. M., Zheng Y. W., *Laser surface cleaning – Basic understanding, engineering efforts and technical barriers,* Proc. SPIE, vol. **4088**, pp. 371-379 (2000e)

Lu Y. F., Song W. D., Zhang Y., Hong M. H., Chong T. C., *Laser removal of particles from solid surfaces,* Riken Review, **32**, pp. 64-70 (2001c)

Luk'yanchuk B. S., Zheng Y. W., Lu Y. F., *Laser cleaning of solid surface: Optical resonance and near-field effects*, Proc. SPIE, vol. **4065**, pp. 576-587 (2000)

Mann K., Wolff-Rottke B., Muller F., *Cleaning of optical surfaces by excimer laser radiation*, Appl. Surf. Sci., **96-98**, pp. 463-468 (1996)

Meja P., Autric M., Alloncle P., Pasquet P., Oltra R., Boquillon J. P., *Laser cleaning of oxidized iron samples: the influence of wavelength and environment*, Appl. Phys. **A 69**, pp. 687-690 (1999a)

Meja P., Autric M., Delaporte P., Alloncle P., *Dry laser cleaning of anodized aluminum*, Appl. Phys. A **69** (Suppl), pp. S343-S346 (1999b)

Meunier M., Wu X., Beaudoin F., Sacher E., Simard-Normandin M., *Excimer laser cleaning for microelectronics: modeling, applications and challenges*, Proc. SPIE, vol. **3618**, pp. 290-301 (1999)

Mosbacher M., Chaoui N., Siegel J., Dobler V., Solis J., Boneberg J., Afonso C. N., Leiderer P., *A comparison of ns and ps steam laser cleaning of Si surfaces*, Appl. Phys. A **69** (Suppl), pp. S331 - S334 (1999)

Mosbacher M., Dobler V., Boneberg J., Leiderer P., *Universal threshold for the steam laser cleaning of submicron spherical particles from silicon*, Appl. Phys. A **70**, pp. 669 - 672 (2000)

Oltra R., Arenholz E., Leiderer P., Kautek W., Fotakis C., Autric M., Afonso C., Wazen P., *Modelling and diagnostic of pulsed laser-solid interactions. Applications to laser cleaning,* Proc. SPIE, vol. **3885**, pp. 499-508 (2000)

Park H. K., Grigoropoulos C. P., Leung W. P., Tam A. C., *A practical excimer laser-based cleaning tool for removal of surface contaminants*, IEEE Transactions on Components, Packaging & Manufacturing Technology - A, **17**, pp. 631-643 (1994)

Psyllaki P., Pasquet P., Meja P., Oltra R., Autric M., *Laser cleaning of steels after high temperature oxidation*, 2000 Conference on Lasers and Electro-Optics Europe, (IEEE Piscataway NJ, USA), IEEE Catalog Number 00TH8505 (2000)

Song W. D., Lu Y. F., Hong M. H., Low T. S., *Laser cleaning of magnetic disks*, Proc. SPIE, vol. **3550**, pp. 19-26 (1998)

Tam A. C., Leung W. P., Zapka W., Ziemlich W., *Laser cleaning techniques for removal of surface particulates*, J. Appl. Phys., **71**, pp. 3515-3523 (1992)

Tam A. C., Park H. K., Grigoropoulos C. P., *Laser cleaning of surface contaminants*, Appl. Surf. Sci., **127-129**, pp. 721-725 (1998)

Vereecke G., Röhr E., Heyns M. M., *Laser-assisted removal of particles on silicon wafers*, J. Appl. Phys., **85**, pp. 3837-3843 (1999)

Vereecke G., Röhr E., Heyns M. M., *Influence of beam incidence angle on dry laser cleaning of surface particles,* Appl. Surf. Sci., **157**, pp. 67-73 (2000)

Wu X., Sacher E., Meunier M., *Excimer laser induced removal of particles from hydrophilic silicon surfaces*, J. Adhesion, **70**, pp. 167-178 (1999a)

Wu X., Sacher E., Meunier M., *The effects of hydrogen bonds on the adhesion of inorganic oxide particles on hydrophilic silicon surfaces*, J. Appl. Phys., **86**, pp. 1744-1748 (1999b)

Wu X., Sacher E., Meunier M., *The modeling of excimer laser particle removal from hydrophilic silicon surfaces*, J. Appl. Phys., **8**, pp. 3618-3627 (2000)

Yavas O., Suzuki N., Takai M., *Laser cleaning of field emitter arrays for enhanced electron emission*, Appl. Phys. Lett., **72**, pp. 2797-2799 (1998)

Zapka W., Ziemlich W., Tam A. C., *Efficient pulsed laser removal of 0.2 μm sized particles from a solid surface*, Appl. Phys. Lett., **58**, pp. 2217 - 2219 (1991)

Zapka W., Ziemlich W., Leung W. P., Tam A. C., *"Laser cleaning" removes particles from surfaces*, Microelectronic Engineering, **20**, pp. 171-183 (1993a)

Zapka W., Ziemlich W., Leung W. P., Tam A. C., *Laser Cleaning: laser induced removal of particles from surfaces*, Advanced Materials for Optics & Electronics, **2**, pp. 63-70 (1993b)

Zapka W., Lilischkis R., Zapka K. F., *Laser cleaning of silicon membrane,* Proc. SPIE, vol. **3996**, pp. 92-96 (2000)

Zheng Y. W., Luk'yanchuk B. S., Lu Y. F., Song W. D., Mai Z. H., *Dry laser cleaning of particles from solid substrates: Experiments and theory,* J. Appl. Phys., **90**, pp. 2135-2142 (2001a)

Zheng Y. W., Lu Y. F., Song W. D., *Angular effect in laser removal of spherical silica particles from silica wafers,* J. Appl. Phys., **90**, pp. 59-63 (2001b)

Chapter 5

LIQUID-ASSISTED PULSED LASER CLEANING WITH NEAR-INFRARED AND ULTRAVIOLET PULSED LASERS

C. P. Grigoropoulos, D. Kim

Nanosecond pulsed laser irradiation is shown to be effective for cleaning contaminant particles as small as 0.3 μm in diameter and a variety of organic/inorganic contaminants from metallic substrates and optical components. In the steam-cleaning approach a micron-thick liquid film is deposited on the surface just before the laser irradiation. Experiments are also done to examine the cleaning of surfaces immersed in liquids. The cleaning threshold and efficiency are investigated for the fundamental, frequency-doubled and frequency-tripled Nd: YAG laser harmonics. The particle removal, rapid phase-change and thin liquid film ablation processes are examined via in-situ non-contact diagnostics.

Keywords: Steam laser cleaning, threshold, non-contact diagnostics

PACS: 42.62.Cf, 81.65.Cf, 68.35.Np, 81.65.-b, 85.40.-e, 81.07.Wx

1. Introduction

As semiconductor and microelectronic device dimensions become smaller in the rapidly advancing microfabrication technology, submicron particulate contamination becomes a critical issue affecting device performance and manufacturing yield loss. Many efforts have been made to develop highly efficient surface cleaning techniques, especially for removing submicron-sized particulates from solid substrates (Lu, 1996; Mann, 1996; Mittal, 1988; Park, 1994; She, 1999; Tam, 1992; Zapka, 1993). Laser cleaning has shown good potential due to its several important advantages over conventional techniques; noteworthy are the non-contact nature and the ability to avoid damage on sensitive parts. In this review, recent advances in our work on liquid-assisted laser cleaning are summarized.

Laser cleaning techniques generally operate in dry or liquid-assisted modes. Several efforts have been devoted to dry laser cleaning (Lu, 1996; Mann, 1996). Experimental evidence shows convincingly that due to the rapid expansion of the laser-irradiated substrate, particulates can be ejected from the surface with a large initial acceleration (nanosecond pulses excite surface accelerations estimated to be of the order of 10^8 m/s^2). However, it has been verified in this laboratory that dry cleaning using a Nd:YAG laser (1064 nm and 355 nm) or a KrF excimer laser (248nm) is not effective for the cleaning of 0.3 μm and smaller aluminum oxide particulates from Si or NiP hard disk surfaces. The possible reason is that the adhesion force between the submicron particulate and the solid surface is beyond the dry laser cleaning capability. In the dry cleaning process, particle removal occurs by the high strain rate associated with rapid thermal expansion of the heated surface. The order of magnitude estimate of the inertia force under the assumption of one-dimensional expansion of the heat affected zone can be shown to be much smaller than predictions of the adhesion forces. On the other hand, it has been reported that laser cleaning efficiency can be enhanced when a micron-thick liquid film is deposited on the surface to be cleaned (Park, 1994; She, 1999; Tam, 1992; Zapka, 1993). While the liquid film is essentially transparent to the incident radiation, the substrate absorbs laser energy, causing a rapid surface temperature rise and substantial liquid superheat. The resulting explosive vaporization of the liquid film can impart a strong enough force to overcome the particulate-substrate surface adhesion.

Van der Waals force is dominant in the adhesion of a submicron particle on a dry solid surface. Electrostatic force is important for particles of diameter larger than ~ 50 μm. Since the magnitudes of Van der Waals and electrostatic forces are much greater than gravitational forces for small particles (see Fig. 1), the effect of gravity is neglected for small particle adhesion. In general, immersion of particles in liquid decreases Van der Waals and electrostatic adhesion because of the shielding effect. However, if liquid bridge is formed at the particle-surface interface by capillary condensation of humidity in the ambient air or by residual liquid from liquid-involving processes, capillary adhesion force, i.e., surface-tension force, can increase the particle bonding force tremendously (Mittal, 1988). The capillary adhesion force, i.e., surface-tension force, depends on the wetting property of the liquid and surface. In the liquid-assisted laser cleaning, the thin liquid film applied to the surface completely covers the contaminant particles, thereby eliminating capillary adhesion force. The thin liquid film disappears by natural evaporation or can be completely removed by additional dry laser pulses at the end of the process. In the case of Si surface cleaning, chemical-bond formation by oxidation is important. Therefore, cleaning processes for the chemical-contamination removal such as HF cleaning are usually employed.

The major adhesion forces are linearly proportional to the particle size (diameter), but the cleaning force is generally proportional to the surface area or volume of the particle. For example, the total drag force in the gas/liquid jet spraying and the effective cleaning force in ultrasonic cleaning are proportional to the frontal surface area normal to the fluid flow and to the cavitation-pressure wave, respectively. Therefore, the removal of a particle becomes more and more difficult as the particle size decreases. Fig. 1 displays predictions of adhesion forces acting on alumina (Al_2O_3) particles on pure Ni surfaces. In the calculation, it has been assumed that the particle-surface spacing is 4 Å on the basis that this value is comparable to typical molecular spacing. In the figure, the adhesion forces have been normalized by gravitational force, i.e., the weight of alumina particles.

In this Chapter, we describe the liquid-assisted laser cleaning process for removing submicron particles using a Nd:YAG or a KrF excimer laser. Particularly, detailed results for removing 0.3 μm-sized alumina particles from NiP hard disk substrates using Nd:YAG laser pulses are reported. In the liquid-assisted laser cleaning, explosive vaporization of thin liquid film on a short time scale plays a central role. Therefore, the dynamics of thin liquid film ablation by a nanosecond laser pulse is examined for elucidating the cleaning mechanism. The acoustic wave generated in the process of rapid liquid evaporation and phase-change kinetics is investigated using photoacoustic-deflection and optical-reflectance probes. Time-resolved images of liquid film ablation obtained by laser flash photography are presented and the liquid-assisted cleaning mechanism is examined. Finally,

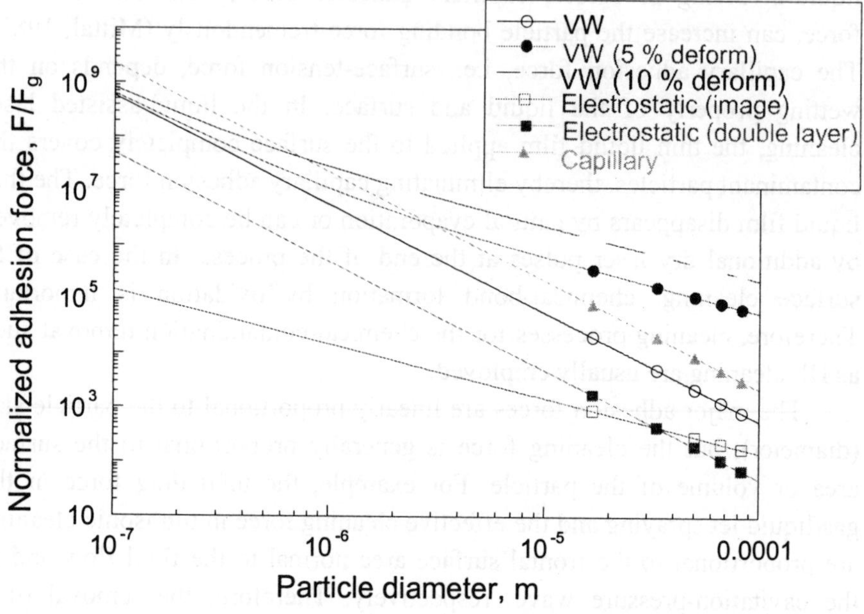

Fig. 1. Adhesion forces between a spherical alumina (Al_2O_3) particle and a Ni surface are normalized by the gravitational force (particle weight). Van der Waals force has been estimated for three different cases, assuming that the particle experiences 0, 5, or 10 % of volume deformation in the contact region.

experiments characterizing the thermodynamics and kinetics of rapid vaporization of bulk liquid in the vicinity of laser-irradiated surface are introduced. The relevance and significance of these studies to the fundamental understanding of the liquid-assisted laser cleaning process is discussed.

2. Experiments

2.1. Laser cleaning system and optical diagnostics

The schematic diagram of the experimental setup is shown in Fig. 2. A Q-switched Nd: YAG laser beam (λ = 1064/355 nm, FWHM = 6 ns) or an excimer laser beam (λ = 248 nm, FWHM = 24 ns) is used to heat the substrate and vaporize the thin liquid film (Park, 1994; She, 1999). For attaining a uniform spatial distribution of laser energy, a tunnel type beam homogenizer is employed in the case of excimer laser cleaning. Meanwhile, the Nd: YAG laser beam having Gaussian intensity distribution is expanded by a concave lens and only the core part with a relatively flat intensity distribution passes through a circular aperture (diameter 3.5 ~ 4 mm). The liquid-puffing system includes a constant-temperature liquid vessel, a vapor-ejection nozzle and pressure-control unit. Once the saturated vapor is ejected from the nozzle upon application of puffing pressure prior to the laser pulse, a thin-liquid film is formed by condensation at the ambient temperature below the saturation temperature. Water-isopropanol mixtures are used as puffing liquid. The sample is mounted on a two-dimensional moving micrometer stage. In this experiment, the laser beam is directed onto the sample at a relatively large incidence angle (~ 40°).

The photoacoustic deflection method is utilized for monitoring the acoustic pulse transmitted from the thin liquid film into the ambient air.

The experimental setup for photoacoustic deflection is displayed in Fig. 2 (b). Pressure waves traversing the probe beam parallel to the target surface induce refractive index gradients deflecting the beam toward the optically denser side. A knife-edge blocks half of the HeNe laser beam for monitoring the deflection of the probe beam with a fast photodetector (rise time < 1 ns). A second HeNe probe beam (λ = 633 nm) is directed onto the center of the clean sample surface at an incidence angle of 10°. The reflectance signal decreases or increases when the HeNe laser light is scattered by bubbles, droplets and vapor leaving the surface. Figure 2 (c) shows the experimental setup for shadowgraph imaging of liquid film ablation by laser flash photography. A N_2 laser-pumped dye laser (λ = 650 nm FWHM = 4 ns) is used as the illumination source for the visualization of the liquid ablation process by laser-flash photography with nanosecond time resolution. A CCD camera equipped with a zoom microscope captures the images exposed by the dye laser with a spatial resolution of 10 μm. The synchronization is regulated by a timing-control unit.

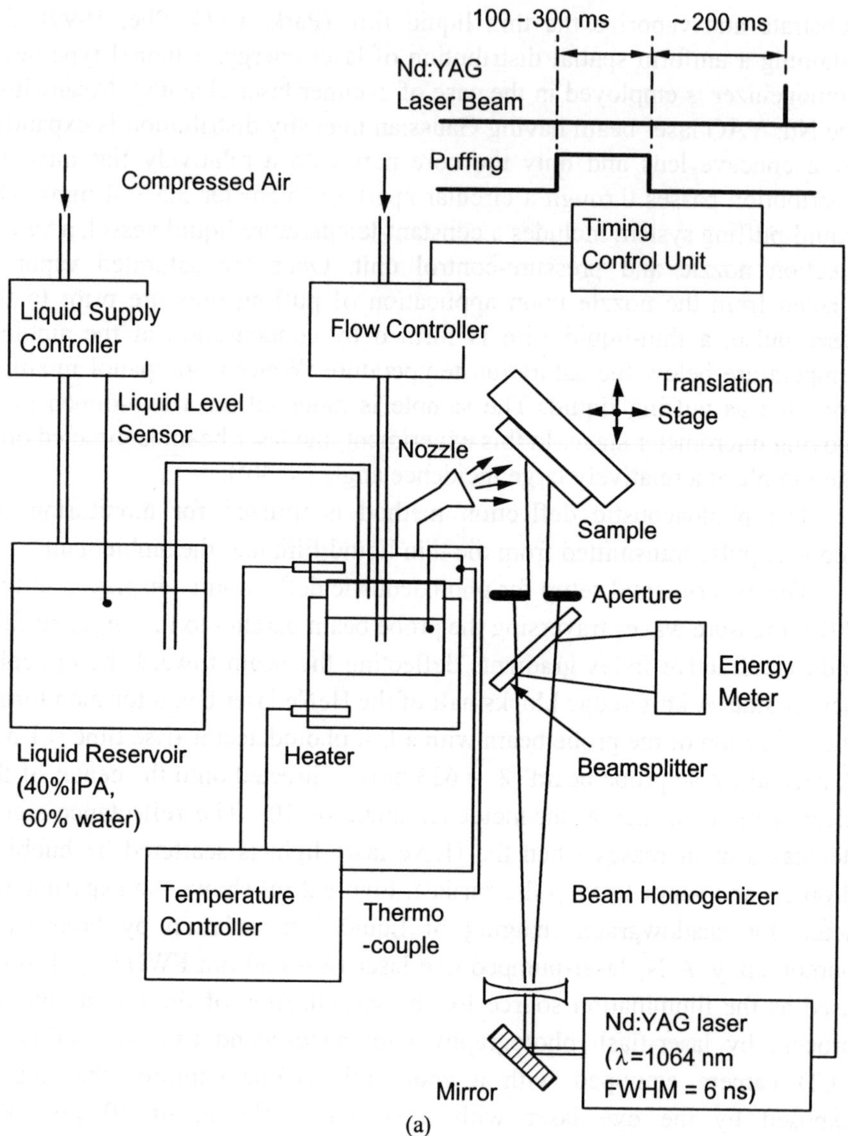

(a)

Fig. 2. Schematic diagram of (a) the Nd: YAG laser (variable wavelength, FWHM = 6 ns) cleaning apparatus, (b) experimental setup for photoacoustic deflection experiment, and (c) experimental setup for shadowgraph imaging of liquid film ablation by laser flash photography.

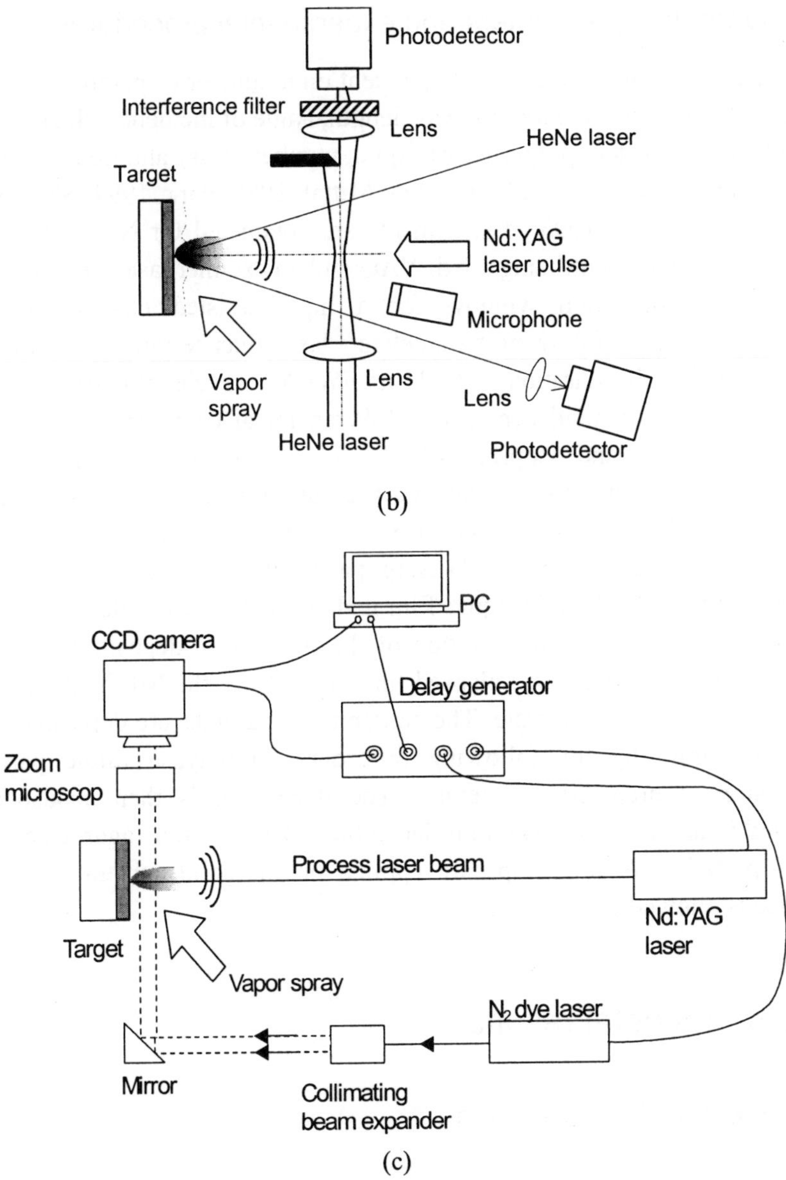

Fig. 2 (continued). Schematic diagram of (a) the Nd: YAG laser (variable wavelength, FWHM=6 ns) cleaning apparatus, (b) experimental setup for photoacoustic deflection experiment, and (c) experimental setup for shadowgraph imaging of liquid film ablation by laser flash photography.

2.2. Operation parameters and experimental procedures

The laser cleaning efficiency is dependent on a number of parameters, such as laser fluence, pulse duration, wavelength, angle of incidence, thickness of liquid film deposited, properties of liquid, number of dry and liquid-assisted (steam) pulses (Lu, 1996; Mann, 1996; Mittal, 1988; Park, 1994; She, 1999; Tam, 1992; Zapka, 1993). Experiments are conducted for NiP and single-crystal Si substrates utilizing a Nd: YAG and an excimer laser, respectively. The typical operation window for a liquid-assisted cleaning system corresponds to the following parameters - laser fluence range: up to about 200 mJ/cm^2, pulse duration: 6 or 24 ns FWHM, wavelength: 1064, 355 or 248 nm, angle of incidence: about 40°, chemical composition of liquid: deionized water and isopropanol (10~50 % by volume), liquid film thickness: a few micrometer (puffing pressure applied for 100 - 300 ms), number of pulses: 1 - 100. In the case of removal of alumina particles from NiP substrates, a mixture of 8% isopropanol and 92% water (by volume) has been utilized. The thin liquid film is deposited by a nozzle driven by a high-pressure N_2 line. The duration of the liquid film deposition is about 200 ms. After an additional elapsed time of 200 ms, the Nd: YAG laser is fired onto the target substrate. The contaminant particles are deposited on a solid substrate by the following procedures. Firstly, a dilute particle dispersion is prepared in acetone. The dispersion is then acoustically agitated in an ultrasonic bath in order to break large particle aggregates into small particles. A few drops are applied on the solid substrate and the surface is allowed to dry in air.

3. Experimental results

3.1. Excimer laser cleaning

A KrF excimer laser-based cleaning tool capable of removing small particulates and organic films has been investigated (Park, 1994). Parametric study has been performed for contaminants composed of 1-μm sized alumina particles and epoxy. The dry cleaning cannot remove small particles but is effective in cleaning off organic films. On the other hand, liquid-assisted cleaning can remove both organic films and particulates. It has been demonstrated that cleaning can best be done with the combination of dry and liquid-assisted cleaning modes for typical contaminants containing both particles and organic compounds. Even in the process of organic-contaminant removal, liquid-assisted cycles need to be incorporated because they remove small broken pieces of the organic film very effectively. The details of the system construction, operation, applicability, cleaning strategy and the probability of thermal damage due to laser cleaning are described in Park *et al.* (1994). Figure 3 illustrates typical cleaning results obtained with the excimer laser cleaning tool. It is observed that dry laser cleaning or ultrasonics removes only the large particles (aggregates) of alumina while two dry and three liquid-assisted cleaning cycles clean the exposed surface completely.

3.2. Nd: YAG laser cleaning

The threshold laser fluences for removing 0.3 μm alumina (Al_2O_3) particles from the NiP substrate have been determined for the $\lambda = 1064$ and 355 nm laser wavelengths by analyzing microscope images of the sample surfaces before and after the laser pulse irradiation. Since the contaminant particles have a certain size distribution with a mean diameter of 0.3 μm, the particle removal threshold is not uniquely determined. The experimental results confirm that large particles or aggregates are removed at much lower laser fluences than those required for complete removal of isolated small particles. The minimum laser fluence for removal of micron-sized or larger contaminants is lower by a factor of about two than the fluence necessary for complete removal of 0.3 μm particles. Accordingly, the laser-cleaning threshold is defined as the laser fluence required for complete removal of the applied contaminants. In this experiment, approximately ten liquid-assisted cleaning cycles have been shown to be enough for complete removal. The measured cleaning threshold based on ten liquid-assisted

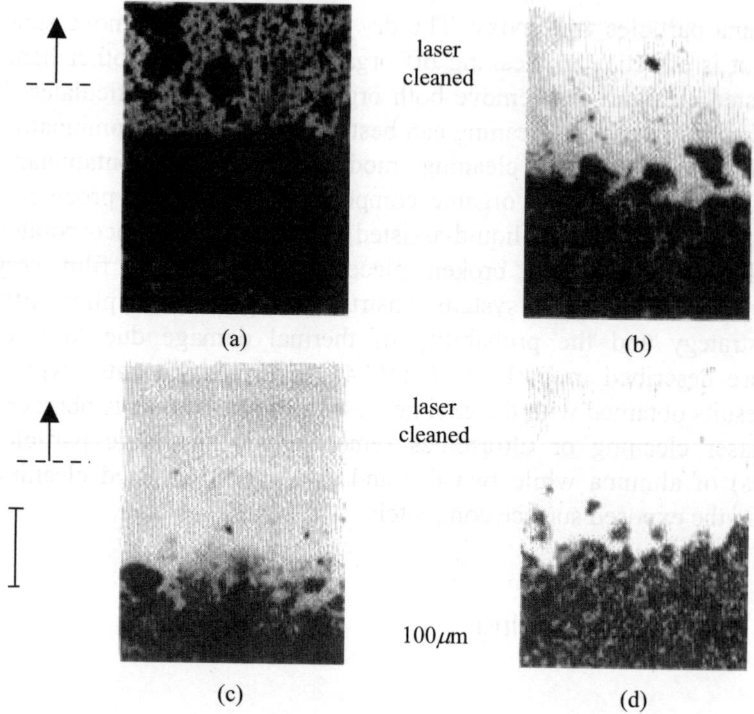

Fig. 3. Microscope photographs of KrF excimer laser (wavelength 248 nm, FWHM 24 ns) cleaning of 1 μm alumina particles with epoxy film on a crystalline silicon surface with the laser fluence of 110 mJ/cm^2. (a) After two dry cleaning cycles. (b) After two dry cleaning cycles followed by three liquid-assisted cleaning cycles. (c) After two times of two dry cleaning followed by three liquid-assisted cleaning cycles. (d) After the ultrasonic cleaning in deionized water for 10min of laser-cleaned surface.

pulses is 86 and 63 mJ/cm^2 for $\lambda = 1064$ and 355 nm, respectively. Typical cleaning results are depicted in Fig. 4 for $\lambda = 1064$ nm. At laser fluences below the cleaning threshold, particles may diminish in size and change their positions but are not completely removed. Similar trends are observed for λ=355 nm, but with a slightly decreased cleaning threshold.

The fact that both the near-IR and the ultraviolet wavelengths are effective for particle removal suggests that photochemical mechanisms play

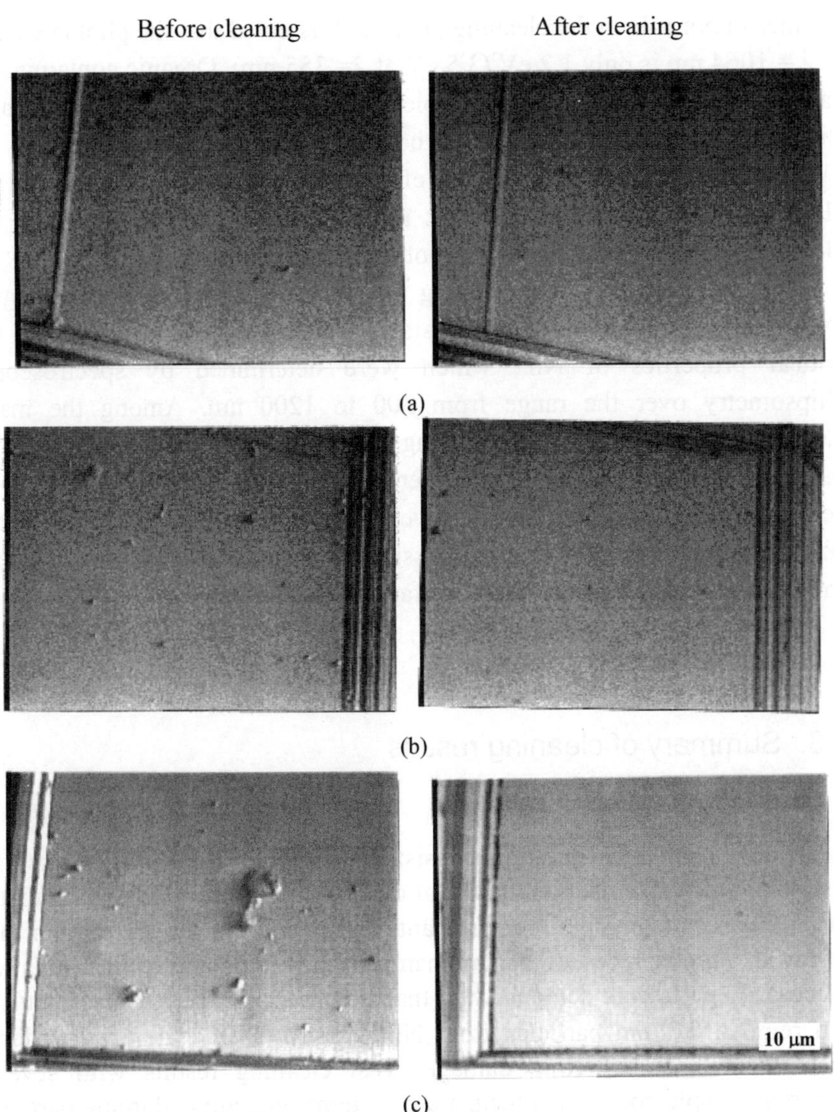

Fig. 4. Optical microscope photographs of a NiP hard disk surface contaminated with 0.3 μm-sized (average) alumina (Al_2O_3) particles and the same spots after 10 Nd: YAG laser pulses with liquid film (λ = 1064 nm) at (a) F = 35 mJ/ cm^2, (b) F = 64 mJ/ cm^2, (c) F = 86 mJ/ cm^2 (θ_i = 40°).

a rather minor role in the cleaning process. It is noted that the photon energy at $\lambda = 1064$ nm is only 1.2 eV (3.5 eV at $\lambda = 355$ nm). Organic contaminants can be ablated directly by ultraviolet laser pulses at relatively low laser fluences; for example, the chemical bonding energy for the C-H bond is 3.5 eV (Lide, 1997). In contrast, the effect of the laser wavelength on the cleaning of transparent particles is not as critical since in this case the cleaning process is not caused by photochemical ablation.

The difference in the cleaning thresholds for the two wavelengths examined can be explained by considering the spectral dependence of the optical properties of NiP, which were determined by spectroscopic ellipsometry over the range from 300 to 1200 nm. Among the many parameters affecting the laser-cleaning performance, the laser beam angle of incidence plays a critical role. Experiments conducted using the near-IR beam at normal incidence angle reveal that complete removal of 0.3 μm-sized alumina particles is not possible and only large (micron-sized) particles begin to detach from the surface at about 53 mJ/cm^2.

3.3. Summary of cleaning results

It has been shown that the liquid-assisted cleaning based on a KrF excimer laser or a Nd: YAG laser is capable of cleaning submicron-sized particulates. In the case of organic contaminants, ultraviolet laser irradiation has removed various types of contaminants by direct photochemical ablation process. Experiments demonstrated that both near-IR and UV laser pulses can remove 0.3 μm particles from NiP substrates at relatively low laser fluences without inflicting damage. Test cleaning results with several different sample contaminant-substrate systems, including alumina particles and epoxy film on NiP, Si, and Cr surfaces, show that the liquid-assisted cleaning with appropriate choice of operation parameters can be very effective in removing a variety of contaminants from solid substrates, especially in cleaning fine particles.

4. Physical mechanisms

4.1. In-situ monitoring of reflectance and visualization

Optical reflectance has been measured for monitoring the kinetics of liquid vaporization in the liquid-assisted cleaning process. In this experiment, a thin liquid film (water-isopropanol mixture) is deposited on clean NiP substrates free of alumina particles and vaporized by a Nd: YAG laser pulse. Figure 5(a) exhibits temporal profiles of the reflectance signal of a continuous-wave HeNe laser (633 nm) at different laser fluences. A Nd: YAG laser pulse at the 355 nm wavelength is irradiated onto the NiP surface at normal angle of incidence to evaporate the liquid film. Figure 5(a), which magnifies a part of Fig. 5(b), displays the details of reflectance transient right after the laser pulse is fired. The change in reflectance signal is caused by light scattering by generated bubbles and vapor plume and can be related with vaporization kinetics. Rapid liquid vaporization on a laser-heated surface immersed in bulk liquid has been investigated utilizing similar reflectance/scattering probes (Yavas, 1993; Yavas, 1994). The reflectance starts showing a sharp drop at about 27 mJ/cm^2, referred to as the "bubble nucleation threshold" (Yavas, 1993; Yavas, 1994). For $\lambda = 1064$ nm, this threshold is determined to be 52 mJ/cm^2. Notable is the fact that this value is very close to the fluence at which detachment of large particles begins for normal incident irradiation. Immediately upon firing the laser at $t = 0$, the signal drops sharply due to scattering through the ejected ablation plume, as shown in the detail depicted in Fig. 5(a). The duration of this reflectance drop is about 400 μs. The long-term reflectance in Fig. 5(b) traces the liquid film formation and depletion, which take place due to liquid-film condensation and natural evaporation (not by the laser pulse) on the time scale of seconds. The oscillations in the reflectance signal correspond to interference fringes created by the changing thickness of the liquid film. The fringe at negative times (i.e. before the laser pulse) in fact corresponds to the liquid film deposition process. It is therefore inferred that the laser-driven ablation is a short time event and that upon its conclusion, the remaining liquid is chiefly removed by evaporation.

The process of liquid-film ablation (on clean NiP substrates) has been visualized by laser flash photography on a nanosecond time scale. The sequence of images of the laser-induced thin film ablation is displayed in Fig. 6. The images capture the acoustic-wave propagation and ablation-plume evolution. The pressure wave traveling into air has a planar form initially, but it gradually attains a spherical shape due to wave diffraction.

The temporal dependence of the pressure wave front position shown in Fig. 6 yields a propagation speed of 350 m/s, which is close to the speed of sound in air. The ablation plume becomes visible at approximately 150 ns after the laser pulse. In the beginning, it forms a very dense region enclosed in a contact discontinuity envelope. Multiple jets are expelled from the liquid film at a later stage. After about 1 μs, the plume front moves with a speed of 300 m/s and produces an acoustic pulse in the ambient air. Studies conducted above the substrate damage threshold showed that the ejecta can attain supersonic speed. As liquid droplets are shed away from the core jet, the plume becomes sparse and after about 100 μs only isolated droplets, decelerated by the viscous action of the ambient air are observed. It is evident that the hydrodynamic response of the liquid film bears a strong influence on the cleaning efficiency.

4.2. Temperature, pressure, and bubble dynamics

The photoacoustic deflection signal measured during the liquid-assisted laser cleaning process is plotted in Fig. 7 for various laser fluences. Pressure waves induced by Nd: YAG laser irradiation onto a clean NiP substrate generate refractive-index gradients in the surrounding air and deflect the probe laser beam in the far field.

Due to the finite size of the probe beam waist, the temporal resolution of this measurement is much longer than the laser pulse width (FWHM = 6 ns). Therefore, the signal cannot be converted to the absolute pressure magnitude or to the exact temporal shape of the acoustic transients. Nevertheless, it is concluded that merely a slight increase in laser fluence above the bubble-nucleation threshold results in significant pressure enhancement. For example, the peak signal amplitude at 27 mJ/cm^2 is almost three times larger than that at 20 mJ/cm^2. These results reveal that

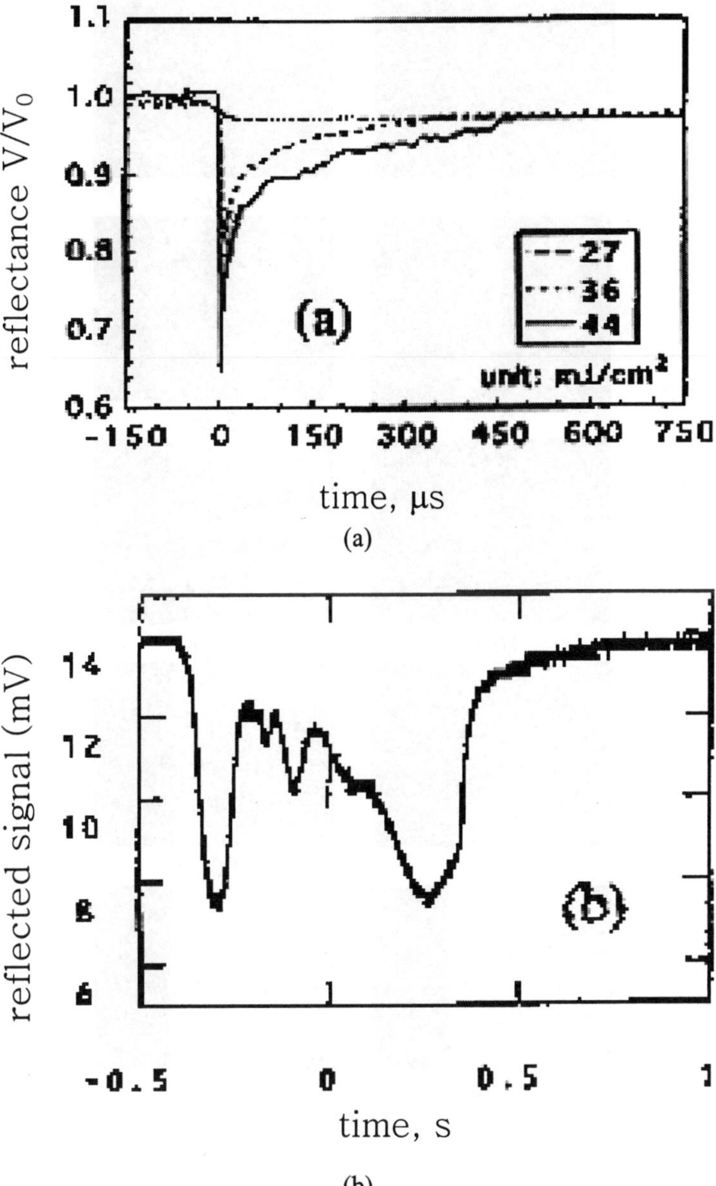

(a)

(b)

Fig. 5. (a) Short-term reflectance signal received by the photodetector at different fluences. (b) Long-term reflectance signal for pulsed laser vaporization at 33 mJ/cm^2. The Nd: YAG laser wavelength $\lambda = 355$ nm.

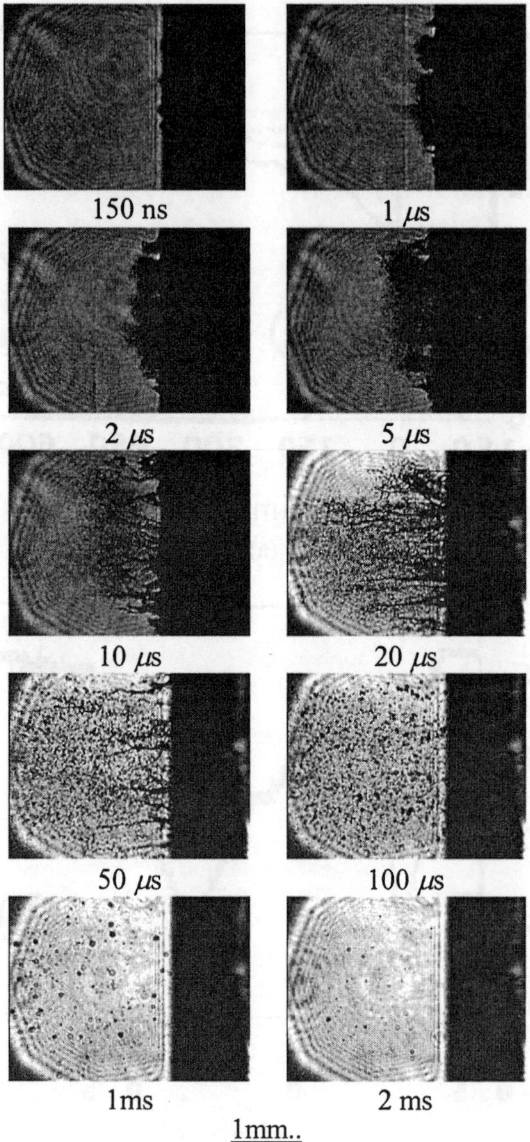

Fig. 6. Sequence of images obtained by laser flash photography in the pulsed Nd: YAG laser vaporization of thin liquid film on clean NiP substrates. The Nd: YAG laser fluence $F = 51.9$ mJ/cm^2. The laser wavelength, $\lambda = 355$ nm.

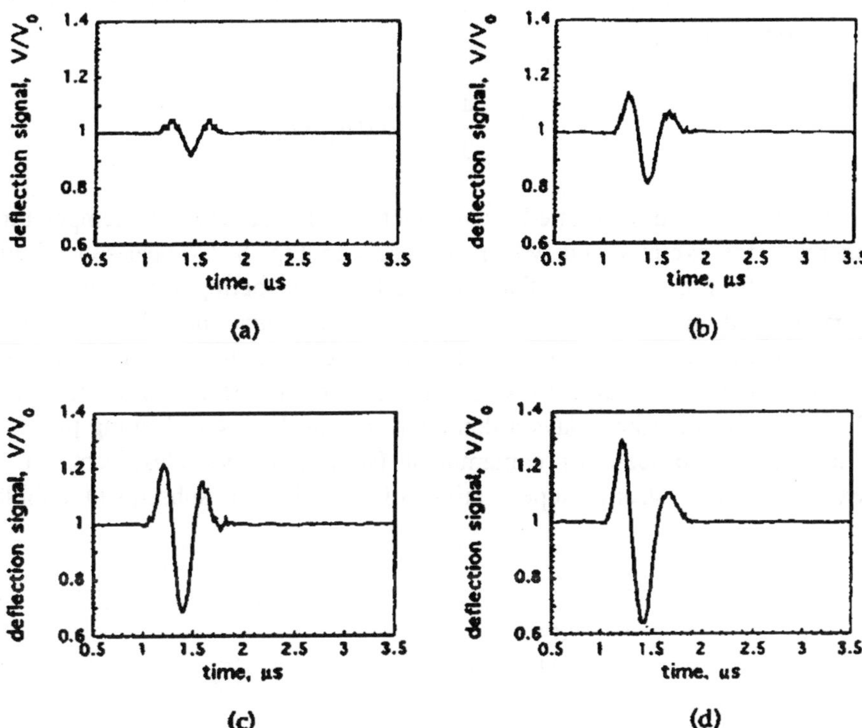

Fig. 7. Temporal deflection signal profiles measured with the photoacoustic deflection method at several Nd:YAG laser fluences: (a) 20, (b) 27, (c) 30, (d) 52 mJ/cm^2. The laser wavelength $\lambda = 355$ nm.

photoacoustic excitation, and enhanced cleaning efficiency in the liquid-assisted cleaning, is produced by vaporization of the applied liquid.

In the laser cleaning process, the laser pulse energy is absorbed by a solid surface during a very short time. Since the temperature increase and the subsequent thermal expansion of the solid material have finite magnitudes at the same time, inertia force of large magnitude is applied to contaminants sitting on the surface. The order of magnitude of this force can be estimated by assuming that the heat affected zone (thermal penetration depth $d_{th} \sim \sqrt{\alpha \tau}$) expands by an amount of $\beta d_{th} \Delta T$. This estimate of the inertia force then becomes (α: thermal diffusivity, τ: laser pulse width, R:

particle radius, ρ: particle density, β: volume expansion coefficient, ΔT: temperature increase)

$$F_i \sim \frac{4}{3} \pi R^3 \rho \beta d_{th} \Delta T / \tau^2 \qquad (1)$$

The inertia force is plotted as a function of characteristic temperature increase for excimer and Nd: YAG lasers in Fig. 8. Comparison of this result with Fig. 1 shows that the inertia force acting on a submicron contaminant may not be large enough to overcome the adhesion force. Superior efficiency of liquid-assisted laser cleaning shown in the previous sections indicates that explosive vaporization or phase explosion at the liquid / solid interface plays a dominant role in the laser cleaning process. The physical process of contaminant (submicron particles) removal is explained by strong superheating of liquid and subsequent bubble

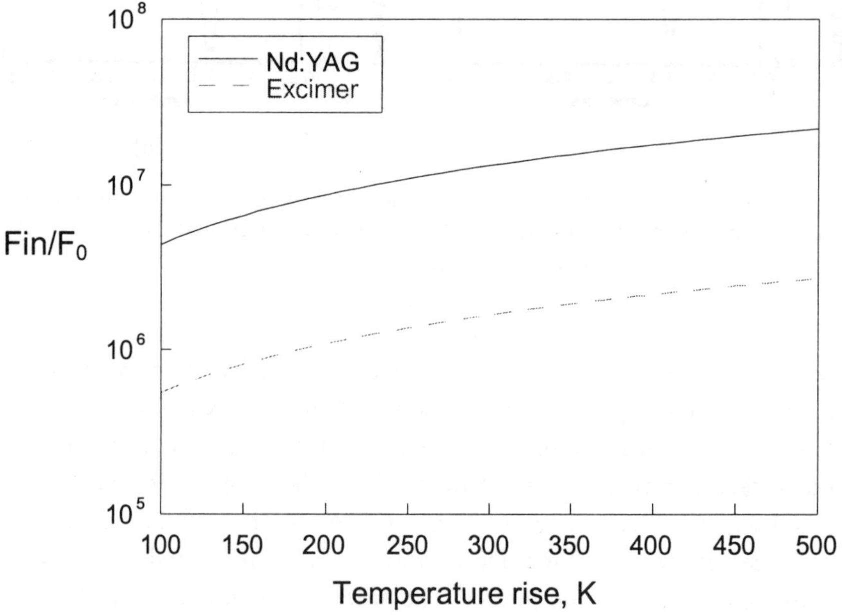

Fig. 8. Normalized inertia force (F_{in}/F_o) acting on a 1 μm-sized alumina particle.

nucleation and growth. When bubble nuclei are formed, and then grow, either at the solid / liquid interface or in the superheated liquid layer, the pressure inside a bubble becomes very high to overcome surface tension force (P_b: pressure inside a bubble, P_∞: ambient pressure, σ: surface tension coefficient, R_b: bubble radius)

$$P_b > P_\infty + \frac{2\sigma}{R_b}. \tag{2}$$

Laser cleaning is achieved by this locally concentrated high pressure around the particle / substrate contact region. The temporal variation of surface temperature has been calculated considering conductive heat transfer, neglecting the phase-change effect in a first order approximation of the actual interface temperature between a solid surface and a superheated liquid. The computation results show that the liquid temperature reaches the nominal boiling temperature, i.e., 100°C at the laser fluence of 26 mJ/cm^2 for the $\lambda = 355$ nm wavelength. The calculated transient temperature at the

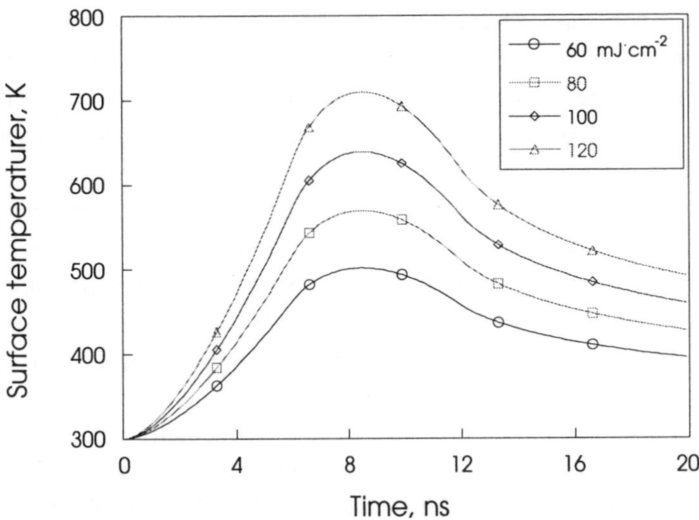

Fig. 9. The temporal variation of the NiP surface temperature irradiated by a Nd: YAG laser pulse for several laser fluences (λ = 1064 nm, FWHM = 6 ns, $\theta_i = 40°$).

interface between the liquid film and the solid surface is displayed in Fig. 9. At 60 mJ/cm^2, superheat close to the critical temperature is predicted, validating the expectation of explosive vaporization. Note, however that for the fluence range examined in the present work, the surface temperature is still far below the NiP melting point of about 1200 K (Chen, 1997).

Another possible effect of introducing a thin liquid film is the reduction of the total adhesion force between contaminants and surfaces by relaxing capillary adhesion as explained in the earlier section.

As the experimental observations indicate, it is essential to investigate the thermophysics of explosive vaporization of liquid induced by nanosecond laser heating in order to understand the mechanisms of liquid-assisted laser cleaning. It is therefore important to quantify the temperature, pressure, and bubble dynamics (size) in the vaporization process, The presence of the thin liquid film whose thickness is not precisely controlled inhibits application and analysis of optical diagnostics. Therefore, a series of optical methods have been employed to probe laser-induced vaporization on a solid surface immersed in bulk liquid (Park, 1996a; Park, 1996b; Park, 1996c; Kim, 2001). In these studies, a KrF excimer laser pulse induces rapid vaporization by heating a thin-film Cr surface immersed in pure water. Pressure waves are generated in the bulk liquid thermoelastically at low laser fluences. Once the laser fluence exceeds the phase-change threshold, bubble nuclei are formed in the heated liquid adjacent to the solid sample and enhanced acoustic emission associated with the bubble dynamics is observed. Evidently, the vaporization process differs from the ablation of thin liquid film upon the laser pulse absorption in the underlying solid surface. The hydrodynamic motion of the liquid film is conspicuously absent in the case of bulk liquid. Nevertheless the thermophysics of phase explosion bear a strong similarity for both cases and the results to be described are helpful in understanding the physical mechanism of liquid-assisted cleaning.

The transient temperature has been experimentally measured using an optical temperature probe made of polysilicon thin films (Park, 1996a; Park, 1996c). The back-surface reflectance of a thin Cr-polysilicon film on a

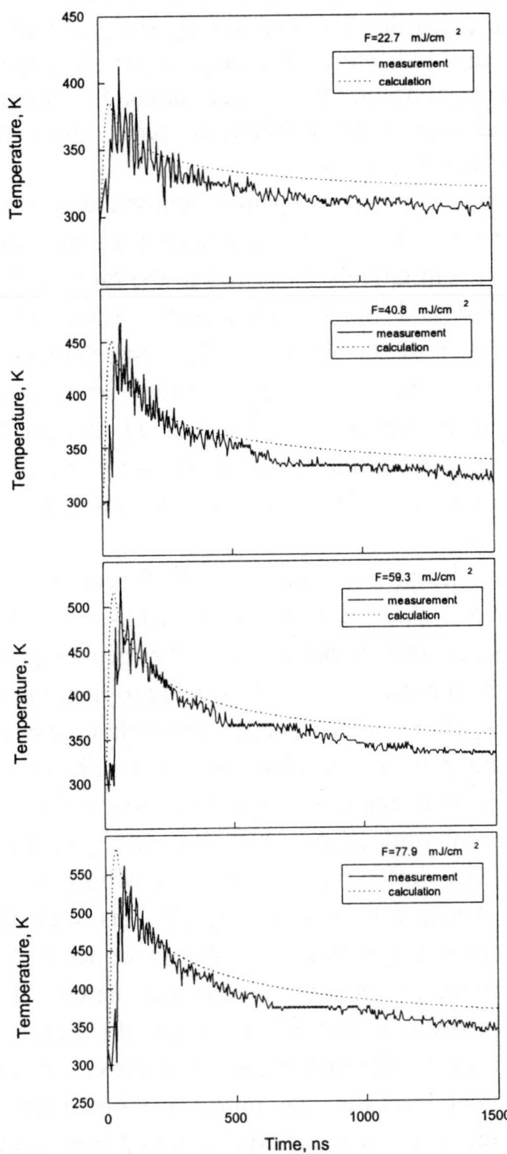

Fig. 10. The temporal variation of the temperature of a Cr surface immersed in water for various excimer laser fluences. The solid and dotted lines represent the results from experiment and calculation, respectively.

quartz substrate has been statically calibrated as a function of temperature for optical measurement of the temperature transient. Figure 10 shows typical temperature profiles obtained in the explosive vaporization process. Experimental results indicate a very large degree of superheating, greatly exceeding the nominal boiling temperature under atmospheric pressure. It is therefore inferred that a non-equilibrium rapid phase transition occurs during the laser cleaning process.

Explosive vaporization accompanies acoustic excitation resulted from the bubble dynamics. The acoustic transient in the rapid vaporization process has been quantitatively measured in the far field by a piezoelectric pressure transducer and by a photoacoustic beam deflection probe as described in Park *et al.* (1996b). A thin Cr-polysilicon film on quartz substrate completely immersed in bulk water has been utilized for this pressure measurement. Results shown in Fig. 11 indicate that the pressure launched into the liquid exceeds the linear pressure caused by thermoelastic expansion of the target for laser fluences exceeding the so-called bubble nucleation threshold.

For disclosing the physical mechanism of acoustic enhancement, the vapor phase kinetics, has been measured semi-quantitatively by optical probing. As a large number of bubble nuclei are generated and exhibit rapid motion on a submicron scale, it is not possible to measure the dynamics of individual bubbles. Therefore, an interference technique has been developed for measuring the optical response of the liquid-solid interface. The interference signal thus represents the collective effect of the produced bubbles. In our recent work (Kim, 2001), the interferometric probe detected the dynamics of an "effective" vapor layer whose growth velocity is estimated at 0.5 - 1 m/s, derived from Fig. 12. This work also revealed that the growth velocity is larger than the collapse velocity. Accordingly, it is inferred that the cavitation pressure emission at the end of bubble collapse stage is not so significant as that due to the initial bubble expansion, which is consistent with the results of the acoustic-transient measurement. Based upon the experimental results, the following conclusions can be drawn concerning the acoustic enhancement due to liquid vaporization:

- Rapid vaporization of liquid yields enhanced acoustic pulse larger than the thermoelastic stress.

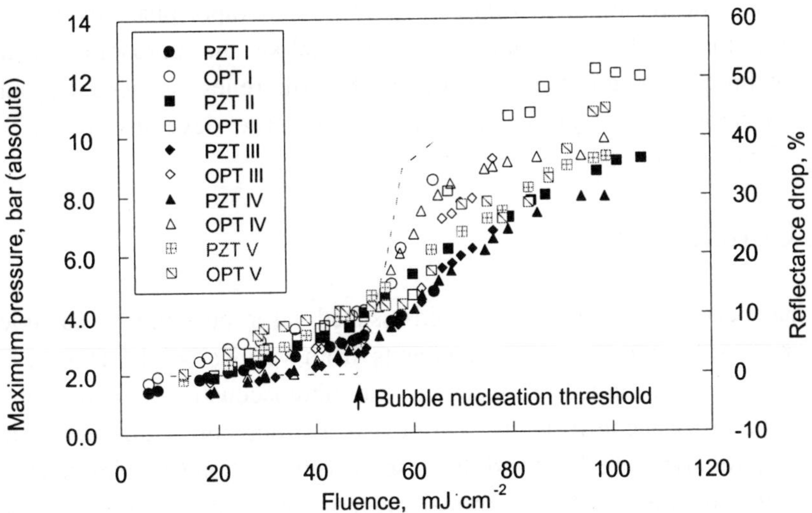

Fig. 11. The pressure pulse amplitudes are plotted as a function of excimer laser fluence. The data produced by the transducer are represented by"PZT" and the data by the deflection probe are labeled by "OPT". The amplitude of the optical specular reflectance drop is also plotted with the dashed line. The arrow marks the bubble nucleation threshold.

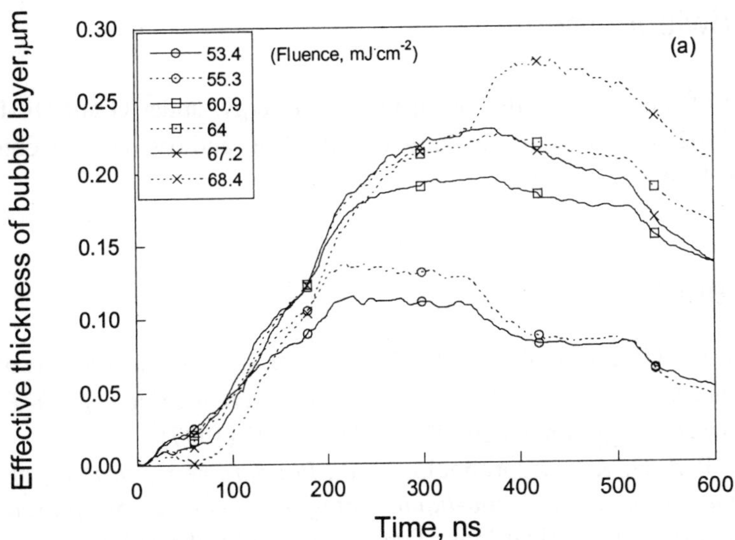

Fig. 12. Effective thickness variation of the bubble layer measured by optical interferometry at different laser fluences.

- The acoustic augmentation comes from rapid vapor phase expansion in the initial stage of bubble growth, unlike the ultrasonic cleaning technique, where the cleaning mechanism relies on acoustic bubble cavitation. No significant cavitation pressure develops in the laser cleaning process.

5. Conclusion

It has been shown that the liquid-assisted laser cleaning is capable of cleaning submicron-sized contaminants from various solid substrates. *In-situ* probing of optical reflectance and photoacoustic deflection in the cleaning process confirms that there exists a strong correlation between the particle-removal threshold and the onset of vaporization (bubble-nucleation). Furthermore, the results of optical diagnostics indicate that the cleaning process is driven by the thermal mechanism involving explosive vaporization of liquid. Accordingly, both near-IR and UV laser pulses have been effective for cleaning submicron particles.

Acknowledgements

Support of this work by the Department of Energy, under Grant DE-FG03-95ER14562 is gratefully acknowledged. The authors are indebted to Andrew Tam's contributions.

References

Chen S. C., Grigoropoulos C. P., *Noncontact nanosecond-time-resolution temperature measurement in excimer laser heating of Ni–P disk substrates*, Appl Phys Lett, **71**, pp. 3191-3193 (1997)

Kim D., Park H. K., Grigoropoulos C. P., *Interferometric probing of rapid vaporization at a solid-liquid interface induced by pulsed-laser irradiation,* Int. J. Heat Mass Transfer, **44,** 3843-3853 (2001)

Lide D. R., Frederikse H. P. R. (Eds.), *The CRC Handbook of Chemistry and Physics*, 78 th Ed., (CRC Press, Boca Raton, 1997)

Lu Y. F., Song W. D., Hong M. H., Chong T. C., Low T. S., *Laser dry cleaning of ZrO_2 particles from air bearing surface of magnetic head sliders*, Mat. Res. Soc. Symp. Proc, vol. **397**, pp. 329-334 (1996)

Mann K., Wolff-Rottke B., Müller F., *Cleaning of optical surfaces by excimer laser radiation*, Appl Surf Sci, **96-98,** pp. 463-468 (1996)

Mittal K. L. (Ed.), *The Particles on Surfaces 1: Detection, Adhesion, and Removal* (Plenum Press, New York, 1988)

Park H. K., Grigoropoulos C. P., Leung W. P., Tam A. C., *A practical excimer laser-based cleaning tool for removal of surface contaminants*, IEEE Trans. Compon. Packag. Manuf. Techn. A, **17**, pp. 631-643 (1994)

Park H. K., Grigoropoulos C. P., Poon C. C., Tam A. C., *Optical probing of the temperature transients during pulsed-laser induced boiling of liquids*, Appl. Phys. Lett., **68**, pp. 596-598 (1996a)

Park H. K., Kim D., Grigoropoulos C. P., Tam A. C., *Pressure generation and measurement in the rapid vaporization of water on a pulsed-laser-heated surface*, J. Appl. Phys., **80,** pp. 4072-4081 (1996b)

Park H. K., Zhang X., Grigoropoulos C. P., Poon C. C., Tam A. C., *Transient temperature during the vaporization of liquid on a pulsed laser-heated solid surface*, J. Heat Transfer., **118**, pp. 702-708 (1996c)

She M., Kim D., Grigoropoulos C. P., *Liquid-assisted pulsed laser cleaning using near-infrared and ultraviolet radiation*, J. Appl. Phys., **86,** pp. 6519- 6524 (1999)

Tam A. C., Leung W. P., Zapka W., Ziemlich W., *Laser-cleaning techniques for removal of surface particulates*, J. Appl. Phys., **71**, pp. 3515-3523 (1992)

Yavas O., Leiderer P., Park H. K., Grigoropoulos C. P., Poon C. C., Leung W. P., Do N., Tam A. C., *Optical reflectance and scattering studies of nucleation and growth of bubbles at a liquid-solid interface induced by pulsed laser heating*, Phys. Rev. Lett., **70**, pp. 1830-1833 (1993)

Yavas O., Leiderer P., Park H. K., Grigoropoulos C. P., Poon C. C., Tam A. C., *Enhanced acoustic cavitation following laser-induced bubble formation: Long-term memory effect*, Phys. Rev. Lett., **72**, pp. 2021-2024 (1994)

Zapka W., Ziemlich W., Leung W. P., Tam A. C., *'Laser cleaning' removes particles from surfaces*, Microelectronic Engineering, **20**, pp. 171-183 (1993)

Chapter 6

STEAM LASER CLEANING OF SILICON SURFACES: LASER-INDUCED GAS BUBBLE NUCLEATION AND EFFICIENCY MEASUREMENTS

P. Leiderer, M. Mosbacher, V. Dobler, A. Schilling,

O. Yavas, B. S. Luk'yanchuk, J. Boneberg

The removal of dust particles from semiconductor surfaces requires new cleaning strategies such as Steam Laser Cleaning (SLC). It is based on laser-induced explosive evaporation of a liquid layer applied on the surface. We have investigated the laser-induced nucleation and growth of gas bubbles at silicon/water, silicon/isopropanol and silver-film/water - interfaces by light scattering and surface plasmon spectroscopy. The achieved superheating of the liquid before bubble nucleation sets in strongly depends on the substrate roughness. On rough metal films it is only about 30 K in water, compared to about 150 K on smooth silicon surfaces. Isopropanol (IPA) on smooth silicon surfaces could be heated to 116°C, corresponding to a superheating of 36 K. In combination with numerical calculations it was possible to determine the heat transfer coefficients silicon-water ($\xi = 3 \cdot 10^7$ W/m^2 K) and silicon – IPA ($\xi = 1 \cdot 10^7$ W/m^2 K). Using optical techniques we have measured the pressure wave created by the growing bubbles and the bubble growth velocities. For a quantitative study of the efficiency of SLC we deposited spherical colloidal particles on industrial silicon wafers. We observed a sharp threshold for particle removal at 110 mJ/cm^2 (laser $\lambda = 532$ nm, FWHM = 8 ns) which is independent of the size (diameter 800 nm down to 60 nm) and material of the particles and efficiencies above 90% for particle removal. On the basis of our results we discuss the validity of the existing SLC models and the perspective of the application of SLC as an industrial cleaning tool.

Keywords: Particle removal; steam laser cleaning; bubble nucleation; heat transfer coefficient, superheating

PACS: 81.65.C; 79.60.Bm; 42.62.Cf

1. Introduction

The interest in nanostructures at surfaces is growing rapidly due to the enormous importance of such structures for basic science and for applications in numerous areas like electronics, optics, data storage, sensor devices etc. As the size of such structures keeps getting smaller, the influence of contaminating dust particles, even if they are tiny, is more and more detrimental. This is in particular true for microelectronics (Hönig, 1988; Hattori, 1990; DeJule, 1998), where the typical width of the structures nowadays is already close to a hundred nanometers, and further reductions are in sight (Sematech Inc., 2000). A single dust particle of the same order of magnitude present on the surface during the chip production process can lead to a malfunction of the whole chip.

Mechanical cleaning techniques like ultrasonics, wiping and brushing, as they are presently used, are either not very efficient in removing very small particles or due to the mechanical contact give rise to new impurities by abrasion. Therefore alternative methods are very desirable (Kohli, 2002). An in principle very promising technique is the so-called "Laser Cleaning" (Bäuerle, 2000): the contaminated surface is heated by a short laser pulse, and the thermal expansion of the irradiated area, if rapid enough, is expected to kick the particle off the surface (Zapka, 1991; Tam, 1992; Engelsberg, 1993). This method is contactless, and the surface region where a dust particle is identified (e.g. by means of light scattering as in commercial wafer inspection tools) can easily be addressed in a controlled way.

It has been demonstrated that pulsed laser radiation is indeed capable of removing dust particles from a surface (Zapka, 1991; Tam, 1992; Engelsberg, 1993; Kelley, 1993; Vereecke, 1999; Halfpenny, 1999; Lu, 2000 a). Yet a closer inspection reveals that there is a great danger to modify the surface locally at the former positions of the particles, e.g. create crater-like structures (Leiderer, 2000; Lu, 2000 b, 2002; Mosbacher, 2001, 2002 a, b; Münzer, 2002). This can be ascribed to the enhancement of the incident light field in the vicinity of the particle (Luk`yanchuk, 2000, 2002; Mosbacher, 2001; Münzer, 2002). For very smooth and delicate surfaces, like those of commercial silicon wafers with an RMS roughness below 0.2 nm, this method is hence not applicable, because the surface defects generated by the cleaning process would be as serious for the further production steps as the presence of the impurity itself.

A slight modification of the cleaning procedure, however, appears to be able to solve this problem: just before the application of the laser pulse a thin liquid film is condensed on the surface, which then is evaporated explosively by the pulse, and the vapor jet carries the particle with it. This "Steam Laser Cleaning" (SLC, in contrast to the "Dry Laser Cleaning", DLC, described before), was introduced by several research groups about a decade ago (Beklemyshev, 1987; Assendel'ft, 1988 a, b; Imen, 1991; Zapka, 1991). In the following years several groups started to investigate the details of the SLC process (Park, 1994 a, b, 1996 a, b; Tam, 1995; Héroux, 1996; Boughaba, 1997; Leiderer, 1998; She, 1999; Mosbacher, 2000, 2002 a, b). The present article gives an account of the recent developments and the underlying physical phenomena of this technique.

2. Experimental

A quantitative analysis of the laser induced bubble nucleation process as well as the determination of cleaning efficiencies of SLC heavily relies on well defined, reproducible experimental boundary conditions. Therefore in this section we will describe the important aspects of our experimental setup.

2.1. Sample preparation

The preparation of samples used in bubble nucleation and SLC efficiency experiments is one of the most critical issues in the described experiments. Slight deviations in the preparation e.g. of contaminated samples may lead to results that are more difficult to interpret or even do not reflect the focused problem. In the following we will describe the preparation of the substrates and liquids used in our experiments.

Metal films on glass: In the experiments that monitored the bubble nucleation process via Surface Plasmon Spectroscopy we used thin silver films as substrate. These films are most suitable for the excitation of surface plasmons. A film of about 50 nm thickness was evaporated in high vacuum either onto the base of a heavy flint glass prism (n = 1.75 for λ = 632.8 nm) or onto a microscope slide (SCHOTT; BK7 glass). A 2 nm thick chromium interlayer was used to increase the adhesion between the silver film and the glass prism. The films prepared in this way are *optically* smooth, yet on the nm scale are rough in comparison with silicon wafers

Industrial silicon wafers: As silicon substrate we used $\langle 100 \rangle$ industrial silicon wafers (Wacker Siltronic, Burghausen, Germany) with an RMS roughness of 0.2 nm. The wafers were cleaned in isopropanol (IPA) in an ultrasonic bath and subsequently have been dried in a nitrogen gas flow.

Water: The water used both in the bubble nucleation and SLC experiments was Millipore water. In the bubble nucleation experiments an additional 200 nm particle filter prevented the contamination of the liquid cell.

Isopropanol (IPA): The Isopropanol (2-Propanol $(CH_3)_2CHOH$), in industrial applications often abbreviated as IPA, was used as bought from the supplier. According to the supplier's specification the purity was at least 99.8%.

Contaminating particles: Most studies on laser cleaning use irregularly shaped particle contaminants (e.g. Al_2O_3, Si_3N_4) to investigate removal efficiencies. Although they may reflect the actual shape and size distribution of contaminants in a real world wafer cleaning process, they are unfavorable in experiments aimed to investigate the underlying basic processes. Therefore, we have used spherical colloidal polystyrene (PS; Interfacial Dynamics Corporation, Portland, OR 97224 USA) and SiO_2 (Bangs Laboratories Inc., Fishers, IL, USA and Duke Scientific Corp., 2463 Faber Place, Palo Alto, CA 94303, USA) particles. Their small size distribution (± 5% for PS, ± 20% for SiO_2) compared to irregular particles and their spherical shape enable studies of removal efficiencies for various well-defined sizes and facilitate future comparisons with theoretical models. Some experiments were also performed using irregularly shaped Al_2O_3 particles (SUMMIT CHEMICALS EUROPE GMBH, Düsseldorf, Germany) as contaminants.

The particles were deposited on the silicon sample by a spin coating process. After dilution with IPA (1:2500) the particle suspension was spun onto the silicon wafer (2000-3000 Rpm). By adjusting the rotation speed, the concentration of particles in the suspension and the total amount of suspension applied onto each sample we were able to prepare samples with more than 95% of isolated spheres at particle densities $\geq 10^7 - 10^8$ (depending on the particles size) per cm^2 (Mosbacher, 2000).

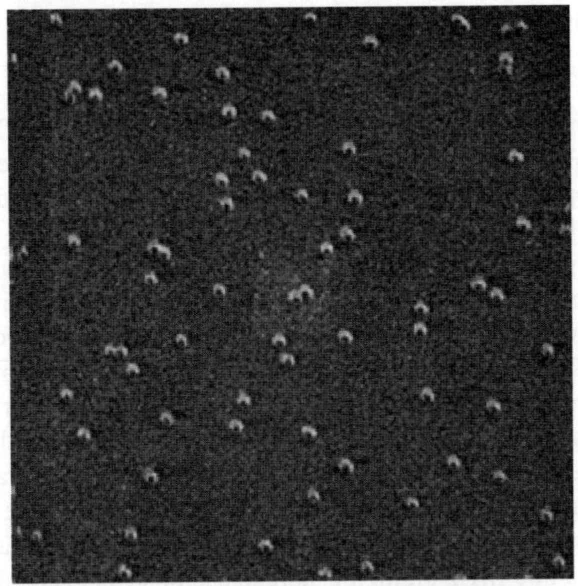

Fig. 1. Typical sample as used in the laser cleaning experiments imaged in an scanning electron microscope. The displayed area is 4.8 mm × 4.8 mm and the particle size is 110 nm.

A typical example can be seen in Fig. 1 where 110 nm sized PS particles were deposited onto a Si wafer. The prevention of particle agglomerates on the samples is important for quantitative experiments, as such agglomerates complicate the interpretation of cleaning results (Halfpenny 1999) and exhibit a different cleaning behavior compared to single particles (Leiderer, 2000; She, 1999).

2.2. Laser sources

In our experiments we applied different laser sources. For the nucleation of gas bubbles on metal substrates we have used both KrF excimer laser pulses (λ = 248 nm, FWHM = 25 ns, spot size 1 × 1 cm^2) and the pulses of a Q-switched Nd: YAG laser (λ = 532 nm, FWHM = 8 ns). As probe lasers we utilized a cw Ar -ion laser (λ = 488 nm, 175 mW) in the SLP - experiments and a HeNe - laser (λ = 633 nm, 5 mW) for the excitation of surface plasmons and the detection of contaminating particles.

2.3. Surface plasmon probe and optical reflectance probe

The experimental setup for the surface plasmon probe (SPP) of the laser-induced bubble nucleation at a liquid-solid interface is depicted in Fig. 2. After the application of the silver film, the prism is mounted on a cuvette filled with Millipore water.

Surface plasmons are excited optically at the silver-water interface via attenuated total reflection (ATR) in the Kretschmann configuration using a 5 mW HeNe probe laser (Kretschmann, 1968). Bubble nucleation takes place at the silver-water interface upon heating by a KrF excimer laser pulse through a quartz window. Any transient change of the surface plasmon resonance angle due to a change of the dielectric function of the water layer on the silver film induced by bubble nucleation, temperature rise, or pressure changes, is monitored by a fast photodiode (rise time $\leq$ 1 ns). The signal is amplified using an ac-coupled 1 GHz bandwidth amplifier, and recorded on a 500 MHz digitizing storage oscilloscope. As indicated in Fig. 2, it is calculated that the appearance of a vapor layer of 1 nm thickness will cause a significant shift of the surface plasmon resonance to a smaller angle of incidence and thus lead to a transient change in the measured intensity, if the angle of incidence of the probe laser is kept fixed. It should be pointed out here that the existence of a vapor layer is just an assumption at this point.

In order to directly compare the results of the novel SPP with the previously used optical reflectance probe (ORP), a second HeNe probe laser, incident from the front side through the liquid and probing the same spot as the SPP, is used to monitor simultaneously the transient reflectance changes caused by light scattering on bubbles.

If not stated otherwise, the SPP measurements presented in the following have been performed by setting the angle of incidence for the probe beam to the middle of the left wing of the surface plasmon resonance at about 53.1° as shown in Fig. 2. In order to verify that no reversible changes of the optical constants of the silver film took place upon laser heating, the complete plasmon resonance curve has been measured prior to and after each series of experiments.

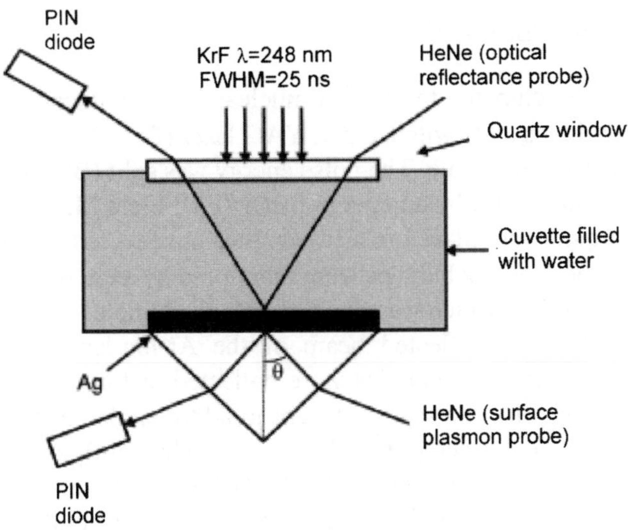

(a) Experimental setup SPP

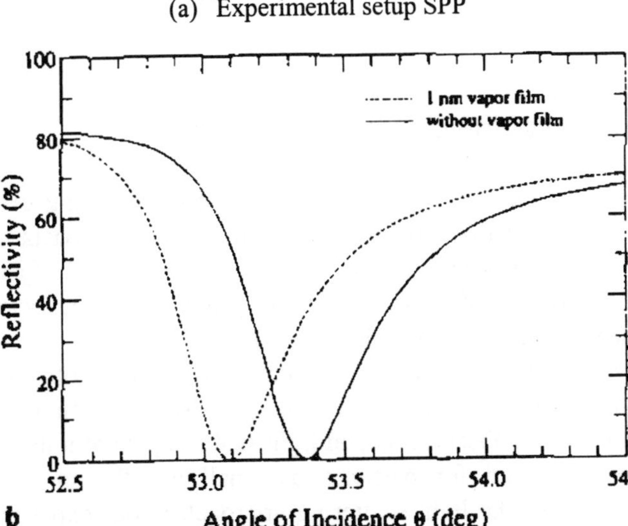

(b) Response of plasmons resonance on bubbles

Fig. 2. (a) Experimental configuration for the ns time resolved study of pulsed laser-induced bubble formation at a liquid-solid interface using the surface plasmon probe (SPP) and optical reflectance probe (ORP) techniques. (b) Calculated shift of the surface plasmon resonance curve caused by a 1 nm thick vapor layer at the water-silver interface.

2.4. Scattered light probe (SLP)

Fig.3 shows the setup for the bubble nucleation experiments. In order to nucleate the bubbles a Q-switched Nd: YAG laser (λ = 532 nm, FWHM = 8 ns) heated the silicon sample. The pulse energy was split (BS) and measured for each individual pulse by an energy meter (FM: Field Master, Coherent). Sample and liquid were placed in a fused silica cuvette that could be heated up to 360 K. The growing bubbles were monitored by a cw Ar-ion laser (λ = 488nm, P = 175 mW), which was focused onto the sample.

Both the specular reflected beam of the Ar-Ion-laser and the light scattered by the nucleated bubbles were collected in forward direction (as shown in the diagram), the scattered light in addition also perpendicular to the incident ray (similar setup, not shown in the diagram). For all the light detection we used fast photodiodes (PD: FND 100, rise time < 1 ns) covered by interference filters (IF). Using a polarizing beam splitter (PBS) the reflected beam was decomposed into its p- and s-polarized constituents that were detected individually.

2.5. Evaluation of the cleaning efficiency

In principle there are two ways of determining the cleaning efficiency of laser cleaning. Either the efficiency is evaluated on the basis of a direct observation in an optical or scanning electron microscope (Halfpenny, 1999; Lu, 2000 a; Fourrier, 2001) or by indirect probe techniques such as light scattering measurements (Mosbacher, 2000).

Recently (Mosbacher, 2000) we have compared both methods in measuring the particle removal efficiency in SLC experiments on bare silicon wafers. Direct observation and light scattering were found to provide consistent results even for particles as small as 100 nm in diameter. Therefore, we chose the light scattering approach in our experiments, as it increases the speed of data analysis.

Using a photomultiplier we were able to detect scattered light from particles with diameters down to 60 nm. The detection of the scattered light from these particles represents the detection limit of our present setup, but might be extended to even smaller sizes by the application of lasers with higher intensity and shorter wavelengths.

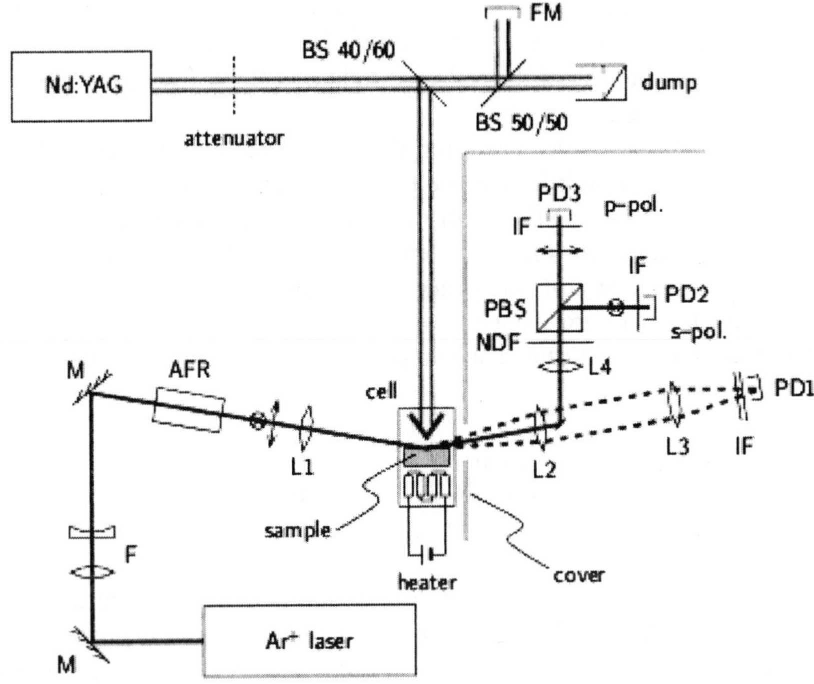

Fig. 3. Optical determination of laser induced bubble nucleation via Scattering light probe (SLP) (details see text).

A 5 mW HeNe laser illuminated a spot with a diameter of < 0.5 mm, which corresponds to several hundred particles at the particle concentration used. This high number of spheres ensured a small statistical error in our experiments, as we were able to monitor several hundred lifts off processes at the same time.

2.6. Determination of laser fluence

The determination of laser fluence in the SLC experiments using the Nd: YAG-laser is described in detail in (Mosbacher, 2000). In brief, the laser fluence was determined relative to the well-known (Kurz, 1983; Lowndes, 1983; Boneberg, 1993) laser fluence of about 310 mJ/cm^2 (at the laser parameters used in the experiments) necessary for melting of a surface layer corresponding to the optical penetration depth of molten Si. In contrast to the solid state molten silicon is metallic, and hence the phase transition is accompanied by an increase in reflectivity. This increase is detected by a time-resolved monitoring of the reflected light of a HeNe laser ($\lambda = 633$ nm) during the experiment.

In the experiments designed for the determination of laser induced bubble nucleation in bulk liquids we chose another, intrinsic laser fluence calibration method. During the experiment we found a distinct change in the reflectivity of the water-silicon-system. As computer simulations of the temperature dependency of this reflectivity show, this change is primarily caused by the temperature change in the water and hence to the applied laser fluence. Based on this, it can be shown (Dobler, 2002) that the integral

$$W = \int_{4\mu s}^{9\mu s} \left(\frac{R_p(t)}{R_{0,p}} - 1 \right) dt \sim \Phi , \tag{1}$$

containing the reflectivity change in p-direction is directly proportional to the applied laser fluence. The upper boundary of the integral is given by the limit of detection time; the lower boundary of 4 μs was chosen in a way that at this time no gas bubbles were present on the surface any more. Hence it is possible to determine from the measured reflectivity changes of the system the applied local laser fluence right at the area where the bubble nucleation is probed.

3. Laser induced bubble nucleation and pressure waves

In SLC the laser induced evaporation of the applied liquid is responsible for the ejection of the contaminating particles from the surface. Hence any deeper understanding of the SLC process requires an investigation of the laser induced bubble nucleation at the substrate - liquid interface. In this section we will both review results on the bubble growth dynamics obtained on rather rough metal films as well as recent results from experiments carried out on industrial silicon wafers.

3.1. Theoretical background

The boiling of liquids is of great relevance for a broad variety of technical applications. Therefore a large amount of literature can be found on this topic. In contrast, there are only a few publications on the early stage of boiling, the nucleation of the gas bubbles. Below we will briefly summarize the relevant theoretical background necessary for the interpretation of our experimental results.

3.1.1. Kinetic limit of superheating

In Fig. 4 we show schematically part of the phase diagram of a one-component liquid. The vapor pressure curve separates the regions where the liquid and the vapor are the thermodynamically stable phases. This curve represents the equilibrium state of a flat liquid-vapor interface. A liquid is called superheated if its temperature (at a given pressure p) is above the temperature $T_{sat}(p)$ on the coexistence curve:

$$T(p) > T_{sat}(p). \tag{2}$$

Such a state can be achieved, e.g., by heating the liquid at constant pressure, as it is mostly done in the experiments described below and schematically indicated by the path $A \rightarrow B$ in Fig. 4. One should add that strictly speaking the pressure will in general increase by some amount due to the effects of thermal expansion. The degree of superheating which can be obtained is affected by two limits:

Thermodynamic limit (spinodal line): In a certain region of the phase diagram the liquid is intrinsically unstable, because the compressibility of the system is negative in this range. Fluctuations in density, however small they are, will therefore grow spontaneously. The boundary of this region is the so-called spinodal or thermodynamic limit of superheating $T_t(p)$ defined by thermodynamic states where

$$\left|\frac{\partial p}{\partial V}\right|_{T,n} = 0. \tag{3}$$

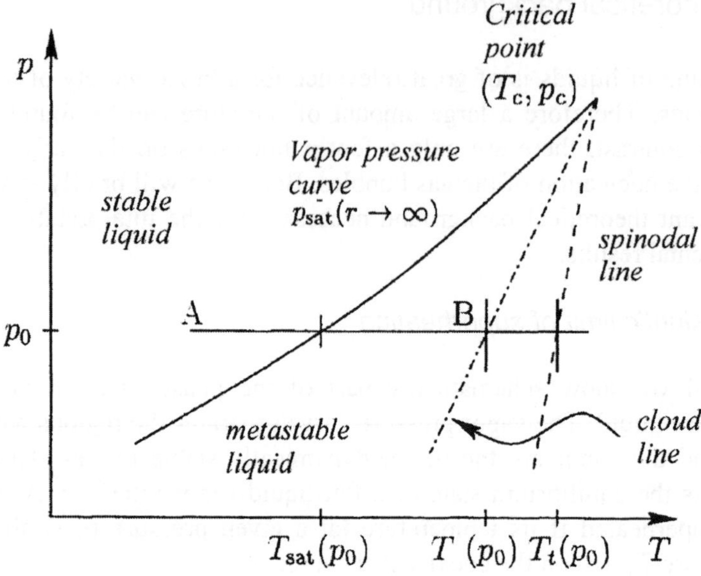

Fig. 4. Schematic p - T - phase diagram of a pure liquid. The path $A \to B$ corresponds to the experimentally realized isobaric heating.

The calculation of the spinodal according to equation (3) is hampered by large uncertainties, because an equation of state is required, whose parameters are not known from experiment in this range. One therefore has to rely on extrapolations (Avesisian, 1985).

Kinetic limit: In the region between the vapor pressure curve and the spinodal the liquid is in a metastable state. In this range a critical nucleus is required for the formation of growing bubbles. If a bubble is smaller than this critical size it will shrink and eventually disappear. The rate J of the formation of critical bubbles grows rapidly in a narrow temperature range, giving rise to a relatively sharp "cloud line" $T_{cl}(p)$ in the metastable region, where the initially transparent liquid becomes cloudy due to the enhanced light scattering from the gas bubbles (or droplets in the case of a supercooled vapor).

A review of the methods used to determine the limits of superheating and relevant data are presented in (Avesisian, 1985). The thermo-dynamics of superheated liquids is discussed, e.g., in (Debenedetti, 1996; Skripov, 1974, 1988).

For the critical nucleation rate the typical value cited in the literature is 10^{12} m^{-3}s^{-1}. This is a convention. Taking into account the conditions of our experiment, with an effective volume of $50 \times 20 \times 0.3$ mm^3 = 300 mm^3 to be probed, and a rate of one nucleus during the time of superheating (roughly 1 μs), a more adequate value is $J_{cr} = 10^{22}$ m^{-3}s^{-1} in our case.

3.1.2. Nucleation theory

Nucleation of a critical nucleus in the volume of the liquid, in the absence of a wall, dust particle etc., is called homogeneous nucleation. By contrast, heterogeneous nucleation is defined as taking place at the interface between the liquid and a wall or other inhomogeneity like a dust particle. In addition there might be gas or vapor bubbles trapped in grooves or cracks of the surface, which can also act as a nucleus for growing vapor bubbles. In the following we discuss these effects and their implications (for more details, see Refs. (Avesisian, 1985; Carey, 1992; Debenedetti, 1996).

Homogeneous nucleation. The pressure p_g in a vapor bubble with equilibrium radius r_{eq} depends on the surface tension σ and the pressure in the surrounding liquid p_l via the Young-Laplace equation. From this equation it can be derived that the equilibrium radius is given by (Carey, 1992)

$$r_{eq} = \frac{2\sigma}{p_{sat}(T_l)\exp\left[v_l\left(p_l - p_{sat}(T_l)\right)/RT_l\right] - p_l}. \tag{4}$$

Here v_l is the specific volume of the liquid (assumed to be incompressible), and T_l and $p_{sat}(T_l)$ are the temperature of the liquid and the corresponding saturated vapor pressure, respectively.

An expansion of the Gibbs free energy G in a Taylor series around the equilibrium radius shows that in a superheated liquid such vapor bubbles are not stable, because $\Delta G(r)$ has a maximum at r_{eq}. This follows from the fact that for $r \to 0$ also $G(r) \to 0$, and $G(r \to \infty) < 0$ (Carey, 1992), leading to a form of $G(r)$ as shown in Fig. 5. The (unstable) equilibrium radius is therefore also the critical radius r_{cr}: bubbles with a radius $r > r_{cr}$ grow spontaneously, whereas smaller bubbles collapse.

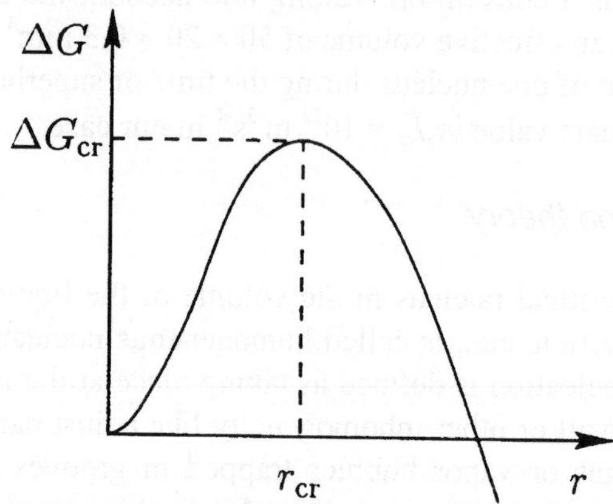

Fig. 5. Schematic representation of the Gibbs free enthalpy ΔG of a bubble as function of its radius. Bubbles nucleated in a superheated liquid grow if their radius exceeds the critical radius r_{cr}.

The number N_n^{hom} of bubbles forming spontaneously from thermally induced fluctuations in a superheated liquid with n molecules via homogeneous nucleation follows a Boltzmann distribution

$$N_n^{\text{hom}} = N_0 \exp\left[-\frac{\Delta G(r)}{k_B T}\right], \tag{5}$$

where N_0 is the number density of the molecules in the liquid. This leads to a rate J (number per volume and time) for the spontaneous formation of bubbles with the critical size r_{cr} (Avesisian, 1985; Carey, 1992) of

$$J = k_f N_0 \exp\left[-\frac{\Delta G(r_{cr})}{k_B T}\right] = k_f N_0 \exp\left[-\frac{16\pi\sigma^3}{3 k_B T_l (\eta\, p_{sat} - p_l)^2}\right]. \tag{6}$$

Here k_f denotes the molecular evaporation rate, and

$$\eta \equiv \exp\left[\frac{v_l}{RT_l}\left(p_l - p_{sat}(T_l)\right)\right]. \tag{7}$$

Hence the rate is a function of both the pressure in the liquid and the temperature T. Since the temperature-dependent quantities show up in the exponent, the nucleation rate varies quite strongly with temperature. Depending on the derivation different values for the prefactor k_f are found (Avesisian, 1985; Carey, 1992; Kagen, 1960; Katz, 1973). A comparison of the various models is given in Debenedetti, 1996. It turns out that the exact value is of minor influence on the experimentally observable nucleation threshold, because the exponential rise of the nucleation rate with superheating leads to an extremely sharp increase over many orders of magnitude within a narrow temperature interval.

Heterogeneous nucleation. Heterogeneous nucleation is the technically more important case, since liquids are most often heated via the container walls, and it is also the relevant case for SLC. The Gibbs free energy of the system is now a function of the contact angle θ of the liquid at the interface. The nucleation rate at a perfectly smooth surface can be derived in a similar way as above. Since the bubble is to be nucleated at an interface, Eq. (5) is modified to (Carey, 1992)

$$N_n^{het} = N_0^{2/3} \exp\left[-\frac{\Delta G(r)}{k_B T}\right], \tag{8}$$

where $N_0^{2/3}$ is the number of molecules per unit area at the interface. The resulting nucleation rate J is:

$$J = \frac{N_0^{2/3}(1+\cos\theta)}{2F}\left(\frac{3F\sigma}{\pi m}\right)^{1/2} \exp\left[-\frac{16\pi F\sigma^3}{3k_B T_l(\eta\, p_{sat} - p_l)^2}\right], \tag{9}$$

where

$$F = F(\theta) = \frac{1}{4}\left(2 + 3\cos\theta - \cos^3\theta\right). \tag{10}$$

Once again one finds an exponential increase over many orders of magnitude in a very narrow temperature interval, giving rise to a relatively sharp nucleation threshold. Results for non-planar geometries, e.g. in and around tubes, are given in the review of Cole, 1974.

The relevant material parameters for water, namely vapor pressure $p_{sat}(T)$, surface tension $\sigma(T)$ and density $\rho(T)$, are plotted in Fig. 6. Using these values the nucleation rate $J(\theta)$ can be calculated. The results for several contact angles are shown in Fig. 7 as a function of temperature. It is obvious that the limit of superheating T_{cl} (the "cloud line" mentioned above) drops significantly for contact angles above about 80°. We note that the non-monotonous behavior of T_{cl} as a function of θ, as described by Carey, 1992, is not reproduced with the material parameters of water shown in Fig. 6.

It should also be mentioned that a description in terms of equilibrium thermodynamics, as it is given here, is strictly speaking only approximate, because nucleation of embryonic gas bubbles is a dynamic process and therefore should also be treated by a dynamic theory. Moreover, in the experiments large temperature gradients appear, which are also not included in this treatment.

Nucleation at gas bubbles. When a surface is immersed in a liquid, air or vapor bubbles can be trapped in grooves at the surface, if the contact angle of the advancing liquid θ_a is larger than the angle γ of the groove (Carey, 1992; Debenedetti, 1996). Such inclusions act as nucleation sites, when the liquid is being superheated, and vapor bubbles start growing from there. The theoretical treatment depends on the details of the groove geometry and therefore can be quite complicated. Since this situation of a rough surface is the "normal" case in technical applications, a large amount of literature exists both with respect to theory and experiment, which is only to be mentioned here (Carey, 1992; Tong, 1997)

3.2. Experiments on metal films

Having summarized the relevant aspects of nucleation theory, we will now discuss experimental findings of laser induced bubble nucleation. First experiments in this field have been carried out on thin silver films, detecting

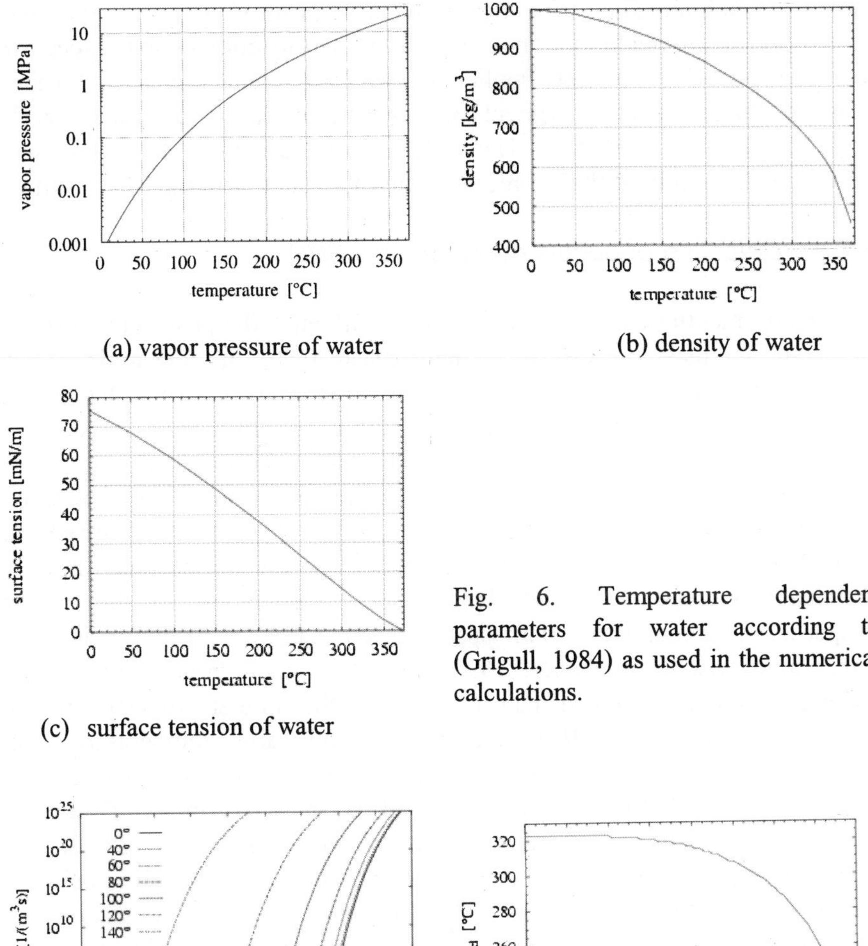

(a) vapor pressure of water

(b) density of water

(c) surface tension of water

Fig. 6. Temperature dependent parameters for water according to (Grigull, 1984) as used in the numerical calculations.

(a)

(b)

Fig. 7. (a) Nucleation rate $J(\theta)$ for bubbles calculated according to Eq. (9) and using the material parameters plotted in Fig. 6. (b) Superheating temperature T_{cl} at the cloud line for a rate of $J = 10^{22}\ m^{-3}s^{-1}$.

the nucleated bubbles by the optical reflectance probe (ORP) technique. The sensitivity of this technique, however, is limited to bubbles that exceed in diameter l/2. Only by the application of evanescent waves in Surface Plasmon Probe (SPP) experiments we have been able to monitor the early stage gas bubble formation. The order of the following discussion will follow the life cycle of the bubbles and therefore start with the SPP results.

3.2.1. Detection of bubble nucleation via SPP

As already mentioned, several effects may influence the position and shape of the surface plasmon resonance (Herminghaus, 1990). In order to determine the contribution of the temperature rise in the silver film, experiments have been conducted first with the bare silver film in the absence of liquid. Only a slight transient decrease in the reflected intensity is observed in this case when the laser pulse heats the film, due to the combined effect of plasmon resonance shift and broadening. However, if the experiment is carried out with a water-filled cuvette, the amplitude of the reflectance drop is drastically increased. As calculations show the change of the optical properties of the dielectric half space (in our case water) due to a given temperature rise (Schiebener, 1990) causes a much larger shift of the surface plasmon resonance than a change of the optical properties of the silver film (Johnson, 1972) induced by the same temperature rise. Consequently, the resulting shape of the surface plasmon resonance curve is predominantly governed by the changes of the refractive index of water as function of the temperature and can be used to probe the bubble growth.

Bubble nucleation threshold. As long as the laser fluence is below a certain value (which corresponds to the threshold for bubble nucleation, see below), the SPP signal exhibits no particular extra features, as shown in the topmost trace of Fig. 8a). When the excimer laser fluence is increased further, however, an additional structure in the SPP signal is observed as seen in the lower traces.

A hump in the reflectance drop starts to appear at an excimer laser fluence of $\Phi = 10.5$ mJ/cm^2, and becomes more pronounced with increasing laser fluence. The appearance of this hump can be interpreted as the onset of bubble nucleation at the water-silver interface. The surface plasmons are effectively scattered by the bubbles, and consequently the surface plasmon resonance is broadened and shifted, resulting in a temporary increase of the

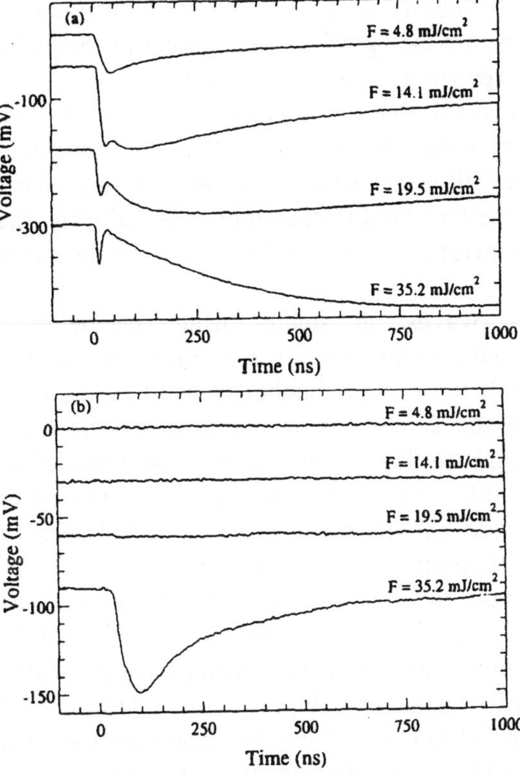

Fig. 8. (a) Surface plasmon probe (SPP) and (b) optical reflectance (ORP) signals for a water-silver interface, irradiated with increasing excimer laser fluence. 10 mV correspond to a reflectance change of 3 %. The curves are offset for clarity.

SPP intensity. When the bubbles collapse, the surface plasmon resonance becomes narrower again and the SPP signal decreases to a level, which is given by the resonance shift due to the actual temperature present at the water-silver interface.

For the laser fluence of $\Phi = 10.5$ mJ/cm^2 numerical computations based on the 1D heat transfer equation yield a peak temperature of $T = 384$ K at the water-silver interface, i.e., a surprisingly low liquid superheating of only about 10 K! This superheating is of the same order as derived for a similar system by Park, 1996. While the hump in the plasmon signal starts to appear at $\Phi = 10.5$ mJ/cm^2 and becomes more pronounced with increasing laser fluence, the simultaneously acquired reflectivity signals for ORP exhibit no particular features up to a fluence level of $\Phi = 19.5$ mJ/cm^2. Above this threshold fluence, for which computations predict a peak temperature of $T = 473$ K at the water-silver interface (liquid superheating » 100 K), a transient decrease in the ORP signal starts to appear as shown in Fig. 8b). This result clearly demonstrates the high sensitivity of the surface plasmon probe to small bubbles, which collapse before they grow up to the order of the probe beam wavelength and therefore remain invisible to the ORP.

The difference in the threshold fluences acquired using these two different methods can be explained by assuming that the threshold temperature obtained using the SPP corresponds to the temperature where bubble nucleation sets in, and the threshold temperature obtained using the ORP corresponds to the temperature where bubble growth becomes effective.

Bubble-induced pressure waves. In studies using piezoelectric transducers we have demonstrated the generation of high intensity acoustic pulses on a nanosecond time scale due to rapid bubble growth (Yavas, 1994 a, b, c). This enhancement of the acoustic pulse intensity by the rapid bubble growth process is thought to play a major role in the removal of submicroscopic particles in SLC. Since the piezoelectric transducer measurements could not provide the absolute pressure amplitudes, a quantitative data analysis was not possible with this method.

Applying the SPP, however, this quantitative information can be obtained on a nanosecond time scale (Schilling, 1996; Yavas, 1997 b). In order to eliminate any temperature effects, the experimental setup presented in Fig. 2 has been slightly modified as shown in Fig. 9. A thin quartz slide is placed in the water at a variable distance from the prism. The quartz slide is coated with 20 nm chromium and 60 nm silver on the surface facing the prism. The irradiation of the quartz slide from the rear side results in a sudden heating of the silver film and in bubble nucleation at the silver-water interface. The acoustic pulse generated during this process propagates toward the prism and can be detected by the SPP. The known relation for the dielectric constant of water as a function of pressure (Schiebener, 1990) is utilized to extract the absolute pressure amplitudes from the shift of the surface plasmon resonance curve. An example for the transient SPP signal resulting from the pulse and acquired using this setup is presented in Fig. 9.

A detailed description of the data acquisition system and the analysis of the SPP signals, i.e., conversion into absolute pressure amplitudes, is given elsewhere (Schilling, 1996; Yavas, 1997 b). The signal exhibits several peaks according to the propagation of the acoustic pulse in water and its repeated reflection between the prism and the quartz slide. The first peak

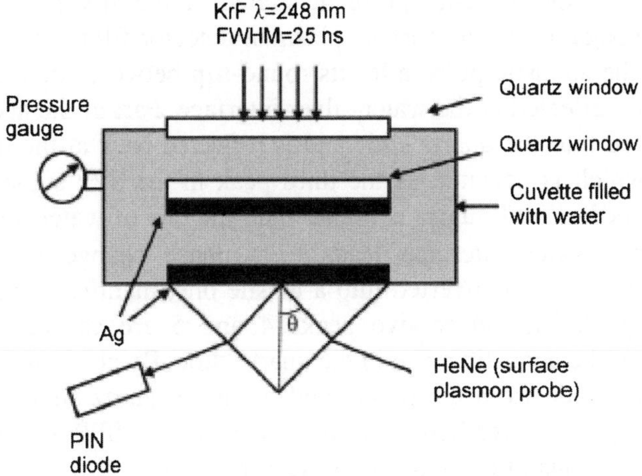

(a) Experimental setup for absolute pressure measurement using SPP

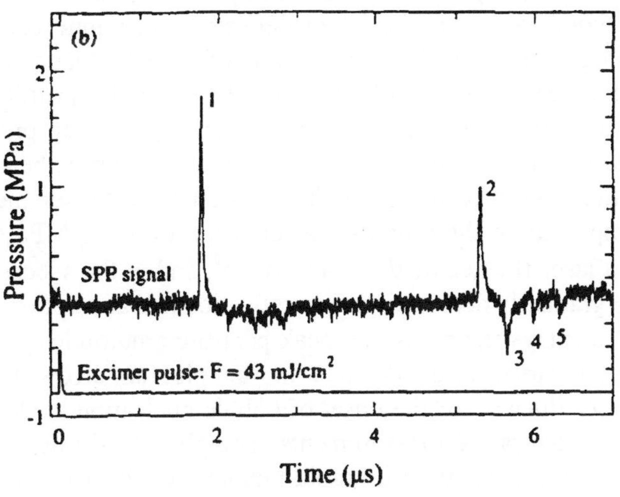

(b) Transient pressure signal

Fig. 9. (a) Experimental configuration for the absolute measurement of acoustic pulse amplitudes on a nanosecond time scale using the surface plasmon technique. In order to eliminate thermal effects on the silver film probed by SPP, the pressure is generated on the quartz slide upon excimer laser irradiation. (b) Transient pressure signal acquired using the surface plasmon probe.

corresponds to the acoustic pulse generated at the silver-water interface after its propagation to the surface plasmon detector film. The second peak represents the acoustic pulse after its round-trip between the prism and the quartz slide, reflected at the water-silver interface. Part of the acoustic pulse is transmitted into the quartz slide and is reflected back at the quartz-water interface, which is detected as the third peak in the SPP signal. Since the acoustic impedance of quartz is higher than the one of water, the reflection at the quartz-water interface leads to a phase change of π, and the compressive wave is converted into a tensile one, manifested in a negative pressure signal. The successive peaks 4 and 5 are caused by multiple reflections of the acoustic pulse in the quartz slide. By changing the distance of the quartz slide from the prism, and by using quartz slides of different thickness, we have verified that the peaks in the SPP signal represent multiple reflections of the acoustic pulse as described.

The width of the acoustic pulse is measured to be about 40 ns, which is much shorter than the previously measured pressure pulse width of » 100 ns using the photoacoustic probe beam deflection technique (Park, 1996 b) demonstrating the improved time resolution of the SPP. Due to the localized probe depth (» 200 nm), the SPP provides the real pulse profiles, while the photoacoustic deflection probe detects generally a larger pressure pulse width because of its integrating nature, i.e., the finite probe beam waist and the interaction length of the probe beam. According to our measurements the peak amplitude of the initial pulse amounts to » 2.5 MPa for the laser irradiation a laser fluence of $\Phi = 43$ mJ/cm^2, and in the successive echoes its intensity gradually decreases due to reflection and absorption losses.

Fig. 10 displays the measured peak pressure amplitudes as function of the incident excimer laser fluence. In the plot an arrow indicates the threshold laser fluence for the onset of bubble nucleation as determined by optical reflectance probe measurements. The plot clearly demonstrates the enhanced pressure generation by rapid bubble growth. An increase in laser fluence results in an increased maximum pressure. This pressure can both result from a faster bubble growth due to a higher superheating of the liquid as well as activation of more nucleation sites and hence an increase in the number of generated bubbles. The onset of a plateau, however, indicates that in the fluence range used saturation sets in for the maximum achievable bubble size, their growth velocity and/or number density of bubbles.

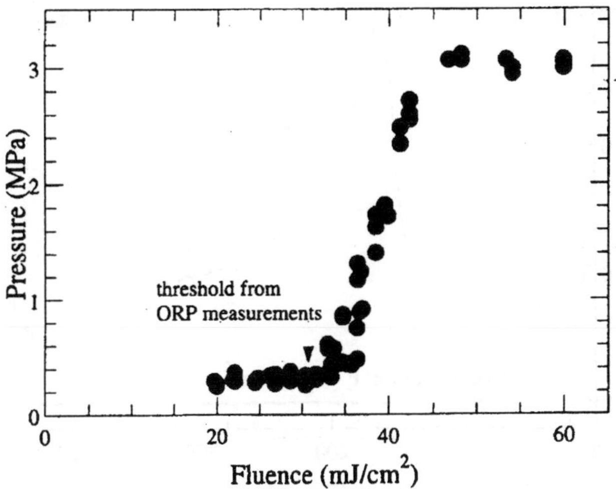

Fig. 10. Peak pressure amplitude as a function of the applied laser fluence measured by surface plasmon probe. The arrow marks the threshold for bubble growth as determined by optical reflectance probe.

Pressure amplitude of the same order of magnitude, but about two to three times lower, has been previously measured for a similar system using photoacoustic probe beam deflection and piezoelectric transducer techniques (Park, 1996 b). The discrepancy can be attributed to different samples and to the difference in temporal resolution of the test methods.

Direct proof for laser-induced bubbles. The fact that the reflection of the acoustic pulse at an acoustically less dense medium leads to a phase change, i.e., to the conversion of the compressive pulse into a tensile one, could be used to directly prove the generation of bubbles by excimer laser heating of the water-silver interface. As long as a bubble layer is present at the water-silver interface on the quartz slide, the acoustic pulse will be reflected as a tensile wave, since the reflection will take place at a water-vapor interface. When the bubbles collapse and the reflection takes place at the water-silver interface, the reflected pulse will be compressive.

This has been verified by placing a quartz slide close to the prism and successively increasing the distance after each excimer laser pulse. Representative results are depicted in Fig. 11. They clearly demonstrate the existence of bubbles up to several hundred ns after the excimer laser pulse,

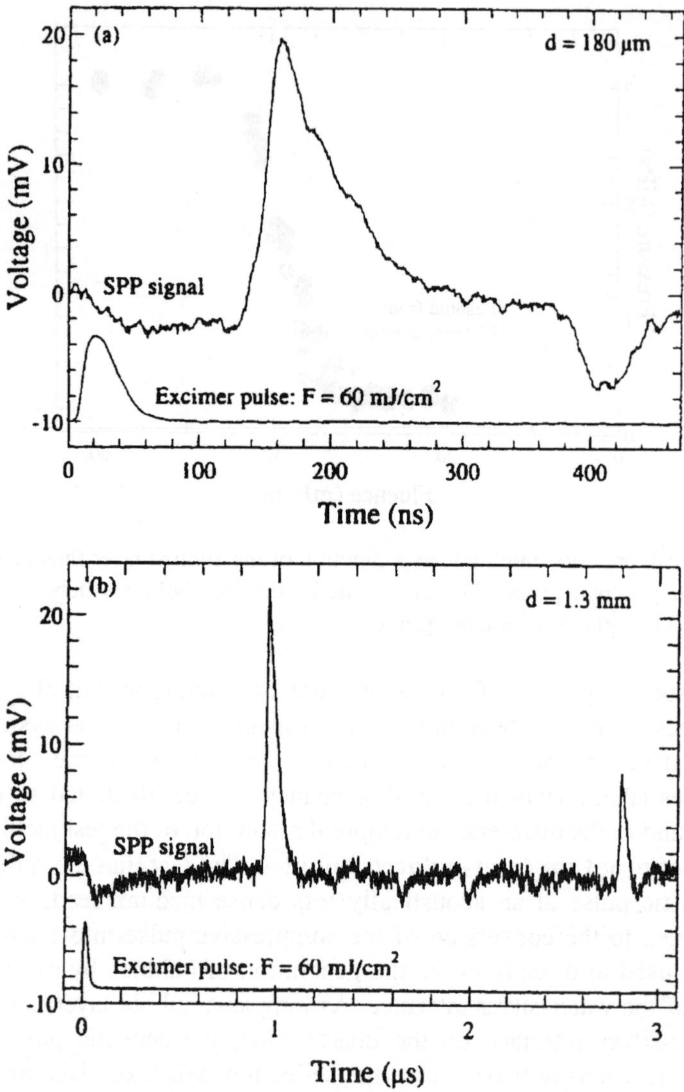

Fig. 11. (a) Direct proof of bubbles by the phase reversal of the acoustic pulse upon reflection at the water-silver interface when bubbles are present. The distance d between the two surfaces is 180 mm, and the delay time between the first and the second (in this case inverted) peak is 240 ns, corresponding to one round-trip of the acoustic pulse. (b) On a longer time scale, when the bubbles are collapsed, reflection takes place without phase reversal (d = 1.3 mm in this case).

manifested by a tensile reflected acoustic pulse. Although our previous studies already had delivered a clear evidence for laser-induced bubbles, the results presented here provide a more direct proof for such bubbles at a liquid-solid interface.

The amplitude of the reflected pressure pulse as a function of the distance d between the quartz slide and the sample surface (first echo) is plotted in Fig. 12. At d = 180 mm (the smallest distance that could be achieved experimentally), corresponding to a delay time of about 240 ns, the pressure amplitude has a negative value, as already seen in Fig. 11. As the distance is increased, the pressure peak drops in magnitude, changes sign around d = 0.5 mm, and then approaches a positive saturation value. We interpret this dependence as a gradual decrease of the bubble size and/or number, which already proceeds after 240 ns, and a complete collapse of the bubbles when the signal reaches saturation after about 1.8 μs.

Bubble growth velocities. Besides the determination of bubble nucleation thresholds the measurement of bubble growth velocities is also of great interest for the understanding of SLC. Models that predict particle as a result of shock waves generated by explosively growing gas bubbles rely on this parameter (Lu, 1999).

It is well known that the transition from Rayleigh to Mie scattering occurs at a bubble radius R_{tr} of about $\lambda_{probe} / 2\pi n_l$ (Born & Wolf, 1999). Accordingly we could use results from the optical reflectance probe experiments to determine the bubble growth velocity in excimer laser heated silver film-water systems (Yavas, 1994 c). Kim, 1996, independently confirmed the obtained value of 3.6 m/s.

Let us compare these experimental value with the one predicted by hydrodynamics. We start from the Rayleigh equation that describes the mechanical energy balance for a bubble growing in an incompressible (ρ_l = *const*) liquid (Carey, 1992):

$$R\frac{d^2R}{dt^2}+\frac{3}{2}\left(\frac{dR}{dt}\right)^2=\frac{1}{\rho_l}\left(p_v-p_\infty-\frac{2\sigma}{R}\right). \tag{11}$$

Here p_v is the pressure of the vapor at the corresponding temperature, p_∞ is the ambient pressure and σ is the interfacial tension. In its early

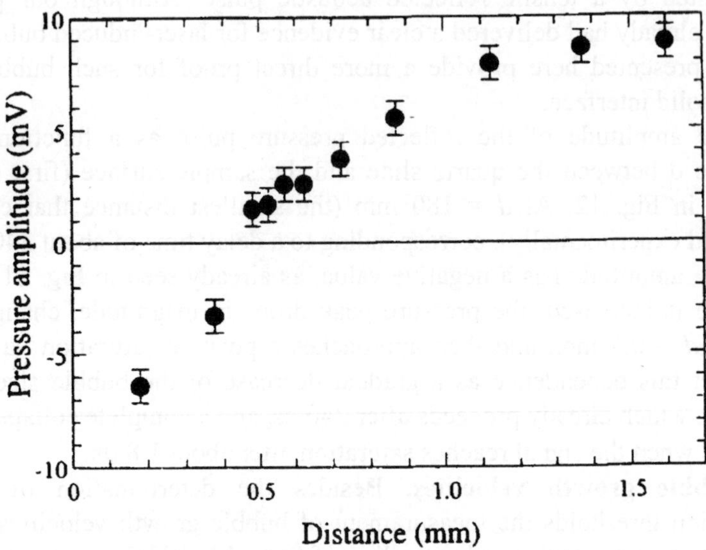

Fig. 12. The amplitude of the reflected pressure pulse as a function of the distance between the quartz slide and the SPP sample surface. Only at a distance $d = 1.3$ mm the signal reaches saturation, indicating that up to the round-trip time corresponding to that distance, 1.8 μs, bubbles are present after their generation by the excimer laser pulse.

stage the bubble growth is inertia-controlled and $2\sigma / R$ is much smaller than $p_v - p_\infty$. Neglecting this expression in Eq. (11) we obtain

$$R\frac{d^2R}{dt^2} + \frac{3}{2}\left(\frac{dR}{dt}\right)^2 = \frac{1}{\rho_l}(p_v - p_\infty). \tag{12}$$

If one furthermore assumes that the inertial force $\propto \ddot{R}$ is negligible, one obtains the expressions for the bubble growth velocity v

$$v = \dot{R} = \sqrt{\frac{2}{3}\frac{p_v - p_\infty}{\rho_l}}, \tag{13}$$

and the maximum growth velocity v_{max}

$$v_{max} = \dot{R} = \sqrt{\frac{2}{3}\frac{p_{max} - p_{\infty}}{\rho_l}}, \tag{14}$$

that are commonly used (Yavas, 1994 b; Lu, 1999) in computations of growth velocities for the modeling of SLC. In our opinion this formula overestimates the actual situation. Inserting the experimental parameters given in (Yavas, 1994 c) results in a bubble growth velocity of about 10 m/s, about a factor of 3 larger than the experimentally determined value.

Considering the gas-bubble expansion process as a reversible one, the relation between the pressure in the vapor and the bubble radius is given by

$$p_v(R) = p_{max}\left(\frac{R_0}{R}\right)^{3\gamma}. \tag{15}$$

However, in reality the requirement of a reversible process will not be fulfilled due to the dissipation of energy via heat conduction, viscosity, etc. A thorough analysis of these dissipation processes requires the understanding of the actual dissipation mechanism and the numerical solution of hydrodynamic equations. Some of these examinations were done in the papers of (Matsumoto, 1990, 1991 a, b; Kameda, 1993). Nevertheless, a simple estimation can be obtained by merely introducing a relaxation time τ, which characterizes the dissipation. Taking into account this relaxation term and additionally considering the inertial force we derive a modified Rayleigh equation

$$R\frac{d^2R}{dt^2} + \frac{3}{2}\left(\frac{dR}{dt}\right)^2 = \frac{1}{\rho_l}\left(p_{max}\left(\frac{R_0}{R}\right)^{3\gamma} exp\left[-\frac{t}{\tau}\right] - p_{\infty}\right). \tag{16}$$

In this equation the adiabatic process corresponds to the limiting case of $\tau \to \infty$, whereas the non-adiabatic case is modeled via a finite value $\tau \neq 0$.

For convenience in the numerical calculations we introduced dimensionless variables into this equation. Numerically we computed the bubble growth velocities resulting from the simplified Eq. (14) and for the cases of adiabatic and nonadiabatic vapor based on the modified Rayleigh Eq. (16). The resulting velocity as function of the normalized time is plotted in Fig. 13. From the comparison it is obvious that by considering the energy dissipation the bubble expansion velocities are reduced by a factor 2-3 compared to the simplified approach and are in good agreement with the experimental values.

3.3. Bubble nucleation on silicon wafers

Since the application of SPP in bubble nucleation studies relies inevitably on a substrate where surface plasmons can be excited, this method is not applicable to studies on silicon wafers. On the other hand, silicon wafers are the key material in semiconductor industry and SLC of these substrates is based on the laser induced bubble nucleation and explosive evaporation of the applied liquid. Hence reliable information on the bubble nucleation thresholds and bubble growth dynamics is needed not only for practical application of SLC but also for its theoretical modeling.

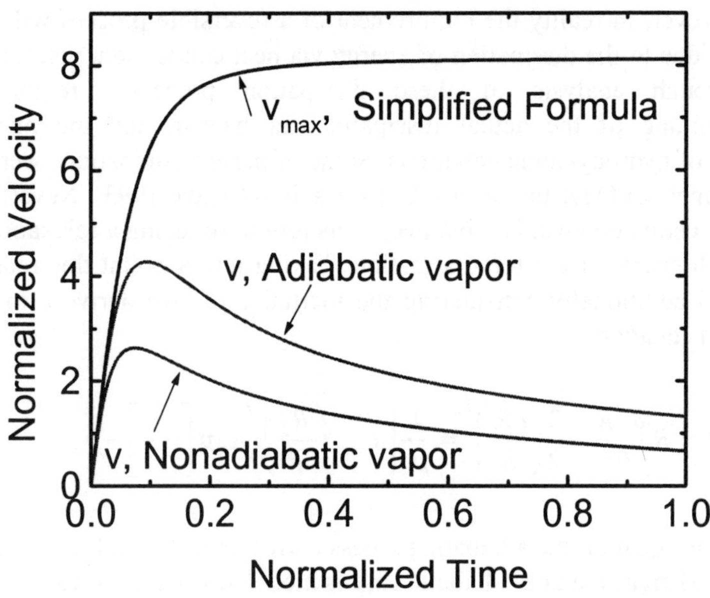

Fig. 13. Computed bubble growth velocities applying different simplifications of the Rayleigh equation.

In order to probe the bubble nucleation on silicon wafers we therefore monitored the light of an Ar-Ion probe laser that was scattered by the bubbles (SLP). The experiments on rough metal films described above revealed only small superheating of the water layer of about 10 K, compared to the theoretically predicted values at a perfectly smooth surface of about 200 K. To verify the interpretation that these results are dominated by surface roughness, we performed a series of bubble nucleation experiments on smooth substrates as well as on substrates with controlled roughness. As in an application of SLC typically mixtures of water and IPA are used, we additionally studied the bubble nucleation for the system IPA-silicon (covered by a native oxide layer).

3.3.1. Water on smooth silicon wafers

As the first system for the bubble nucleation studies we chose water (starting temperature 23 °C) on a silicon wafer (RMS roughness 0.2 nm) (Mosbacher, 2002 a).

Detecting nucleated bubbles by scattered light. Increasing step by step the pump energy delivered by the Nd: YAG laser, we first detected an increase in the system's reflectivity R for the Ar-Ion laser both for the *s*- and *p*- polarization. This can be clearly seen in Fig.14 a, where we have plotted the relative reflectivity change for both polarizations as a function of time. Within a few hundred nanoseconds the reflectivity increases and afterwards decreases during about 10 μs. Although qualitatively the signals for both polarizations do not differ, the reflectivity increase for *p*-polarization is much higher when compared to *s*-polarization. This can be explained by taking into account the initial reflectivity for the two polarizations at an angle of incidence of 82 °of the probe laser and the temperature dependence of the reflectivity values.

At moderate laser fluence no bubble nucleation took place and hence no scattered light was detected (Fig. 14 b). A slight increase in laser fluence changes this scenario. Above a certain, well-defined threshold we observed scattered light both in the plane of incidence of the probe laser (denoted by $\|$) and perpendicular to this plane ($\perp$). The scattered signals, shown in Fig. 15 b, indicate the presence of nucleated gas bubbles on the silicon surface (Yavas, 1993). After a time span of about 800 ns the scattered light disappears, which shows that the nucleated bubbles exist for a well-defined lifetime.

Compared to the situation when laser fluence below the nucleation threshold was applied, the shape of the reflected intensity signal changes only a little. A small dip in the s-polarized intensity is the only hint on the nucleated bubbles. From this comparison it is clear that the detection of scattered light is a much more sensitive tool to monitor the dynamics of laser induced bubble nucleation and especially determine the threshold laser fluence for nucleation. Only a further increase in the applied pump laser fluence creates enough bubbles to allow bubble detection in the reflected light.

The nucleation temperature. We repeated these experiments for different starting temperatures of the water. Typical examples for the maximum of the scattered intensity as function of the applied pump laser fluence are shown in Fig. 16 for starting temperatures T_0 of 23 °C and 70 °C. Remarkable for both cases is the well-defined, sharp threshold in laser fluence, below which no bubble nucleation is detected. A comparison of this threshold for the two starting temperatures shows that for T_0 = 70 °C much less laser fluence is needed to nucleate bubbles.

This result can be interpreted in the following way: the bubble nucleation rate J increases by several orders of magnitude in a narrow temperature interval. Hence below some sharp temperature threshold of the water layer adjacent to the substrate surface no bubbles will be nucleated, whereas above this threshold many bubbles form. During the heating the liquid temperature is raised at constant pressure p_0 up to the kinetic limit of superheating $T_{cl}(p_0)$. An increase in the starting temperature of the water results in less energy ΔE that must be deposited in the liquid for heating it up to T_{cl}. In our case, this energy is deposited via heat transfer from the silicon substrate, which is heated by the pump laser. Assuming a constant heat capacity of water and silicon at the temperature intervals considered here 1 one can write for the threshold laser fluence Φ_{th}:

$$T_{cl} - T_0 \propto \Delta E \;\sim T_{Si} - T_0 \propto \Phi_{th} . \tag{17}$$

In order to verify this linear dependence of Φ_{th} on the starting temperature T_0 we repeated the detection of threshold fluences for a total of

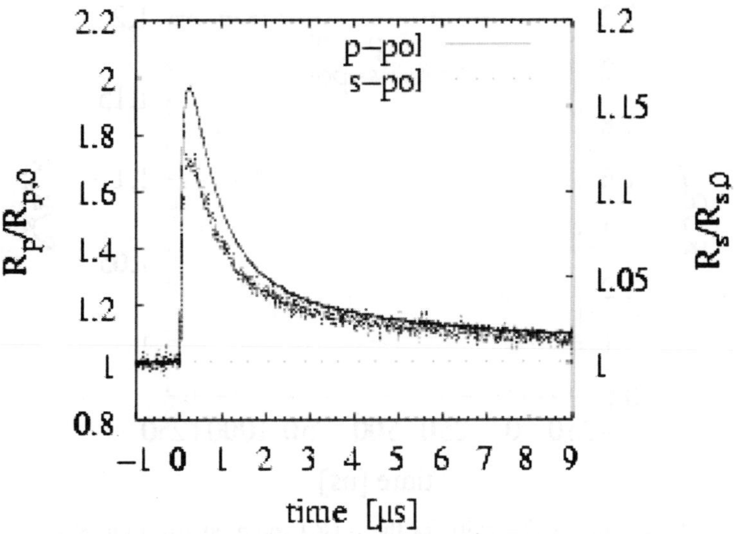

(a) Relative reflectivity for the *s*- (R_s) and *p*-polarization (R_p).

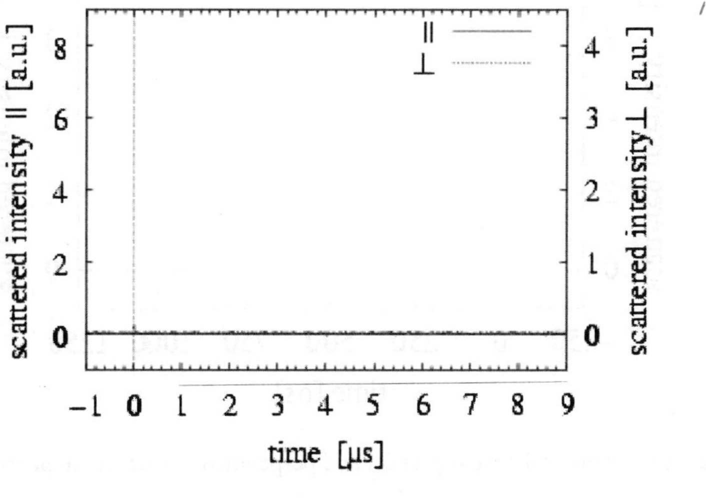

(b) Scattered intensity.

Fig. 14. Detection of the reflected and scattered light intensity from a Si-H_2O interface at a laser fluence of 71 mJ/cm^2, which is below the bubble nucleation threshold.

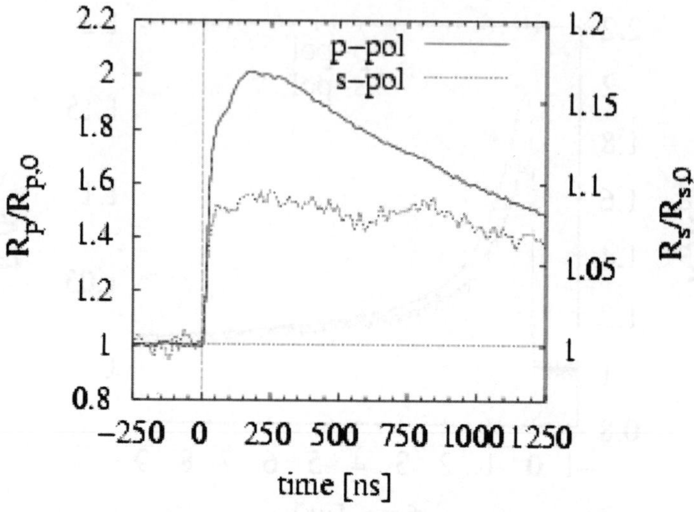

(a) Relative reflectivity for the *s*- (R_s) and *p*-polarization (R_p).

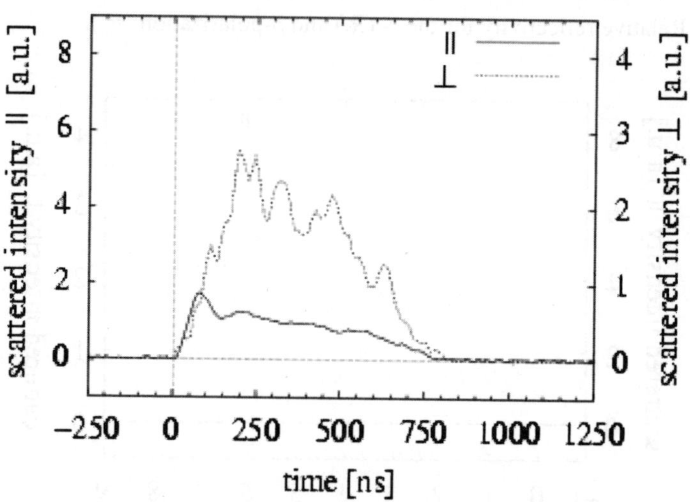

(b) Scattered intensity detected parallel and perpendicular to incident probe beam.

Fig. 15. Detection of the reflected and scattered light intensity from a Si- H_2O interface at a laser fluence of 75 mJ/cm^2, which is above the bubble nucleation threshold.

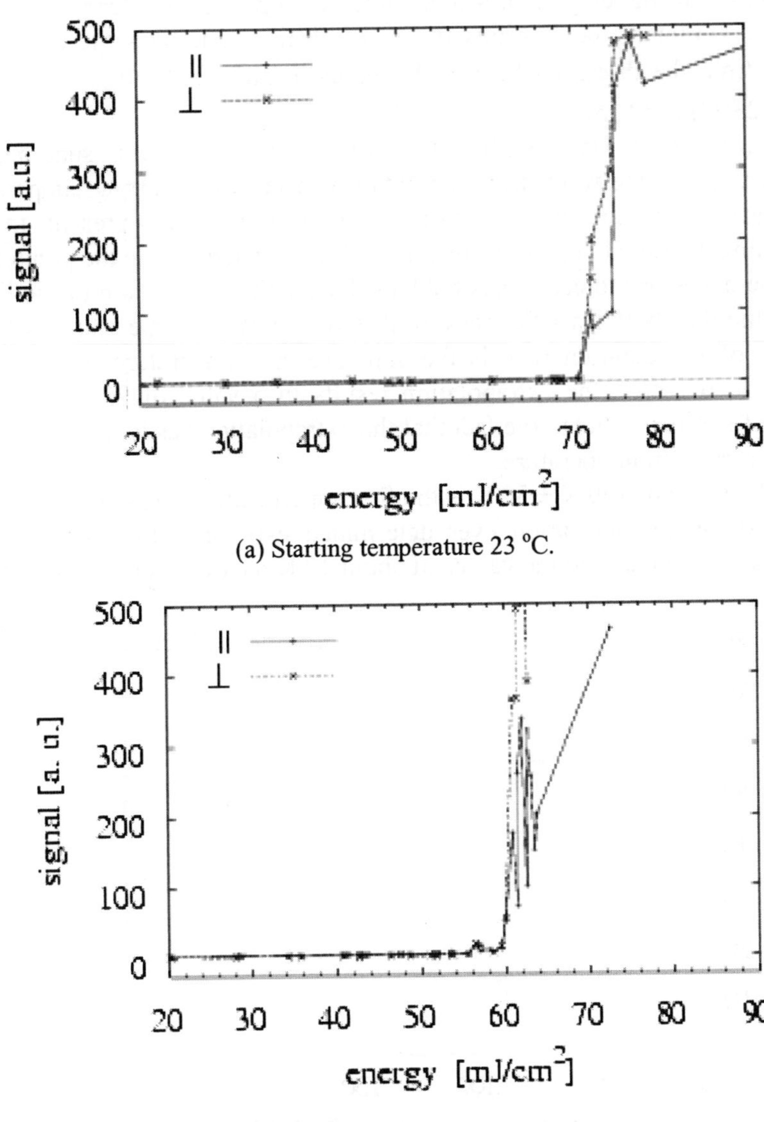

(a) Starting temperature 23 °C.

(b) Starting temperature 70 °C.

Fig. 16. Maximum scattered intensity of the probe laser as function of the applied pump laser fluence for different starting temperatures T_0 of the water. Note the well-defined, sharp threshold of bubble nucleation in laser fluence.

six more starting temperatures in the interval between room temperature and 87 °C. A plot of these results in Fig.17 clearly underlines the theoretical prediction; the threshold fluence decreases linearly from Φ_{th} = 72 mJ/cm^2 (T_0 = 23°C) to 51 mJ/cm^2 (T_0 = 87°C).

Moreover it is possible to extract the maximum superheating temperature of the water from the obtained data. If the starting temperature of water were to reach the superheating temperature present in our experiments, Eq. (17) predicts that no additional energy from the pump laser will be necessary to create gas bubbles. Hence the intersection of the linear fit of the threshold laser fluences for different starting temperatures with the x-axis of the diagram reveals the superheating temperature of the water layer. In our case the extrapolation results in a value of 250°C ± 30°C, where the error is due to the fact that the extrapolation has to be done over a large interval in temperature.

The value of 150 K ± 30 K is the first superheating temperature for laser induced bubble nucleation ever determined experimentally on a smooth substrate. Compared to the values of about 10 K measured on metal films by

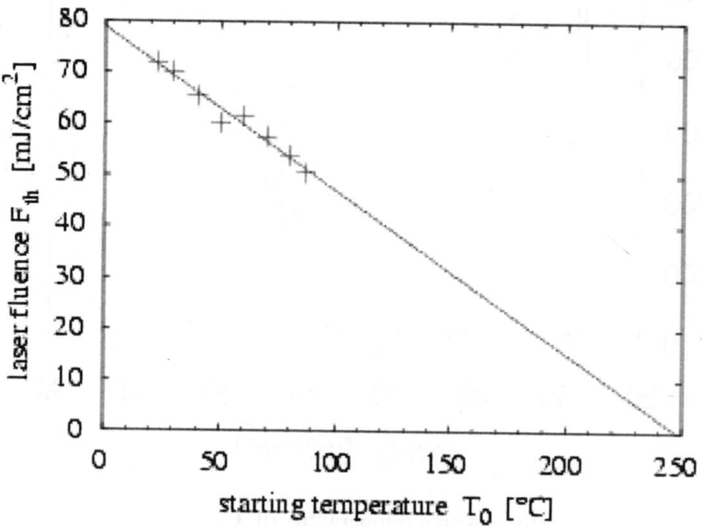

Fig. 17. Threshold fluences for laser induced bubble nucleation at a silicon/ water interface as function of the starting temperature T_0 of the water layer. An extrapolation of the linear dependence yields a superheating temperature of 250°C ± 30°C of the water.

SPP and 30 K ± 30 K determined on silver films in a control experiment using SLP, our value is much closer to the theoretical prediction of 205 K (see Fig. 7), especially if the experimental error is considered.

3.3.2. Water on structured silicon substrates

The observed high superheating values for water on smooth silicon substrates support the interpretation that the small superheating found previously for water-metal film-systems is due to the surface roughness. In order to corroborate this assertion we have performed nucleation experiments on surfaces with controlled roughness.

Silicon samples with controlled surface roughness. A controlled modification of the surface roughness of the silicon was achieved by creating well-defined holes with a diameter of a few hundred nanometers in the surface (Lu, 2000 b, 2002; Mosbacher, 2001; Münzer, 2002). For this purpose we utilized a method described in detail in (Münzer, 2002).

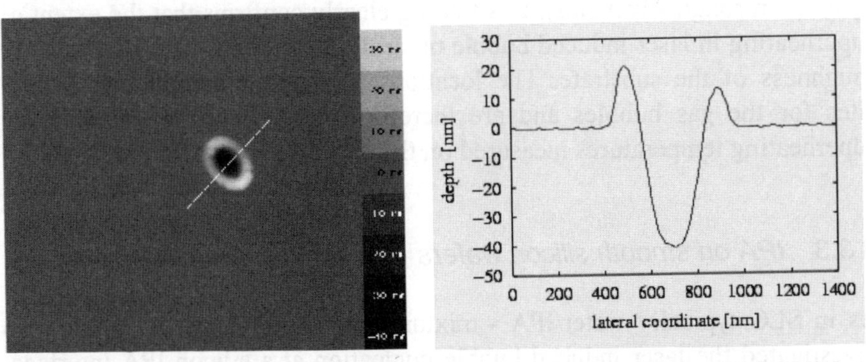

Fig. 18. A typical hole on a silicon wafer structured by nano near filed drilling as imaged by an atomic force microscope. The size of the imaged area is 3 × 3 mm^2, the white line indicates the cross section displayed in the right image.

In brief, we applied spherical PS colloidal particles 810 nm in diameter onto the substrate as described in Section 2. Illumination of the samples prepared in this way with a fs laser pulse (λ = 800 nm, FWHM = 150 fs) results in particle removal (Mosbacher, 1999, 2001) in a dry laser cleaning (DLC) process. Since for the applied laser parameters the dominating cleaning mechanism in DLC is local substrate ablation caused by the enhancement of the applied laser fluence in the near field of the particles

(Mosbacher, 2001, 2002 a) holes are created at the former position of the particles. A typical example of such a hole is shown in Fig.18. By adjusting the number density of the colloidal particles and hence the hole density on the surface we could control the roughness of the surface.

Nucleation thresholds. In the same way as discussed above for smooth silicon wafers we measured the nucleation thresholds in laser fluence for the structured substrates. A typical result obtained for T_0 = 70 ° C is shown in Fig. 19. In comparison to the results obtained on smooth surfaces the nucleation threshold is not as pronounced: the slope of the signal increase is much smaller and rather similar to the results obtained on silver films.

Nucleation temperature. Repeating the above experiment for different starting temperatures T_0 of the water, we again obtained a linear dependence of the threshold laser fluence on T_0 (see Fig. 20). Extrapolating the obtained results to vanishing fluence yields a nucleation temperature of (160 ± 13) °C. Consequently, the superheating is reduced to 60 K on these rough surfaces in comparison to the value of 150 K measured on smooth surfaces. Calling to mind the superheating of about 30 K measured by SLP on rough silver films this finding clearly confirms that the extent of superheating in laser induced bubble nucleation is dominated by the surface roughness of the substrate. The locations of roughness act as nucleation sites for the gas bubbles and are therefore responsible for the different superheating temperatures measured on the various substrates.

3.3.3. IPA on smooth silicon wafers

As in SLC typically water/IPA - mixtures are used (Tam, 1992), we also investigated the laser induced bubble nucleation at a silicon-IPA interface. Qualitatively the results are very similar to the ones obtained for the silicon-water system. Fig. 21 shows the extrapolation of the threshold fluences in order to obtain the bubble nucleation temperature. Compared to the boiling point of 82°C the nucleation temperature of (116 ± 5) °C exhibits only a small superheating of about 30 K.

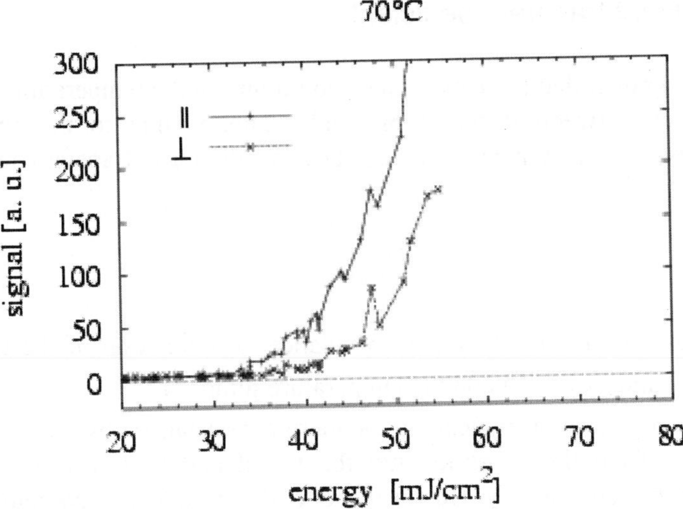

Fig. 19. Intensity of the light scattered by bubbles nucleated in water (T_0 = 70°C) on a rough silicon surface. Note that the threshold is not as sharp as for smooth surfaces.

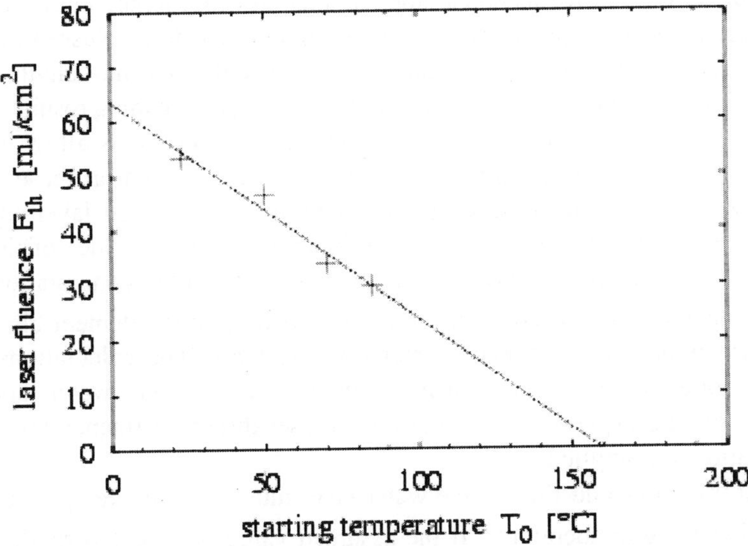

Fig. 20. Threshold fluences for bubble nucleation on rough silicon surfaces as function of the start temperature T_0. The extrapolation to vanishing fluence yields a nucleation temperature of (160 ± 13) °C.

3.4. Heat transfer coefficient

It is well known that there exists a discontinuity in the temperature profile at boundaries between different materials. This is due to the finite heat conductivity of the boundary region. The so-called heat transfer coefficient

$$\xi = \frac{\dot{Q}}{A\Delta T}, \tag{18}$$

quantifies this thermal boundary resistance as a function of the heat flow $\dot{Q}$, the boundary area A and the temperature jump ΔT.

With regard to laser cleaning the thermal resistance obviously limits the heat flow from the substrate into the liquid and thus lowers the liquid temperature considerably. However, none of the published computations of temperature profiles in laser cleaning (Lu, 1999; Wu, 2000) incorporates this fact. Probably one reason is that although the phenomenon is well investigated for low temperatures below 50 K (Kapitza resistance, see Swartz, 1989) and for technical applications at room temperature, for long time scales (several seconds) and macroscopic dimensions (Nusselt-number, see Cerbe, 1999), there are no data at ns time scales and nm length scales, which would be needed for the interpretation of laser cleaning results.

The data obtained in the bubble nucleation experiments allow us now for the first time to determine the heat transfer coefficient at these scales. Fig. 22 shows computed maximum temperatures of the water layer adjacent to the silicon surface as a function of different assumed values of the heat transfer coefficient ξ between silicon and water. Our computations are based on the 1D heat equation implemented in a finite element algorithm and use temperature dependent material constants. The calculations have been done for different starting temperatures of the water and the corresponding experimentally determined laser threshold fluence for bubble nucleation was applied.

At this threshold fluence the water layer must reach the temperature T_{cl} for all starting temperatures. If the assumed value of ξ is too small or too large, however, the water layer will reach maximum temperatures that are different for each starting temperature. It can be seen from these computations that for all starting temperatures of the water the graphs of the

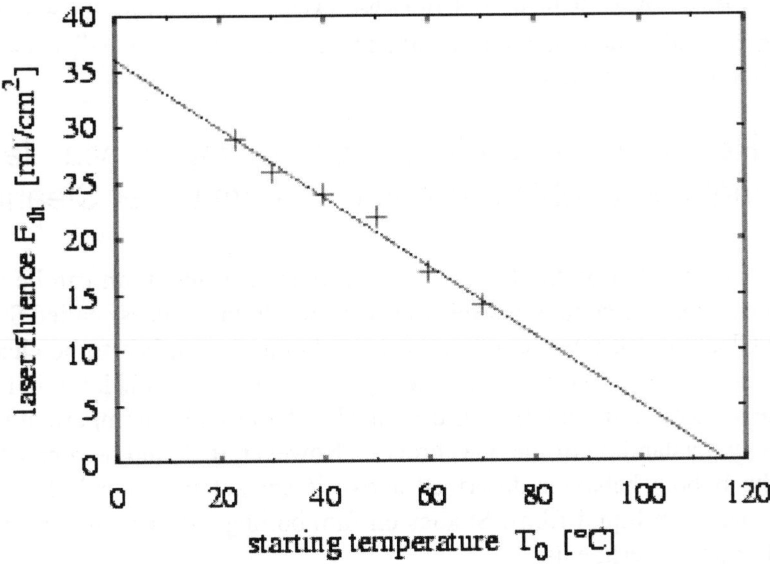

Fig. 21. Threshold fluences for bubble nucleation in IPA on smooth silicon surfaces as function of T_0. The extrapolation to vanishing fluence yields a nucleation temperature of (116 ± 5) °C.

calculated maximum temperatures as function of the assumed heat transfer coefficient intersect at one single point. This point is given by a value of $\xi_{H_2O} = 3 \cdot 10^7$ W/m^{2}K and a maximum water temperature of 250 °C - just as determined experimentally. Therefore this z-value represents the heat transfer coefficient in the studied system. Interestingly the computations exclude values of $\xi < 3 \cdot 10^6$ W/m^{2}K, as in this case the equilibrium boiling temperature of 100 °C is not reached for all starting temperatures.

In the same way we can determine the heat transfer coefficient for IPA on a smooth silicon wafer (see Fig. 23). Here the computations reveal a value of $\xi_{IPA} = 1 \cdot 10^7$ W/m^{2}K. Again, values smaller than $\xi_{IPA} = = 5 \cdot 10^6$ W/m^{2}K can be excluded as the boiling temperature is not reached for all T_0.

The ratio $\xi_{IPA}/\xi_{H_2O} = 0.33$ can be understood qualitatively in the frame of the theory of acoustic impedances. A comparison of the acoustic impedances of IPA ($Z_{IPA} = \rho c = 0.947 \cdot 10^6$ kg/m^{2}s) and water ($Z_{H_2O} = 1.49$

$\cdot 106$ kg/m^{2}s) shows that the transmission probability $\alpha_{Si \to IPA}$ for a phonon at perpendicular incidence on the silicon - IPA interface is smaller than for the silicon - water system.

4. Removal of particles on surfaces via laser induced bubble nucleation: Steam Laser Cleaning

The forces exerted by the laser induced explosive evaporation can be used to remove particle contamination from surfaces. In this process called Steam Laser Cleaning (SLC) a liquid is condensed onto the surface to be cleaned and irradiated by a laser pulse. Clearly the experiments on laser induced bubble nucleation in bulk liquids described so far provide an important basis for the understanding of the SLC process. However, it should be pointed out that the bubble behavior described above is characteristic for bulk liquid rather than thin liquid films. Studies on film boiling still have to be carried out in future experiments.

4.1. Efficiency measurements

4.1.1. Dependence on the number of applied laser pulses

Clearly, with view to applications, the most interesting question in a surface cleaning process is its efficiency. More precisely, one is interested in the minimum particle size that can be removed, the percentage of particles of different size that can be removed, and in the process parameters necessary for particle removal.

As described in Section 2 we have used spherical colloidal particles made of PS and SiO_2 as model contaminants in our cleaning studies. For these particles we have determined cleaning efficiencies for a wide range of particle sizes. While varying the process parameters regarding the applied Nd: YAG laser (number of pulses, laser fluence) we have up to now not varied the liquid film properties. The film thickness was kept at a constant value in between 200 nm and 400 nm as determined by reflectometry. We also did not vary the composition of the liquid; a mixture of water (90%)

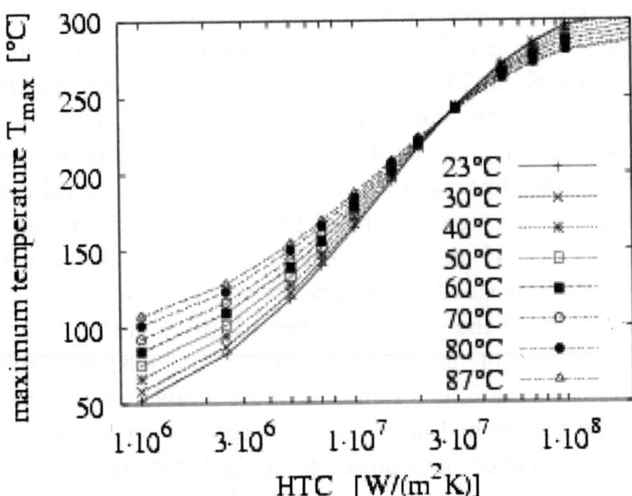

Fig. 22. Computed maximum temperatures in water on a silicon surface as function of the heat transfer coefficient (HTC) ξ_{H_2O} for different starting temperatures T_0. As laser fluence we used the experimentally determined threshold fluence for bubble nucleation.

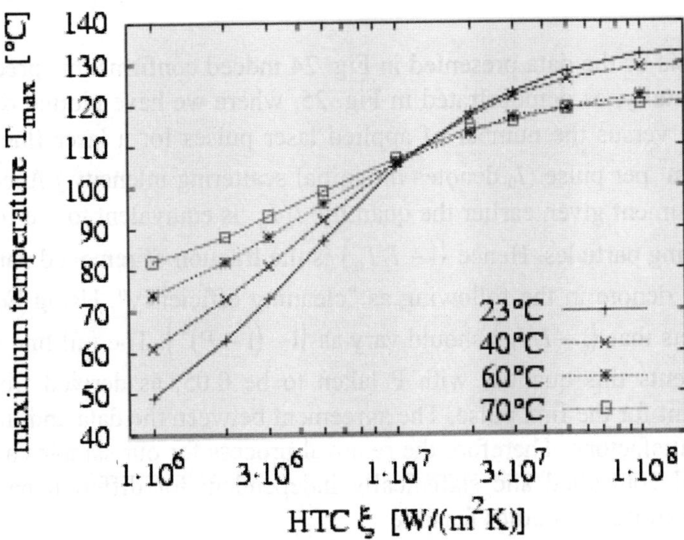

Fig. 23. Computed maximum temperatures in IPA on silicon as function of the heat transfer coefficient ξ_{IPA} for different starting temperatures T_0. As laser fluence we used the threshold laser fluence.

and IPA (10%) was evaporated in all experiments.

When preparing samples with a high density of particle contaminants (average distance about 10 mm) of course one must make sure that they do not influence each other during the cleaning process. Fig. 24 shows the removal efficiency for 800 nm silica spheres as a function of laser fluence for 1, 2, 10, 20 cleaning steps. At a laser fluence of 115 mJ/cm^2 the first pulse removes only 5 percent of the particles, whereas at the somewhat increased level of 140 mJ/cm^2 already nearly half of the particles are detached under otherwise identical conditions.

If the majority of the particles is bound to the surface with the same energy, as one might expect for almost monodisperse spheres, the probability P that a particle is removed by an individual laser pulse should only depend on the applied laser fluence F and not on the previous pulses. Hence when a sequence of n pulses is applied, the fraction of remaining particles N_r / N_0 (where N_0 is the initial particle number) should be given by

$$\frac{N_r}{N_0} = (1 - P)^n . \tag{19}$$

An analysis of the data presented in Fig. 24 indeed confirms this prediction experimentally, as demonstrated in Fig. 25, where we have plotted data for $(1 - I/I_0)$ versus the number of applied laser pulses for a laser fluence of 115 mJ/cm^2 per pulse (I_0 denotes the initial scattering intensity.) According to the argument given earlier the quantity I/I_0 is equivalent to the fraction of remaining particles. Hence $(1 - I/I_0)$ is the fraction of removed particles, which we denote in the following as "cleaning efficiency". Using Eq. (19) one expects that $(1 - I/I_0)$ should vary as $(1 - (1 - P)^n)$. The full line in Fig. 25 represents this quantity with P taken to be 0.05, as derived from the datum point for the first pulse. The agreement between the data and this line is quite satisfactory. Therefore the removal process for our samples appears to be well controlled and statistically independent for different particles, making quantitative studies feasible.

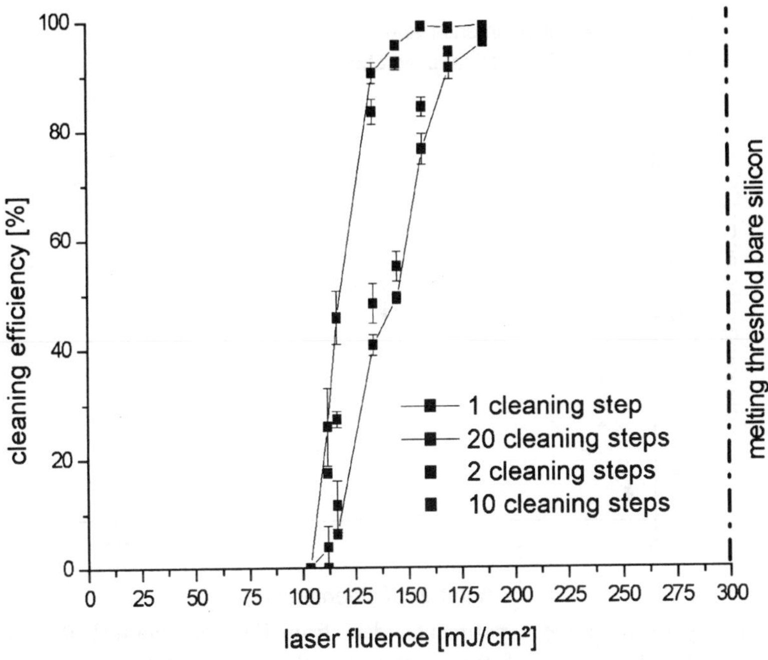

Fig. 24. Cleaning efficiency for 800 nm silica spheres as a function of laser fluence for 1, 2, 10, 20 cleaning steps.

4.1.2. Dependence on the laser fluence and variation of particle size

Results of the energy dependence of particle removal by the steam laser cleaning process for PS spheres with diameters of 800 nm, 500 nm, and 60 nm are plotted in Fig. 26. With regard to a better illustration of the cleaning behavior the results for 20 cleaning steps are plotted, which is justified by the statistically independent cleaning behavior.

For all sizes we obtain a similar behavior of the efficiency as a function of laser fluence: a steep increase in the cleaning efficiency at of $\Phi = 110$ mJ/cm^2. This absolute value of the threshold fluence was determined as described in Section 2. A comparison of the laser fluence needed for cleaning to the one necessary for melting the optical penetration depth of the silicon substrate resulted in a value of 0.35, which corresponds to above quoted threshold fluence.

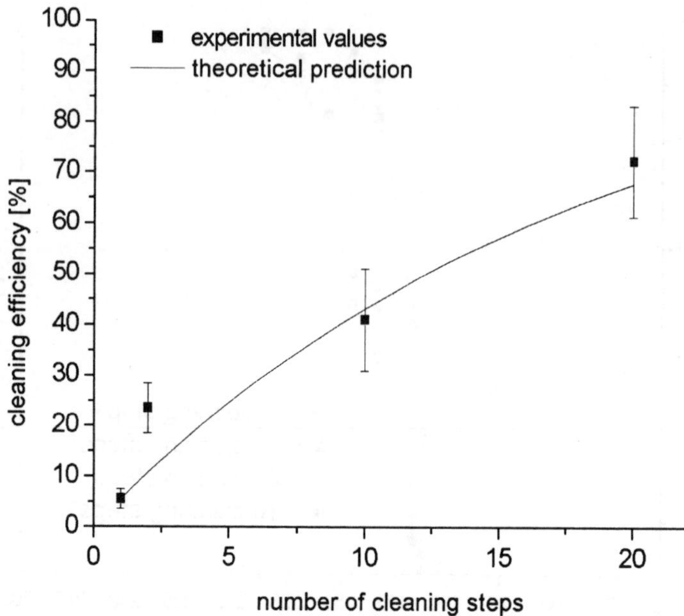

Fig. 25. Comparison of the experimental data from Fig. 24 obtained at a laser fluence of 115 mJ/cm^2 and of the theoretical value provided the particles are detached statistically independent.

The same cleaning threshold was found for PS spheres with diameters of 235 nm and 300 nm. These results are not shown in the graph for the sake of clarity. At fluences above 170 mJ/cm^2 more than 90% of the particles were removed after 20 cleaning steps (steam pulse plus laser pulse). For comparison the threshold where a (bare) Si surface would start to melt at the laser parameters used here is 270 mJ/cm^2.

In order to study the influence of the particle material on the cleaning efficiency we used silica spheres with diameters of 800 nm and 500 nm and polydisperse alumina particles with a mean diameter of 300 nm. Once again we obtain the same threshold as for PS spheres (Fig. 27) and the same dependence of the cleaning efficiency on the laser fluence. Thus the cleaning threshold is independent of both particle size and material for the investigated particles. Experiments using slightly different laser parameters (λ = 583 nm, FWHM = 2.5 ns and 7 ns as well as λ = 532 nm and FWHM = 2.5 ns) revealed again the same cleaning threshold as reported above (Mosbacher, 2000).

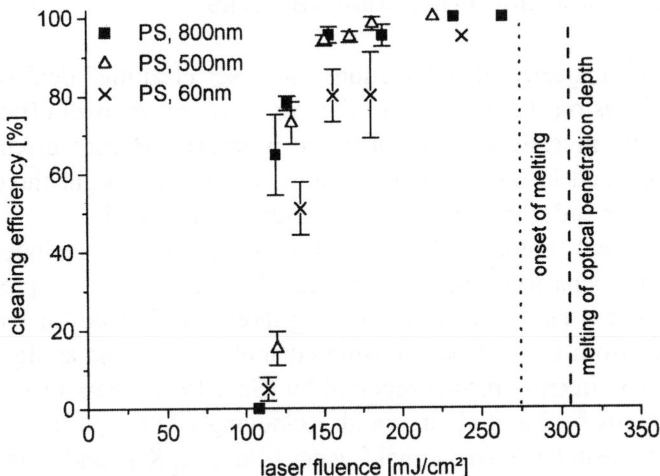

Fig. 26. Experimentally determined cleaning efficiencies as a function of applied laser fluence for various PS spheres. The cleaning threshold is size independent and the cleaning efficiency exhibits a steep increase above the onset of cleaning.

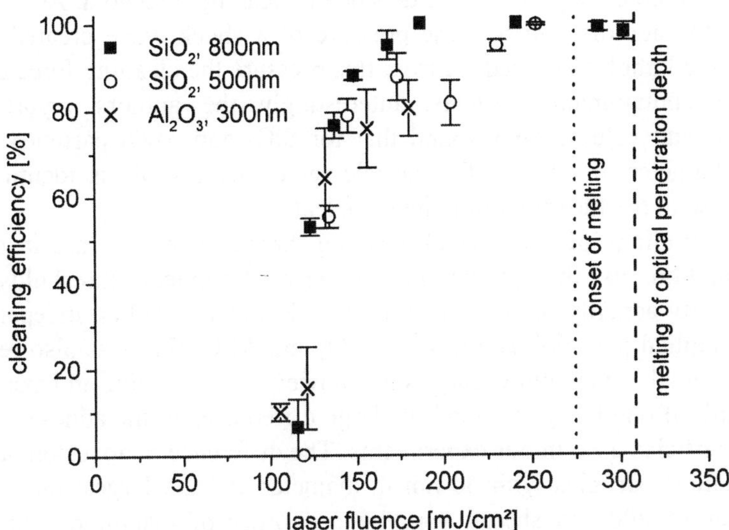

Fig. 27. Experimentally determined cleaning efficiencies as a function of applied laser fluence for particles of different geometry, material and size. A material- and size independent threshold and a steep increase of the cleaning efficiency was monitored.

4.2. Discussion and concluding remarks

Whereas the majority of publications on "laser cleaning" deal with DLC, only few focus on the quantitative determination of removal efficiencies in SLC and the investigation of the basic processes. Before discussing our results described in the previous sections, we will highlight three of these SLC studies in order to provide the necessary background.

In a recent theoretical study (Lu, 1999) have developed a first approximate scenario for the SLC process. The most important prediction of this treatment is size dependent cleaning threshold fluence for the removal of particles from surfaces when using ethanol and acetone as liquids in the process. Experimental results reported by (She, 1999) seem to indicate that the predictions by Lu et al. are valid. Removing Al_2O_3 - particles from NiP surfaces and using a "micron thick" water film, She, Kim and Grigoropoulos found that "the minimum laser fluence for removal of micron-sized or larger contaminants was lower by a factor of about 2 than the fluence necessary for complete removal of 0.3 μm particles".

The second theoretical description of SLC by Wu, Sacher and Meunier, 2000 also predicts a particle size dependent cleaning threshold. As in the Lu model, the authors compute the pressure of a shock wave created by the nucleating bubbles and deduce from this pressure the cleaning force exerted on the contaminant particles. Interestingly the authors report some experimental values. They found that for SiO_2 and Al_2O_3 particles with a mean diameter of 100 nm the particle removal threshold is identical and very close to the threshold of bubble nucleation.

Our findings of a universal cleaning threshold in SLC are in serious contradiction to the predictions of size dependent thresholds. The universality actually refers to two aspects. First we noted an independence of the applied particle materials PS, SiO_2 and Al_2O_3 that was also reported by Wu et al. Even more surprising, however, is the size independence, especially if one brings to mind the large differences in the adhesion forces of the particles used in our experiments. The difference in adhesion between the smallest particles with 60 nm in diameter and the largest ones with a diameter of 800 nm should vary by one order of magnitude, since the adhesion force is proportional to the particle radius (Johnson, 1971; Heim, 1999).

Although the investigations presented here are the first systematic quantitative studies of cleaning thresholds in SLC, many open questions still have to be answered prior to a full understanding of the processes underlying particle removal in SLC. Most of these refer to the dependence of the cleaning threshold and maybe its uniformity on process parameters such as liquid film thickness, liquid composition or laser wavelength. For instance one possible explanation for the different findings of the Grigoropoulos group may be their film thickness of a few micron compared to our 200-400 nm thick films.

The importance of controlling the liquid film properties and understanding its influence on the SLC process becomes even more obvious taking into account that the mechanism considered to be responsible for particle removal - the explosive liquid evaporation - should depend critically on the liquid film properties. As already mentioned above Wu et al. point out a close agreement of the particle removal threshold and the bubble nucleation threshold in their experiments. Indeed such an agreement provides one possible explanation for the universality of the cleaning threshold for particles completely embedded in the liquid film (Mosbacher, 2000). Should SLC be solely governed by an explosion of the liquid film, the process consequently will be determined by parameters of the film rather than particle properties.

Although this is a possible and plausible interpretation of our results, still a lot of research has to be carried out in the field of laser induced bubble nucleation and explosive film evaporation in order to unequivocally clarify the relevant SLC processes. The experiments reported in the first part of this paper were also motivated by this requirement. Improving techniques such as Scattered Light Probes (SLP) and developing powerful tools like Surface Plasmon Probe (SPP) we have been able to elucidate crucial details of laser induced bubble nucleation on surfaces. These include the recording of the pressure waves and detecting the associated pressure amplitudes in the MPa regime. From the experiments we were also able to deduce the gas bubbles growth velocities, which were found to be in the range of a few m/s.

Probably the most important results, however, are related to the measurement of the laser threshold fluences necessary for bubble nucleation and especially the related superheating temperatures of the liquids. Superheating of the liquid can occur in SLC due to the deposition of the laser energy during a very short time interval of a few ns. Since the pressure amplitudes in SLC depend very strongly on the liquid temperature, the liquid superheating may be the key parameter of the process. One major finding of our experiments in this field is the strong dependence of the extent of superheating on the surface roughness. On thin metal films only small amounts of superheating in the order of 10-30 K have been measured. In contrast on smooth silicon wafers we measured superheating of about 150 K! Moreover, from these superheating temperatures it was possible to show that the heat does not flow freely from the substrate to the liquid, but is inhibited by the solid-liquid heat resistance described in the case of a silicon-water interface by a heat transfer coefficient of $3 \cdot 10^7$ W/m^{2}K.

These two results - high superheating temperatures on smooth silicon wafers and the existence of a finite temperature jump between substrate and liquid - severely question the predictions of the existing SLC models (Lu, 1999; Wu, 2000). Both models strongly rely on several assumptions whose validity must be challenged against the experimental findings. In reference (Lu, 1999), e.g., the bubble growth velocities are computed using the oversimplified formula (14) and our results on bubble growth and superheating (Yavas, 1997 a) obtained on rough metal films are transferred to the silicon-liquid system, which is clearly invalid. Additionally both groups obtain the temperature of the liquid layer by a numerical computation assuming a perfect heat flow from the substrate into the liquid. This is invalid as well, which can be illustrated by the results of numerical computations we carried out on our own. Inserting not only the correct heat transfer coefficient but also temperature dependent material properties for the liquid and the substrate, our calculations based on the 1D heat equation (Bischof, 1996; Dobler, 2002) show temperature differences of liquid and substrate of about 100 K!

Besides the above comments some even more fundamental questions remain. Both models rely on the assumption that shock or pressure waves produced by the growing bubbles/ growing vapor layer exert the removal force on the particles. However, the details of this scenario are still open.

Is it really a vapor layer, as suggested by Wu, which is responsible for the removal force? Or are a few gas bubbles enough, as it is incorporated in the Lu model? Where are these gas bubbles nucleated - all over the bare silicon surface as Lu suggests? In this case as well as in the assumption of a vapor layer - why is it possible to perform "steam" cleaning just by adsorbing a small amount of liquid at the particle-surface interstice via capillary condensation (Fourrier, 2001; Mosbacher, 2002 a)? Here the above approaches to explain the cleaning mechanism fail. And - related to the last comment - what is the influence of the film parameters, especially the film thickness on the cleaning process and its theoretical description. Can the results obtained so far on bulk liquid be transferred on liquid films? From our point of view these open questions clearly show that still a lot of experimental work on SLC and its fundamentals has to be carried out and that all models proposed so far should be regarded as a first approximate approach.

Even the experimental results pose questions: The threshold for bubble nucleation by a single Nd: YAG-pulse measured for the silicon/bulk water system is about 80 mJ/cm^2. When compared to the universal cleaning threshold in SLC of 110 mJ/cm^2 there is a difference, which will be the subject of further investigations.

Of course all the results on basic processes like laser induced bubble nucleation are very interesting on their own, however, in the end SLC is intended to become a cleaning method applicable in the semiconductor industry. Considering the unsolved questions raised above one might doubt that there is any hope of developing SLC to be a reliable tool in the near future.

We do not agree and believe, that SLC is extremely promising for the process parameters we used. The existence of the universal cleaning threshold at a laser fluence a factor of three below the melting threshold of bare silicon shows, that it is possible to choose one fixed laser fluence well below the substrate melting fluence to remove particles of a broad size range. Moreover, this is also a considerable advantage in terms of possible applications over DLC, where the threshold is size and material dependent, the removal efficiencies are lower and particles smaller than 100 nm could not be removed (Zapka, 1991; Mosbacher, 1998, 2002 a, b). Thus SLC allows particle removal at lower and therefore safer laser fluences, which is essential for avoiding surface modifications.

In the field of liquid parameters still a lot of systematic research has to be carried out, however, our results with removal efficiencies of 90% for particles 60 nm in diameter show that efficient cleaning is already possible. Thus the basic prerequisite for a real world application of SLC is fulfilled: the process works and suitable process parameters are known. Nevertheless it is not clear whether these parameters are the optimum ones. This optimization will be triggered by the growing knowledge on the fundamentals of SLC and will, together with the investigation of different substrate materials, provide the field of future research on this exciting topic.

Acknowledgements

This review is result of our scientific collaboration between the University of Konstanz, Germany and Data Storage Institute, Singapore. We thank Dr. Nikita Arnold (Johannes-Kepler-University, Linz, Austria), Dr. Bernd-Uwe Runge, Johannes Graf and Florian Lang (all University of Konstanz) for constructive discussions of the findings of our experiments. Financial support by EU TMR project "Laser Cleaning" (no ERBFMRXCT98 0188) and the Konstanz Center for Modern Optics is gratefully acknowledged. Wacker Siltronic supplied the industrial silicon wafers.

References

Assendel'ft E. Y., Beklemyshev V. I., Makhonin I. I., Petrov Yu. N., Prokhorov A. M., Pustovoy V. I., *Optoacoustic effect on the desorption of microscopic particles from a solid surface into a liquid.* Sov. Tech. Phys. Lett., vol. **14**(6), pp. 444-445 (1988a)

Assendel'ft E. Y., Beklemyshev V. I., Makhonin I. I., Petrov Yu. N., Prokhorov A. M., Pustovoy V. I., *Photodesorption of microscopic particles from a semiconductor surfaces into a liquid*, Sov. Tech. Phys. Lett., vol. **14**(8), pp. 650-654 (1988b)

Avesisian C. T., *The homogeneous nucleation limit of liquids*, J. Phys. Chem. Ref. Data, **14**, pp. 695-729 (1985)

Bäuerle D., *Laser Processing and Chemistry*, 3^{d} ed., (Springer-Verlag, Berlin, 2000)

Beklemyshev V. I., Makarov V. V., Makhonin I. I., Petrov Yu. N., Prokhorov A. M., Pustovoy V. I., *Photodesorption of metal ions in a semiconductor-water system*, JETP Letters, vol. **46**, pp. 347-350 (1987)

Bischof J., *Metallische Dünnfilmschmelzen nach Pulslaserbestrahlung: Phasenumwandlungen und Instabilitäten*, PhD thesis (Universität Konstanz, 1996)

Boneberg J., *Dynamische Verfestigung von Halbleiterschichten nach ns-Laser-Annealing*, PhD thesis (University of Konstanz, 1993)

Born M., Wolf E., *Principles of Optics*, 7-th Edition, (Cambridge University Press, 1999)

Boughaba S., Wu X., Sacher E., Meunier M., *Liquid explosive evaporative removal of submicron particles from hydrophilic oxidized silicon surface,* J. Adhesion, **61**, pp. 293-307 (1997)

Carey V. P., *Liquid-Vapor Phase Change Phenomena* (Hemisphere Publishing Cooperation, Washington, 1992)

Cerbe G., Hoffmann H. J., *Einführung in die Thermodynamik* (Carl Hanser Verlag, München, 1999)

Cole R., *Boiling nucleation*, Adv. Heat Transfer, **10**, pp. 85-167 (1974)

Debenedetti P. G., *Metastable liquids: Concepts and Principles* (Princeton University Press, Princeton, New Jersey, 1996)

DeJule R., *Trends in wafer cleaning*. Semiconductor International, **8**, pp. 65-68 (1998)

Dobler V., *Dynamische Messungen zum Laser Cleaning*, PhD thesis (University of Konstanz, 2002)

Engelsberg A. C., In: *Surface Chemical Cleaning and Passivation for Semiconductor Processing*, Ed. by G. Higashi, E. Irene, T. Ohmi, Proc. MRS, vol. **315**, p. 255 (1993)

Fourrier T., Schrems G., Mühlberger T., Arnold N., Heitz J., Bäuerle D., Mosbacher M., Boneberg J., Leiderer P., *Laser cleaning of polymer surfaces*, Appl. Phys. A **72**, pp. 1-6 (2001)

Grigull U., Starub J., Schiebner P., *Steam tables in SI-units* (Springer-Verlag, Berlin, Heidelberg, 1984)

Halfpenny D. R., Kane D. M., *A quantitative analysis of single pulse UV dry laser cleaning*, J. Appl. Phys., **86**, pp. 6641-6646 (1999)

Hattori T., Solid State Tech., **8**, July 1990

Heim L.-O., Blum J., Preuss M., Butt H.-J., *Adhesion and friction forces between spherical micrometer-sized particles*, Phys. Rev. Lett, **83** (16), pp. 3328-3331 (1999)

Herminghaus S., Leiderer P., *Surface plasmon enhanced transient thermoreflectance*, Appl. Phys. A **51**, p. 350 (1990)

Hoenig S. A., In: *Particles on Surfaces: Detection, Adhesion and Removal*, Ed. by K. L. Mittal, vol. **1**, p. 3 (Plenum Press, New York, NY, 1988)

Héroux J. B., Boughaba S., Ressejac I., Sacher E., Meunier M., *CO_2 laser-assisted removal of submicron particles from solid surfaces*, J. Appl. Phys., **79**, pp. 2857-2862 (1996)

Imen K., Lee S. J., Allen S. D., *Laser-assisted micron scale particle removal*, Appl. Phys. Lett., **58**, pp. 203-205 (1991)

Johnson K. L., Kendall K., Roberts A. D., *Surface energy and the contact of elastic solids*, Proc. R. Soc. Lond. A, **324**, pp. 301-313 (1971)

Johnson P. B., Christy R. W., Phys. Rev. B, **6**, p. 4370 (1972)

Kagen Y., *The kinetics of boiling of a pure liquid*, Russian J. Phys. Chem., 34, pp. 42-46 (1960)

Kameda M., Yamada M., Matsumoto Y., *Nonlinear oscillation of a small gas bubble*, In: Advances in nolinear acoustics, Ed. by H. Hobaek (World Scientific, 1993)

Katz J. L., Blander M., *Condensation and boiling: Correction to homogeneous nucleation theory for nonideal gases*, J. Colloid Interface Sci., **42**, pp. 496-502 (1973)

Kelley J. D., Hovis F. E., *A thermal detachment mechsnism for particle removal from surfaces by pulsed laser irradiation*, Microelectronic Engineering, **20**, 159-170 (1993)

Kim D., Park H. K., Grigoropoulos C. P., *National Heat Trans. Conf. Proc.*, **4**, p. 69 (1996)

Kohli R., In: *Particles on Surfaces*, Ed. by K. L. Mittal (VSP Publishing, 2002)

Kretschmann E., Raether H., *Z. Naturf. A*, **23**, p. 2135 (1968)

Kurz H., Lompre L. A., Liu J. M., J. de Physique, **10** (44), C5-23 (1983)

Leiderer P., Boneberg J., Dobler V., Mosbacher M., Münzer H.-J., Chaoui N., Siegel J., Solis J., Afonso C. N., Fourrier T., Schrems G., Bäuerle D., *Laser-induced particle removal from silicon wafers*, Proc. SPIE, vol. **4065**, pp. 249-259 (2000)

Leiderer P., Boneberg J., Mosbacher M., Schilling A., Yavas O., *Laser cleaning of silicon surfaces*, Proc. SPIE, vol. **3274**, pp. 68-77 (1998)

Lowndes D. H., Wood R. F., Westbrook R. D., *Pulsed neodymium: yttrium aluminum garnet laser (532 nm) melting of crystalline silicon: Experiment and theory*, Appl. Phys. Lett., **43** (3), pp. 258-260 (1983)

Lu Y. F., Zhang Y., Wan Y. H., Song W. D., *Laser cleaning of silicon surfaces with deposition of different liquid films*, Appl. Surf. Sci., **138-139**, pp. 140-144 (1999)

Lu Y. F., Zheng Y. W., Song W. D., *Laser induced removal of spherical particles from silicon wafers*, J. Appl. Phys. **87**, pp. 1534-1539 (2000a)

Lu Y. F., Zhang L., Song W. D., Zheng Y. W., Luk'yanchuk B. S., *Laser writing of sub-wavelength structure on silicon (100) surfaces with particle enhanced optical irradiation,* JETP Letters, vol. **72**, pp. 457-459 (2000 b)

Lu Y. F., Zhang L., Song W. D., Zheng Y.W., Luk`yanchuk B. S., *Particle-Enhanced Near-Field Optical Effect and Laser Writing for Nanostructure Fabrication,* Proc. SPIE, vol. **4426**, pp.143-145 (2002)

Luk'yanchuk B. S., Zheng Y. W., Lu Y. F., *Laser Cleaning of the surface: Optical resonance and near-field effects*, Proc. SPIE, vol. **4065**, pp. 576-587 (2000)

Luk'yanchuk B. S., Zheng Y. W., Lu Y. F., *Basic physical problems related to dry laser cleaning*, RIKEN Review No. 43, pp. 47-65 (2002)

Matsumoto Y., Takemura F., *Numerical analysis on a bubble motion with full equations*, In: M. F. Hamilton, D. T. Blackstock (Eds.), Frontiers of nonlinear acoustics: Proceedings of the 12th ISNA, (Elsevier Science Publishers, London, 1990)

Matsumoto Y., Kameda M., Takemura F., Ohashi H., *Numerical simulations of pressure wave behaviors in bubbly liquids*, In: G. Matsui, A. Serizawa, Y. Tsuji (Eds.), Proceedings of the International Conference on Multiphase Flows, pp. 327-330 (1991 a)

Matsumoto Y., Takemura F., Kameda M., *Bubble motion in an oscillatory pressure field*, In: H. Kato, O. Furuya (Eds.), "Cavitation' 91", The first ASME-JSME Fluid Engineering Conference, vol. **116**, pp. 33-38 (The American Society of Mechanical Engineers, 1991 b)

Mosbacher M., Dobler V., Boneberg J., Leiderer P., CPD 2.12. In CLEO/Europe - EQEC, (1998)

Mosbacher M., Chaoui N., Siegel J., Dobler V., Solis J., Boneberg J., Afonso C. N., Leiderer P., *A comparison of ns and ps steam laser cleaning of Si surfaces*, Appl. Phys. A, **69** [Suppl.], pp. 331-334, (1999)

Mosbacher M., Dobler V., Boneberg J., Leiderer P., *Universal threshold for the steam laser cleaning of submicron spherical particles from silicon,* Appl. Phys. A, **70**, pp. 669-672 (2000)

Mosbacher M., Münzer H.-J., Zimmermann J., Solis J., Boneberg J., Leiderer P., *Optical field enhancement effects in laser-assisted particle removal*, Appl. Phys. A, **72**, (2001)

Mosbacher M., Bertsch M., Münzer H.-J., Dobler V., Runge B.-U., Bäuerle D., Boneberg J., Leiderer P., *Laser cleaning of silicon wafers - mechanisms and efficiencies*, Proc. SPIE, vol. **4426**, pp. 308-314 (2002 a)

Mosbacher M., Münzer H.-J., Bertsch M., Dobler V., Chaoui N., Siegel J., Oltra R., Bäuerle D., Boneberg J., Leiderer P., *Laser assisted particle removal from silicon wafers*, In: "Particles on Surfaces", Ed. by K. L. Mittal (VSP Publishing, 2002 b)

Münzer H.-J., Mosbacher M., Bertsch M., Dubbers O., Burmeister F., Pack A., Wannemacher R., Runge B.-U., Bäuerle D., Boneberg J., Leiderer P., *Optical near field effects in surface nanostructuring and laser cleaning, Proc. SPIE*, vol. **4426**, pp. 180-183 (2002)

Park H. K., Grigoropoulos C. P., Leung W. P., Tam A. C., *A practical excimer laser-based cleaning tool for removal of surface contaminants*, IEEE Transactions on Components, Packaging, and Manufacturing Technology A, **17** (4), pp. 631-643 (1994 a)

Park H. K., Grigoropoulos C. P., Poon C. C., Tam A. C., Yavas O., Leiderer P., *Optical probing of the temperature and pressure transients at a liquid/solid interface due to pulsed laser-induced vaporization*, Proc. SPIE, vol. **2498**, pp. 32-40 (1994 b)

Park H. K., Grigoropoulos C. P., Poon C. C., Tam A. C., *Optical probing of the temperature transients during pulsed laser-induced boiling of liquids*, Appl. Phys. Lett., **68**, pp. 596-598 (1996 a)

Park H. K., Kim D., Grigoropoulos C. P., *Pressure generation and measurement in the rapid vaporization of water on a pulsed-laser-heated substrate*, J. Appl. Phys., **80**, pp. 4072-4081 (1996 b)

Schiebener P., Straub J., Levelt-Sengers J. M. H., Gallagher J. S., J. Phys. Chem. Ref. Data, **19**, p. 677 (1990)

Schilling A., Yavas O., Bischof J., Boneberg J., Leiderer P., *Absolute pressure measurements on a nanosecond time scale using surface plasmons*, Appl. Phys. Lett., **69**, pp. 4159-4161 (1996)

Sematech Inc., *International technology roadmap for semiconductors* (Technical report, 2000), http://public.itrs.net/

She M., Kim D. S., Grigoropoulos C. P., *Liquid-assisted pulsed laser cleaning using near-infrared and ultraviolet radiation*, J. Appl. Phys., **86** (11), pp. 6519-6524 (1999)

Skripov V. P., *Metastable liquids* (Halsted Press, John Wiley & Sons, New York, 1974)

Skripov V. P., Sinitsyn E. N., Pavlov P. A., Ermakov G. V., Muratov G. N., Bulanov N. V., Badakov V. G., *Thermophysical properties of liquids in the metastable (superheated) state* (Gordon and Breach, New York, 1988)

Swartz E. T., Pohl R. O., *Thermal boundary resistance*, Rev. Mod. Phys., **61**, pp. 605-668 (1989)

Tam A. C., Leung W. P., Zapka W., Ziemlich W., *Laser-cleaning techniques for removal of surface particulates*, J. Appl. Phys., **71** (7), pp. 3515-3523 (1992)

Tam A. C., Leung W. P., Zapka W., In: *Particles on surfaces*, Ed. by K. L. Mittal, pp. 405-418 (Marcel Dekker, New York, 1995)

Tam A. C., Park H. K., Grigoropoulos C. P., *Laser cleaning of surface contaminants*, Appl. Surf. Sci., **127-129**, pp. 721-725 (1998)

Tong L. S., Tang Y. S., *Boiling heat transfer and two-phase flow* (Taylor & Francis, Washington, 1997)

Vereecke G., Röhr E., Heyns M. M., *Laser-assisted removal of particles on silicon wafers*, J. Appl. Phys., **85** (7), pp. 3837-3843 (1999)

Wagner W., Cooper J. R., Dittmer A., Kijima, Kretzschmar H.-J., Kruse A., Mares R., Oguchi K., Sato H., Stöcker I., Sifner O., Takaishi Y., Tanishita I., Trübach J., Willkommen T., *The IAPWS industrial formulation 1997 for the thermodynamic properties of water and steam*, ASME Trans., **122**, pp. 150-182 (2000)

Wu X., Sacher E., Meunier M., *The modeling of excimer laser particle removal from hydrophilic silicon surfaces*, J. Appl. Phys., **87** (8), pp. 3618-3627 (2000)

Yavas O., Leiderer P., Park H. K., Grigoropoulos C. P., Poon C. C., Leung W. P., Do N., Tam A. C., *Optical reflectance and scattering studies of nucleation and growth of bubbles at a liquid-solid interface induced by pulsed laser heating*, Phys. Rev. Lett., **70** (12), pp. 1830-1833 (1993)

Yavas O., *Laserinduzierte Gasblasennukleation*, PhD thesis (University of Konstanz, 1994 a)

Yavas O., Leiderer P., Park H. K., Grigoropoulos C. P., Poon C. C., Leung W. P., Do N., Tam A. C., *Optical and acoustic study of nucleation and growth of bubbles at a liquid-solid interface induced by nanosecond-pulsed-laser heating*, Appl. Phys. A, **58**, pp. 407-415 (1994 b)

Yavas O., Leiderer P., Park H. K., Grigoropoulos C. P., Poon C. C., Tam. A.C. , *Enhanced acoustic cavitation following laser-induced bubble formation: long-term memory effect*, Phys. Rev. Lett., **72**, pp. 2021-2024 (1994 c)

Yavas O., Schilling A., Bischof J., Boneberg J., Leiderer P., *Bubble nucleation and pressure generation during laser cleaning of surfaces*, Appl. Phys. A, **64**, pp. 331-339 (1997 a)

Yavas O., Schilling A., Bischof J., Boneberg J., Leiderer P., *Study of nucleation process during laser cleaning of surfaces*, Laser Physics, **7**, pp. 343-348 (1997 b)

Zapka W., Ziemlich W., Tam A.C., *Efficient pulsed laser removal of 0.2 μm sized particles from a solid surface*, Appl. Phys. Lett., **58** (20), pp. 2217-2219 (1991)

Chapter 7

PHYSICAL MECHANISMS OF LASER CLEANING

V. P. Veiko, E. A. Shakhno

Physical mechanisms of laser cleaning of solid surface from soiling particles and films are discussed. Dry laser cleaning of surface from particles is considered to be a result of inertial force, appearing due to thermal expansion of absorbing particles and/or substrate (shaking-off mechanism). Steam laser cleaning is considered for the cases of absorbing particles and absorbing substrate. In the latter case, peculiarities of bubble formation in the liquid layer under the conditions of ns laser action are analyzed. Laser cleaning of surfaces from films by shaking-off and buckling mechanisms (due to thermal tension) is considered. Conditions of the action of each of them are defined. Other physical mechanisms, blasting and evaporation, are briefly discussed.

Keywords: particle, film, substrate, bubble, shaking-off, buckling, thermal expansion

PACS: 42.62.Cf, 81.65.Cf, 68.35.Np,81.65.-b, 85.40.-e, 81.07.Wx

1. Introduction

Laser cleaning of solid surface attracts attention of many investigators, e.g. Lu (1998); Boneberg (1999); Luk'yanchuk (2001), Mosbacher (1999). The problem of cleaning of the surface from different admixtures and contaminants in the form of particles and films of different substances arises in many processes: in industry, construction, art restoration, medicine etc. The typical example is from microelectronics, where cleaning of the surface from particles of nanometer sizes becomes actual due to continuous increasing of integration degree. Another example is cleaning of oil–paintings, icons and other art works from soiling films, paint layers of later versions, etc., and their return to the original condition. We don't aim here at a description of all possible kinds of soiling where methods of cleaning are useful, but we have to mention investigations of A. Tam devoted to laser cleaning of memory devices (Tam, 1992), sculptures and buildings from graffiti, etc.

The field of laser cleaning applications is continuously extending. Some applications are based not only on sufficient power of laser radiation, but also on its ability to penetrate into the scaled vacuum volumes and interior of complicated constructions (for instance, cleaning of pipes from corrosion without their disjoining, cleaning of metal surfaces from radioactive soiling, etc).

This brief list of real and potential applications of laser cleaning shows that they can be divided into three groups, from the point of view of underlying physical processes:

- cleaning of the surface from soiling particles,
- cleaning of the surface from heterogeneous films,
- cleaning of the surface from films of its own material, modified during the operation: oxide films, radioactive layers etc.

Physical mechanisms of the first and second groups of cleaning processes are considered in this paper. The third group is mainly based, for our opinion, on laser evaporation of film material, and its physical processes are known well enough. Laser cleaning is used for the removal of particles of different sizes (and materials) as well as for the removal of films of soiling substances from solid surfaces. There are dry and steam laser cleaning methods. They are based on pulse laser heating of solid surface, dry and in the presence of a liquid layer on it, correspondingly. The laser cleaning methods are free from shortcomings, which are peculiar to other methods, such as technological complication, necessity of washing and drying, ecological concerns. In the cases, when dry laser cleaning is not effective enough or cannot be used for other reasons, steam laser cleaning is used.

2. Laser cleaning of the solid surface from particles

The physical mechanism of dry laser cleaning is supposed to be connected with the fast thermal expansion of particles and/or substrate under the action of nanosecond laser pulses. Thermal expansion creates the cleaning force overcoming the adhesion force, which binds the particles and the substrate surface (Lu, 1998).

Our investigations of sub–evaporation regimes of the laser induced forward transfer (LIFT) (Veyko, 1997) allow us to use the physical models of solid film shaking–off from the underlying substrate for the description of the dry laser cleaning process.

The advantages of steam laser cleaning are a low threshold and the absence of damage of the surface by cleaning. Therefore steam laser cleaning has been intensively investigated recently. Mainly, there are experimental investigations of threshold characteristics of cleaning (for instance Lu et al., 1998; Mosbacher et al., 1999): superheating of the absorbing substrate surface with respect to the boiling point of liquid, which is necessary for boiling beginning, and threshold intensity, which depends on superheating value. Experimental investigations reveal the most important phenomena and quantitatively define the parameters of the process, but they cannot make a sufficiently complete physical picture of the process, that is necessary for the choice of optimal cleaning regime and prognosis of the results of cleaning.

2.1. Dry laser cleaning

2.1.1. The cleaning force

During laser action, the substrate and/or particles absorb radiation. Substrate or particles heating can be caused (depending on their optical properties) by direct light absorption, by heat transfer from the absorbing substrate to the particles or from the absorbing particles to the substrate, see in Fig.1. It should be noted that in the last case heating of the substrate zone under the particle is not effective because of radial thermal diffusion. If the particles are transparent or semi–transparent, the substrate absorption depends on additional optical effects: focusing action of a particle, multiple–reflection, size dependent effects etc.

Due to particle and/or substrate heating the particle mass center displacement takes place. In the following analysis we shall consider particles and substrate incompressible*. It is a good approximation for small particles. For big particles, the deformation has to be taken into consideration (see Luk'yanchuk, 2001). By accepted approximation, the particle mass center displacement is $\delta = \delta_S + \delta_p$:

* This assumption of incompressibility strongly simplified consideration but it is not clear with which size of the particles one can use this simplification. In any case this 1D model has the general problem – it yields typically too big threshold fluences (comment of Editor).

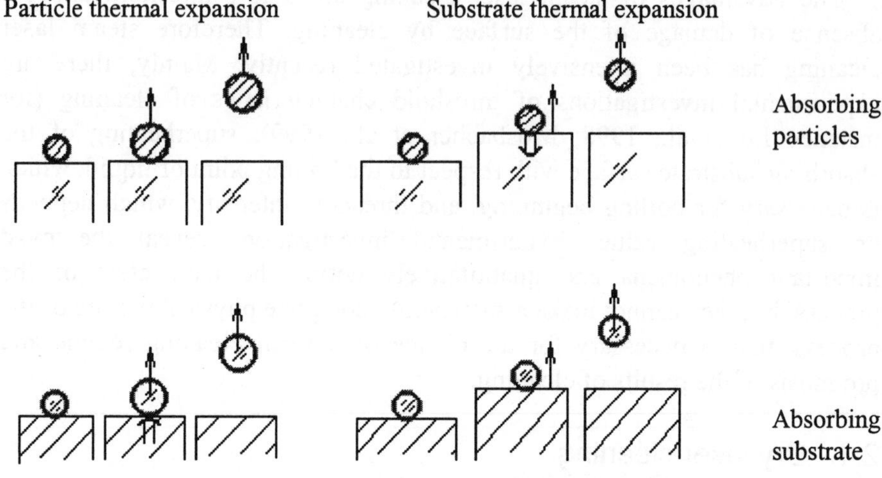

Fig. 1. Illustration of the physical mechanisms of dry laser cleaning.

$$\delta_{S(p)} = k_{S(p)} \int_{0}^{h_{S(p)}} \alpha_{S(p)} T_{S(p)} dy_{S(p)} , \quad (1)$$

δ_S and δ_p are partial displacements caused by substrate (*s*) and particle (*p*) expansion correspondingly, α_S, α_p, h_s, h_p, T_S, T_p are linear coefficient of thermal expansion, thickness and temperature referred to the initial value of the substrate and particle correspondingly. Coordinate axes y_S and y_p are directed from the substrate and particle surfaces into them normally to the substrate surface. The coefficient $k_s = 1$. The coefficient k_p defines the connection between thermal enlargement of the whole particle and its mass center displacement. It changes from 0 to 1. Thermally small particles (height $h_p \le \sqrt{a_p \tau}$, a_p is particle thermal diffusivity, τ is pulse duration), i.e., up to micron size metal particles or smaller dielectric particles, are heated almost uniformly over their volume and $k_p \approx 0.5$. For thermally large absorbing particles, thermal energy concentrates in their upper part and mass center displacement is almost absent, i.e. $k_p \approx 0$. By thermally large transparent particles heated from the substrate, thermal energy concentrates in their lower part and $k_p \approx 1$. Value of k_p is also influenced by the geometry of the particle.

Acceleration of the particle mass center bound with its thermal displacement causes the inertial force $F_0 = m\, d^2\delta/dt^2$, m is the particle mass. The value of this force can be obtained by comparison of equations (1) with energy balance equations:

$$A_{S(p)} q dt = \int_0^{h_{S(p)}} \rho_{S(p)} c_{S(p)} dT_{S(p)} dy_{S(p)}, \tag{2}$$

where q is incident intensity, A_S, A_p, ρ_S, ρ_p, c_S, c_p are effective absorption coefficient, density and specific heat of the substrate and particle. Effective absorption coefficient of some object is defined here as ratio of power received by the object due to radiation absorption and also due to thermal transfer, to power of the incident radiation at it.

The obtained value of inertial force is:

$$F_0 = m \frac{dq}{dt}\left(\frac{\alpha_S A_S}{\rho_S c_S} + k_p \frac{\alpha_p A_p}{\rho_p c_p} \right). \tag{3}$$

During the pulse front edge, $dq/dt > 0$ and the force F_0 ($F_0 > 0$) presses the particle to the substrate surface. During the back edge, $dq/dt < 0$ and the force $|F_0|$ ($F_0 < 0$) acts as a cleaning force, shaking–off the particle from the substrate surface. Particle shaking–off occurs, if the cleaning force exceeds the adhesion force F_a.

Therefore laser fluence value, which is necessary for particle shaking–off (for triangular pulse) is

$$\varepsilon > \frac{1}{4} \frac{F_a}{m} \frac{b\tau^2}{\dfrac{\alpha_S A_S}{\rho_S c_S} + k_p \dfrac{\alpha_p A_p}{\rho_p c_p}}, \tag{4}$$

where $b = 2\tau_1/\tau$ is a coefficient defining the temporal shape of the pulse, τ is pulse duration, τ_1 is pulse back edge duration.

2.1.2. Surface cleaning condition

The cleaning force disappears after the pulse, but the adhesion force remains. The shaken–off particle will not return to the substrate, if its kinetic energy (after finishing its and substrate's expansion) exceeds the work necessary for the particle moving to infinity against adhesion force:

$$\frac{mu_*^2}{2} > \int_{z*}^{\infty} F_a dz\,, \tag{5}$$

where z is the distance between the particle lower surface and the substrate surface, u is particle mass center velocity, z_* and u_* are z and u values in the moment of the particle and surface expansion finishing.

Let us consider some peculiar cases of dry cleaning.

a) Absorbing particles and/or absorbing substrate expansion

Particle motion from the substrate starts at the moment of pulse back edge beginning $t = 0$ (when dq/dt becomes negative). The equation of motion is:

$$m\frac{d^2x}{dt^2} = -F_a(z), \tag{6}$$

where x is the distance between the particle mass center and the initial position of the substrate surface. The correlation between coordinates x and z is:

$$\frac{dx}{dt} = \frac{dz}{dt} + \frac{d\delta}{dt}. \tag{7}$$

The initial conditions: $z = z_0$, $dz/dt = 0$ at $t = 0$. The initial distance between the particle and the substrate z_0 is defined (see Lu et al., 1998) as atomic separation.

Combining equation (6) with cleaning condition (5), one obtains the value of the cleaning force $|F_0|$, which is necessary for complete particle removal. For large enough values of $F_a(z_0)/m$ (particle size is $d << 10$ μm) and triangular pulse it is:

$$|F_0| > F_a(z_0) + 2\frac{z_0}{\bar{n}-1}\frac{m}{\tau_1^2}. \qquad (8)$$

The number $\bar{n}$ depends on the kind of adhesion force: $F_a \sim 1/z^{\bar{n}}$. For Van der Waals forces $\bar{n} = 2-3$ (Lu et al., 1998).

Correspondingly, the cleaning condition for laser fluence ε is:

$$\varepsilon > \frac{1}{4}\frac{F_a(z_0)}{m}\frac{b\tau^2}{\frac{\alpha_S A_S}{\rho_S c_S} + k_p\frac{\alpha_p A_p}{\rho_p c_p}} + 2\frac{\frac{z_0}{(\bar{n}-1)b}}{\frac{\alpha_S A_S}{\rho_S c_S} + k_p\frac{\alpha_p A_p}{\rho_p c_p}}. \qquad (9)$$

The second term of correlation (9) is usually relatively small, except small values of $F_a(z_0)/m$ (large particles and weak adhesion). Therefore cleaning conditions (4) and (9) are usually approximately equal.

b) Transparent particles expansion

If the particles are transparent and the substrate is an absorbing one, but the thermal expansion coefficient of the particle is much larger than that of the substrate ($\alpha_p >> \alpha_S$), the cleaning force is caused by particle expansion due to its heating by the substrate. In this case substrate expansion is almost absent. Particle expansion is stopped due to thermal contact interruption by particle separation from the substrate. Therefore the values u_* and z_* in the cleaning condition (5) are defined in the moment $t = 0$. The calculated fluence, which is necessary for complete cleaning is:

$$\varepsilon > \frac{\rho_p c_p}{k_p \alpha_p A_p}\tau\sqrt{\frac{z_0}{2(\bar{n}-1)}\frac{F_a(z_0)}{m}}. \qquad (10)$$

A comparison of cleaning conditions (4) and (10) shows that the value of ε given by (4) is usually significantly larger than the value given by (10), except the cases of small $F_a(z_0)/m$.

2.1.3. Dry laser cleaning in the multipulse regime

The exact value of adhesion force is difficult to define due to indefinites of particle shape, its real contact area with the substrate and other factors. Therefore we shall use the experimentally defined value of the adhesion force per unit area of the particles section: $P = \rho_p d F_a / m$ (d is particle size). The P value depends on material properties, particle shape and a number of other factors, and is about $\sim 10^6$ N/m^2. Analyzing the cleaning conditions (Veiko, 1997), one can see that the calculated values of P are much less than the real values. In practice, when the multiple pulse regime is used, the number of pulses is about $N \approx 100$ (Lu, 1998). The cleaning conditions can be evaluated in this case using the following consideration. The partial particle shaking–off, i.e. decreasing of its contact area with the substrate, occurs in each pulse. Adhesion force decreases each time by some value ΔF_a. The partial relaxation occurs between the pulses. So, adhesion force decreases by value $(1-\bar{\beta})N\Delta F_a$ after the N–th pulse ($\bar{\beta} < 1$ is relaxation coefficient, it supposedly increases with decreasing laser repetition rate f, increasing particle size d and depends on the materials of the substrate and particles and particles shape). If $(1-\bar{\beta})N\Delta F_a = F_a$, the particle is removed from the substrate after the N–th pulse*. From (4) one can obtain values of P and d for which cleaning takes place:

$$\frac{P}{d} < 4(1-\bar{\beta})\frac{N\rho_p}{b\tau^2}\left(\frac{\alpha_S A_S}{\rho_S c_S} + k_p \frac{\alpha_p A_p}{\rho_p c_p}\right)\varepsilon. \tag{11}$$

---

* This idea, in fact, assumes the reduction of adhesion (like bond-breaking). Another explanation of the multiple pulse effects is related to the spread of adhesion conditions, particle size distributions and variation of the laser pulse parameters (comment of Editor).

2.1.4. Discussion

The worked out physical mechanisms of dry and steam laser cleaning of the surface show the following features.

Threshold fluence depends on the physical properties of the particle and substrate material, particle size and temporal shape of the pulse. Dry cleaning effectiveness depends on optical and thermal properties of the substrate, particle and the coefficients of their thermal expansion. In particular, the effective cleaning of the substrate from thermally large particles is possible, if radiation is absorbed by the substrate. For thermally small particles radiation absorption by the substrate, as well as by the particles can be used. If radiation is absorbed by small particles ($\leq 0.1\mu$m), cleaning effectiveness increases by particle size increasing; by larger sizes of the particles it decreases, since the thermal energy received by the lower part of the particle is insufficient.

2.2. Steam laser cleaning

In steam laser cleaning the substrate surface is covered by a thin layer of liquid. The particles and the liquid layer are removed from the substrate surface by the action of laser pulse. The physical mechanism of the cleaning process is defined by the type of heat transfer to the evaporating liquid (Fig.2).

2.2.1. Absorbing particles at the transparent substrate

The upper part of the particle heats mainly via light absorption. Liquid above the particle's top also heats and evaporates due to heat transfer from the heated particle top, its evaporation is like micro–explosion. In this way, the particle top becomes free from the liquid. Then the particle heats and a thin layer of liquid around the particle heats and evaporates. The resulting vapor pressure pushes off the liquid from the particle surface near the evaporation region. As a result, the heat transfer from the heated particle to the liquid decreases and evaporation decreases or ends. Therefore effective evaporation occurs only in the narrow region at the particle surface near the isotherm of liquid boiling temperature T_b (which depends on the conditions of evaporation, particularly on vapor pressure). So, the vapor channel forms

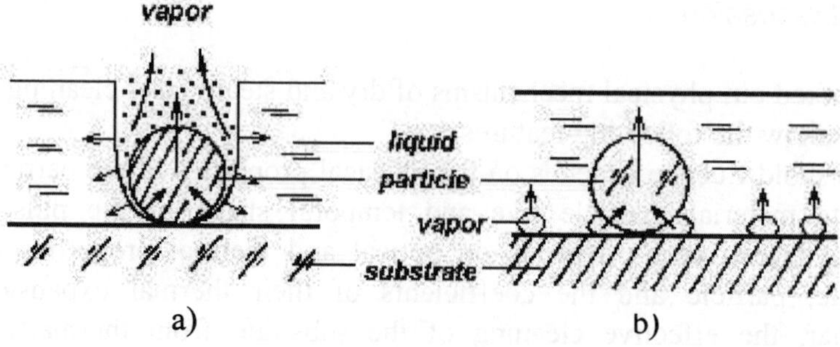

Fig. 2. Illustration of the physical mechanisms of steam laser cleaning: a - absorbing particle, b - absorbing substrate.

around the particle surface. It broadens up and vapor pressure is maximal in its lower part (Fig.2a).

The action of vapor pressure on the particle depends on the position of the effective evaporation region. When it is situated near the particle's upper part, the vertical projection of the force of vapor pressure F_n (normal to the substrate surface) is directed down, to the substrate surface, and presses the particle to it. During vapor channel development, the effective evaporation region transfers to the lower part of the particle. By this, F_n becomes directed up, and if $F_n > F_a$, it pushes the particle up from the substrate surface.

Threshold fluence ε_{th} of steam laser cleaning can be estimated as the fluence necessary for heating of the whole particle to the temperature T_b (neglecting latent heat of evaporation and thermal losses due to thermal conductivity):

$$\varepsilon_{th} = \frac{\rho_p c_p h_p (T_b - T_{in})}{\overline{A}_p}, \tag{12}$$

where $\overline{A}_p$ is average absorption coefficient of the particle which includes the influence of the angle of incidence, T_{in} is the initial temperature.

The shape of the particles and their position relatively to the substrate surface considerably influence the effectiveness and possibility of cleaning. Naturally, removal of the particles is difficult, if its maximum cross–section is near the substrate surface (like the hemisphere lying at the plane surface). Then, the force F_n is always directed down. The presence of some particles of such kind explains the observed effect of incomplete cleaning (Lu, 1998).

2.2.2. Transparent particles at the absorbing substrate

In laser cleaning, the surface layer of absorbing substrate is heated by irradiation: the adjacent layer of liquid is also heated due to thermal flow from the substrate, thus the conditions of liquid boiling are created (Fig. 2b). Experiments (Yavas, 1994) show, that liquid boiling and particle removal take place if the substrate surface and the liquid layer are heated to the temperature, which is higher than the boiling temperature at normal pressure, that is the evidence of bubble mechanism of cleaning.

Superheating is necessary for the formation and growth of supercritical bubbles. Therefore the regime of laser cleaning is defined by the condition of formation of bubbles of size larger than some critical value $h_{\min}$, which is defined by equation

$$P_T - P_0 = 2\sigma H_{th}, \tag{13}$$

where P_T is vapor pressure in the bubble, P_0 is ambient pressure (atmospheric pressure), σ is surface tension of liquid, H is mean curvature of bubble surface, which depends on its size and form, H_{th} is H value for a bubble of critical size $h_{\min}$. In particular, for a spherical bubble $H = \frac{1}{r}$, r is its radius.

Bubble formation has some peculiarities in nanosecond regime of substrate heating. First of all, it is large temperature gradients in the liquid near the substrate surface. As a result surface tension is different in different parts of the vapor bubble surface. Surplus pressure in the bubble $P_T - P_0$ (left part of the equation (13)) does not depend on coordinate, since vapor pressure in the bubble is nearly constant in its volume under considered conditions. Therefore, the product σH_{th} is also constant along the bubble surface. But surface tension σ is a function of temperature (and consequently of coordinate), therefore the form of the bubble is not spherical. This fact complicates the definition of the critical (minimum) size of a stable bubble and correspondingly the threshold characteristics of steam laser cleaning.

On the other hand, the bubble size (its height) cannot exceed the liquid layer thickness h_{max} heated to the boiling temperature by normal pressure. The value of h_{max} is defined by the substrate surface temperature and thermal properties of the liquid and substrate.

Typical dependencies of minimum h_{min} and maximum h_{max} sizes of the stable bubble on substrate surface temperature T_m are shown in Fig. 3. It is obvious, that substrate surface cleaning can be carried out only by $T_m > T_{th}$ (T_{th} is the threshold value of the substrate surface, by which $h_{max} = h_{min}$).

To calculate the value of T_{th} the dependences $h_{min}(T_m)$ and $h_{max}(T_m)$ should be defined. As the liquid temperature depends on coordinate, the bubble is not spherical and its form is determined by the distribution of surface tension pressure at the bubble surface. To evaluate h_{min} the following approximate method is used. We define the mean value of surface tension $\overline{\sigma}$, taking into account surface roughness (as bubbles arise generally in micro hollows):

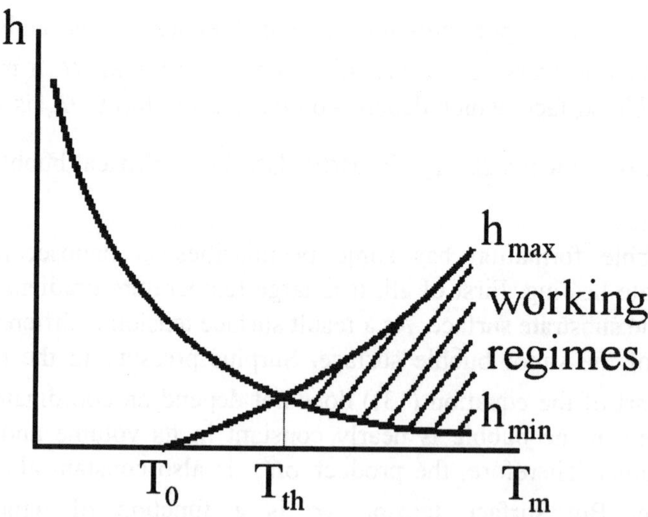

Fig. 3. The area of working regimes of steam laser cleaning and the threshold value of substrate surface temperature.

$$\bar{\sigma} = \sigma(T_l = T_m)\frac{R}{h_{min}} + \frac{\sigma(T_l = T_m) + \sigma(T_l = T_0)}{2}\frac{h_{min} - R}{h_{min}}, \tag{14}$$

where R is height of unevenness of rough surface, T_l is liquid temperature.

Taking into account, that the lower part of the spherical surface is absent (due to a certain angle between the surface of the bubble and the substrate), one can use relation between $h_{\min}$ and the mean value of bubble radius r:

$$h_{\min} = r(1 + \cos\alpha). \tag{15}$$

For rough surface the angle α is equal to the sum of the angle of wetting and the angle of element surface inclination. Surface tension of liquid σ according to experimental data decreases approximately linearly with increasing temperature of the liquid T_l (Kikoin, 1976) therefore we use approximation:

$$\sigma(T_l) = \sigma_0 \frac{T_c - T_l}{T_c - T_0}, \tag{16}$$

where σ_0 is surface tension at the boiling temperature T_0 (and normal pressure), T_c is the critical temperature of the liquid. Approximation (16) gives values $\sigma = \sigma_0$ for $T_l = T_0$ and $\sigma = 0$ by $T_l = T_c$.

As bubbles arise generally in the microhollows of a rough surface, the vapor pressure in the bubble P_T can be defined as the pressure of saturated vapor P_s calculated by the temperature of substrate surface T_m: $P_T = P_s(T = T_m)$. Dependence of saturated vapor pressure on temperature can be obtained from the Clapeiron equation:

$$P_T = P_0 \exp\left[\frac{L_v \mu}{R_g T_0}\left(1 - \frac{T_0}{T}\right)\right], \tag{17}$$

where L_v is evaporation heat of the liquid, μ is its molecular mass, R_g is gas constant. The dependence (17) was verified by experiment long ago and reliably (see Kikoin, 1976).

The maximum size of the bubble $h_{\max}$ is defined by the thickness of the liquid layer heated to the boiling temperature T_0 and the height of unevenness of a rough surface of substrate R.

Dependence of the temperature of transparent liquid T_l on distance x from absorbing substrate surface is defined by thermal properties of the liquid, maximum temperature of the substrate surface and temporal dependence of substrate surface temperature (that depends on character of light absorption). In any case spatial distribution of temperature in the liquid near the substrate surface (for $x < \sqrt{a\tau}$, a is thermal diffusivity of the liquid, τ is pulse duration) in the moment of pulse finishing is

$$T_l \approx \left(T_m - T_{in}\right)\left(1 - \frac{\gamma x}{\sqrt{a\tau}}\right) + T_{in}, \tag{18}$$

where T_m is substrate surface temperature in the moment of pulse finishing, T_{in} - initial temperature, coefficient $\gamma = -\left.\frac{\partial T_l}{\partial x}\right|_{x=0} \frac{\sqrt{a\tau}}{T_m - T_{in}} \approx 1$. If laser power during the pulse is constant, $\gamma = \sqrt{\pi}/2$ for high–absorbing substrate (depth of light penetration $\delta << \sqrt{a_s \tau}$, a_s is thermal diffusivity of the substrate), $\gamma = 2/\sqrt{\pi}$ for week–absorbing substrate ($\delta >> \sqrt{a_s \tau}$) and for a substrate covered by metal film, which is used in model experiments, for instance Yavas (1994). So, the maximum size of the bubble $h_{\max}$ is:

$$h_{\max} = \frac{\sqrt{a\tau}}{\gamma} \frac{T_m - T_0}{T_m - T_{in}} + R. \tag{19}$$

The threshold value of the surface temperature $T_m = T_{th}$ is defined by equating of $h_{\max}$ and $h_{\min}$.

Superheating of the substrate $\Delta T = T_{th} - T_0$ versus unevenness of a rough surface is given in Fig.4 for water and ethanol for different values of angle α and conditions of laser radiation absorbance γ. One can see, that the superheating required for bubble evaporation beginning is less for ethanol, than for water, and it decreases by roughness increasing.

Analysis of equations (14)–(19) shows, that value T_{th}/T_0 is defined also by the following combinations: $\frac{1+\cos\alpha}{P_0}\sigma_0 \frac{\gamma}{\sqrt{a\tau}}$, $\frac{L_v\mu}{R_g T_0}$, $\frac{T_{in}}{T_0}$ and $\frac{T_0}{T_c}$. For the liquids, used by steam laser cleaning, values of coefficients $\frac{L_v\mu}{R_g T_0}$ and T_0/T_c vary insignificantly, so value T_{th}/T_0 varies in the boundaries of some percents. A calculation of dependence of T_{th}/T_0 on value T_{in}/T_0 showed, that T_{th}/T_0 varies also slightly (for ethanol in boundaries of 3%) by change of T_{in} from 0^0C to T_0. That is evidence that preliminary heating of the pattern does not considerably influence the cleaning threshold. Value T_{th}/T_0 mainly depends on the parameter related to surface tension of the liquid (Fig. 5).

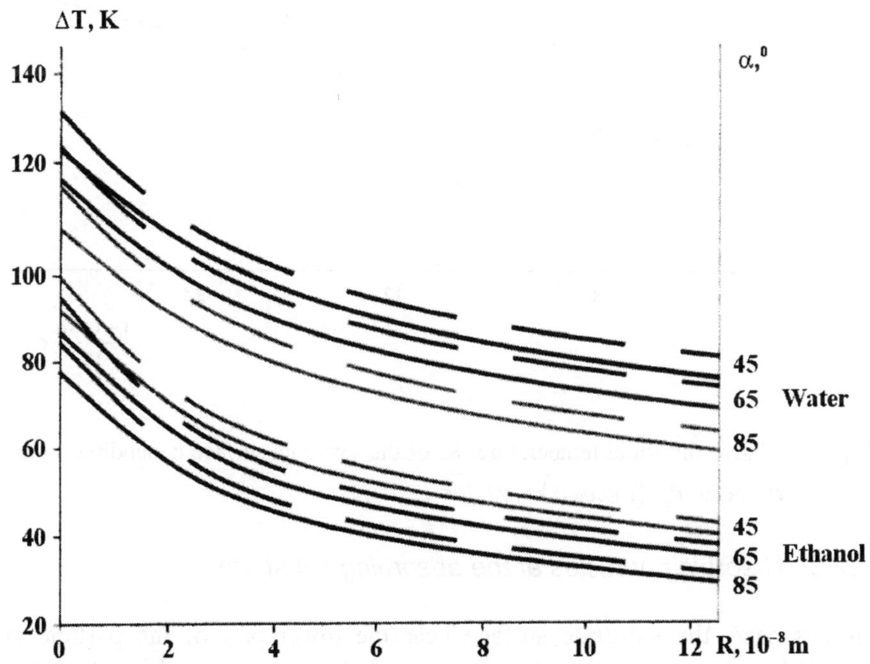

Fig. 4. Superheating of the substrate surface depending on its roughness: solid lines $\gamma = \sqrt{\pi}/2$, dash lines $\gamma = 2/\sqrt{\pi}$.

So, threshold temperature is mainly defined by boiling temperature and surface tension of the liquid and the roughness of the substrate surface (calculated for ethanol, high–absorbing substrate $\gamma = \sqrt{\pi}/2$, $R = 10^{-8}$ m, $\alpha = 65^0$). It should be noted, that the threshold of cleaning may be some less, than the calculated one, in the case of weak–absorbing substrates ($\delta >> \sqrt{a_s \tau}$). It is caused by h_{max} increasing due to thermal flow from the substrate to the liquid layer after pulse finishing. The surface temperature decreases slightly and h_{min} change is not significant. This problem is not analyzed in this paper because of the absence of reliable data on optical properties of the substrate, but it will be investigated somewhere.

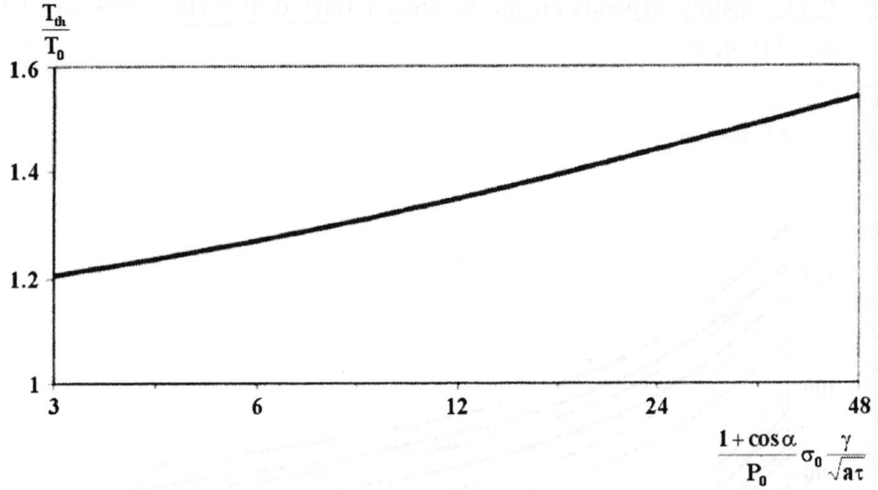

Fig. 5. Relative threshold temperature, K, of the substrate surface depending on the parameter $P_0^{-1}(1+\cos\alpha)\sigma_0\gamma/\sqrt{a\tau}$.

2.2.3. Absorbing particles at the absorbing substrate

In this case, the substrate surface near the lower part of the particle is situated in its shadow and does not absorb laser radiation. Liquid in this region can heated by heat transfer both from the particle top and from irradiated substrate zone. The criterion of the prevailing thermal transfer way is $\varphi = \overline{A}_p\sqrt{a_p}/\left(A_S\sqrt{a_S}\right)$. If $\varphi > 1$ the liquid in the shadow region heats basically from the particle, the cleaning mechanism is like that considered in p.2.2.1. If $\varphi < 1$ from the substrate (the cleaning mechanism is like that considered in p.2.2.2).

Let us calculate threshold intensity for steam cleaning of NiP substrate from 1 μm Al particles by using isopropanol alcohol as liquid for cleaning ($T_0 = 355$ K) and wavelength $\lambda = 248$. In this case $\varphi > 1$ and assuming $\overline{A}_p = (1 - R_p)/2$ (R_p is reflectivity of the particle material by normal incidence) and $T_0 = 20^0$ C, one can obtain: $\varepsilon_{th} = 32.6$ mJ/cm^2. Experimentally measured value for the same conditions (Boneberg et al., 1999) is 30 mJ/cm^2.

2.2.4. Discussion

The physical mechanism of steam cleaning is connected with prevalent action of vapor pressure to lower part of the particle surface. The laws of steam laser cleaning are different for the substrate and particles materials with different thermal and optical properties. Depending on their correlation, liquid evaporation is caused either by the heating of the particles or of the substrate. There are threshold intensities in both cases. In the case of absorbing particle, it is proportional to the particle size. For absorbing substrate, the threshold value of the substrate surface temperature for laser cleaning mainly depends on the boiling point of the liquid, its surface tension and roughness of the substrate surface. The investigations of the threshold characteristics of bubble boiling by steam laser cleaning carried out in this paper give an opportunity of optimum choice of the liquid and machining regime and prognosis of the results of cleaning.

3. Laser cleaning of the solid surface from films

Laser cleaning of solid surface from films is usually carried out in low–energetical (before–evaporation) regimes, which except destruction of surface layers of substrate, undesirable chemical and structural modifications and thermal tension in it. For this reason the most promising methods of film removal are the following:

– film shaking–off from the substrate surface under the action of the inertial force arising due to thermal expansion of the film and/or substrate,
– film buckling caused by thermal stresses within the film.

3.1. Laser cleaning by buckling mechanism

Buckling of the film (Fig. 6) seems to be the most probable physical mechanism of before–evaporation ablation of elastic films (for example, metal films). Ablation of thin metal films in solid and liquid phase by the action of 10^{-8} s excimer laser pulses has been experimentally investigated (Tóth et al., 1995). The velocity of film fragment movement from the substrate depending on laser fluence was measured. The main features of before– evaporation ablation of W film turned out to be the following. Film ablation begins at fluence $\varepsilon \approx 50\,\text{mJ/cm}^2$, which is much less than the melting threshold and about the threshold of substrate surface gasification.

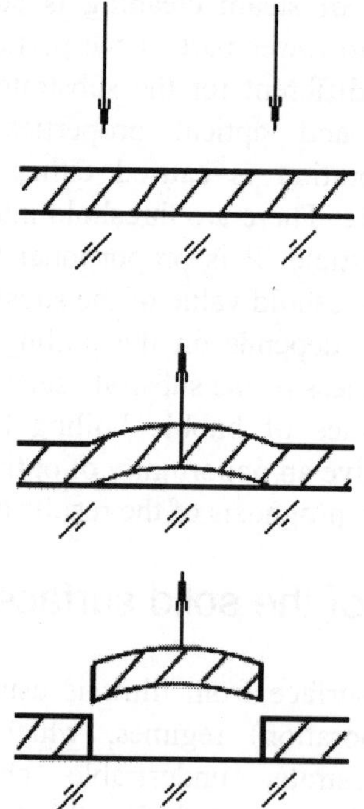

Fig. 6. Buckling mechanism of film removal.

The velocity of the film fragment movement increases with fluence up to $\varepsilon \approx 300\,\text{mJ/cm}^2$ (which lies between the thresholds of film melting and evaporation) and reaches 100 m/s. For higher fluences the velocity of film fragments decreases with fluence. Such dependence of the velocity on laser fluence is not obvious.

The most probable mechanism of before–threshold ablation of the film in such conditions is the following. By laser heating of solid film above the temperature of substrate gasification, the pressure of gas in the cavities at the film–substrate interface and sizes of the cavities increase. Therefore adhesion is partially or completely eliminated in the irradiated zone. For other materials of the film and the substrate, adhesion elimination can occur under the action of other factors, for instance evaporation of the film at the interface or desorption of before–adsorbed gases. But calculations show, that gas pressure is not sufficient for the achievement of such velocity of film fragments. On the other hand, film thermal expansion by heating causes its curving in the irradiated zone, i.e. its movement from the substrate surface. This process can be considered as a process of generation of small thermal compression stress and its transformation into film displacement at once. We shall consider this phenomenon integrally: as a single transformation of compression energy to kinetic energy. The additional argument for this model is a good agreement of calculation results with experimental data (Tóth et al., 1995).

3.1.1. The main regularities and regimes of film buckling

We shall consider the case of a large size of the irradiated zone in one–dimensional approach. According to the suggested model integral thermal stress is

$$\sigma_T = E\alpha_f T\,, \tag{20}$$

where E is elasticity (Young) modulus, α_f is linear thermal expansion coefficient, T is film temperature referred to its initial value.

The density of the compression energy to volume unit is

$$Q = \frac{\sigma_T^2}{2E} = \frac{E}{2}\left(\alpha_f T\right)^2. \tag{21}$$

It transforms to the kinetic energy of film fragment movement from the substrate surface. By this the following regimes of film ablation in solid or liquid phase are possible:

1. Laser fluence ε is larger than the threshold of adhesion elimination ε_0 and such that the film temperature T does not reach the value of melting temperature T_{ml} during the laser pulse and the compression energy and correspondingly the film fragment velocity V are small enough. By this, the complete tearing–off of the film fragment, occurring by its transfer to the distance that is about film thickness h_f happens after laser pulse finishing, i.e. $\int_0^{\tau} V dt < h_f$. In this case, compression energy increases during the whole pulse. The maximum value of fragment velocity corresponds to the pulse finishing.

2. By fluence increasing (film temperature is less than its melting temperature T_{ml}, as in the first regime), fragment velocity increases and it transfers at the distance h_f in time t_*, which is less than pulse duration, i.e. $\int_0^{\tau} V dt > h_f$. By this, film fragment velocity reaches its maximum value at the moment t_* ($t_* < \tau$).

3. Laser fluence is larger, than the film melting threshold, but the time of complete separation of the film fragment t_* is less than the time of its heating to the melting temperature t_H. By this film separation occurs when it is solid still and melting begins after it. Therefore, regularities of film ablation in this regime and the previous one are the same.

4. By fluence increasing, film melting begins before it transfers to the distance h_f (i.e. $t_H < t_*$). By this two processes take place additionally:

- film enlargement due to increasing of its volume by melting, that is analogous to thermal enlargement and causes increasing of the compression energy and film fragment velocity;
- melted regions sizes decreasing due to melt transfer under the action of compression tension, causing decreasing of the compression energy and film fragment velocity.

Film fragment velocity is defined by the combined action of the mentioned factors at the moment of complete separation t_*.

All the considered before–evaporation regimes of film ablation are present in the case of refractory films such as W thickness of which is about 10^{-7} m. For films with lower melting temperature or larger thickness, some of the considered regimes can be absent.

Film ablation in these regimes is investigated in detail below. Thermal transfer from the film to the substrate is not taken into account, as it is considerable only before elimination of adhesion and therefore does not influence the main course of events.

3.1.2. Film ablation without melting before separation of the film fragment

Let us consider simultaneously two possible extreme cases of film ablation: when adhesion is eliminated only due to compression energy and only for some another reason (for instance due to substrate surface layer gasification, or due to gasification of adsorbed molecules). After adhesion elimination the film compression energy is transformed to kinetic energy

$$Q = \frac{\rho_f V^2}{2} + Q_0, \tag{22}$$

where Q_0 is the energy of adhesion elimination per film volume unit (it is zero in the second case). One can define the velocity of fragment movement in both cases for the 1st, 2nd and 3rd regimes, taking into account the dependence of the film temperature on laser fluence $T = A_f\,\varepsilon/\rho_f c_f h_f$ (ρ_f is film density, c_f is film heat capacity, A_f is absorption coefficient of film). The results of calculation are given in Table 1.

Formulae for the threshold fluences separating the regimes are presented in Table 2.

3.1.3. Ablation of the melting film

The film fragment begins to melt during its movement before complete separation in the IV regime. Its velocity depending on laser fluence can be estimated by the following way. Let us look through the change of the film tension caused by film enlargement by its melting and by melt transfer by

Table 1. Velocity of film fragment movement V without melting before separation.

Regime I	Regimes II, III
Adhesion elimination due to compression energy	
$V=\sqrt{\frac{E}{\rho_f}\frac{\alpha_f A_f \varepsilon}{\rho_f c_f h_f}}$	$V=\sqrt{\frac{E}{\rho_f}\frac{\alpha_f A_f \varepsilon}{\rho_f c_f h_f}}\left(\sqrt{\sqrt{\frac{\rho_f}{E}}\frac{8\rho_f c_f h_f \varepsilon}{\alpha_f A_f \varepsilon_0^2 \tau}+1}-1\right)$
Compression energy is not spent for adhesion elimination	
$V=\sqrt{\frac{E}{\rho_f}\frac{\alpha_f A_f \varepsilon}{\rho_f c_f h_f}}\sqrt{1-\left(\frac{\varepsilon_0}{\varepsilon}\right)^2}$	$\varepsilon=\frac{\tau\varepsilon_0 V}{2h_f}\left(\sqrt{\left(\sqrt{\frac{\rho_f}{E}}\frac{8\rho_f c_f h_f V}{\alpha_f A_f \varepsilon_0}\right)^2+1}-1\right)$

Table 2. Threshold fluences.

Between the I and II regimes, ε_1	Between the III and IV regimes, ε_2
Adhesion elimination due to compression energy	
$\varepsilon_1=\varepsilon_0+\frac{2h_f}{\tau}\sqrt{\frac{\rho_f}{E}\frac{\rho_f c_f h_f}{\alpha_f A_f}}$	$\varepsilon_2=\sqrt{\frac{E}{\rho_f}}\frac{\tau\alpha_f T_m \varepsilon_0}{2h_f}\left(\frac{\rho_f c_f h_f T_m}{A_f \varepsilon_0}-1\right)$
Compression energy is not spent for adhesion elimination	
$\frac{\varepsilon_1/\varepsilon_0}{(\varepsilon_1/\varepsilon_0-1)\sqrt{(\varepsilon_1/\varepsilon_0)^2-1}}=\frac{\tau}{2h_f}\sqrt{\frac{E}{\rho_f}\frac{\alpha_f A_f \varepsilon_0}{\rho_f c_f h_f}}$	$\varepsilon_2=\sqrt{\frac{E}{\rho_f}}\frac{\tau\alpha_f T_m \varepsilon_0}{2h_f}\left[\sqrt{\left(\frac{\rho_f c_f h_f T_m}{A_f \varepsilon_0}\right)^2+1}-1\right]$

the action of compression tension. There the one-dimensional case will be considered (if irradiated zone at the film surface has a form of a long strip).

Let melting begin in one or several "hot points", forming one or several melting regions in the irradiated zone of the film. We shall consider, that the melted region has initial thickness h_f and half–width $l|_{t=0}=h_f$, and then the width of the melted region increases by absorbing laser energy. The work of the compression tension can be defined as following:

$$dA=2\sigma_T h_f U dt, \tag{23}$$

where U is the speed of movement of the melted region boundary by the action of compression tension σ_T. On the other hand, compression tension causes a change of the kinetic energy in the melt. Neglecting melt viscosity:

$$dA = d\left(\frac{2\rho_f h_f \ell \overline{U_l^2}}{2}\right), \tag{24}$$

where ℓ is half–width of the melting region, $\sqrt{\overline{U_l^2}}$ is squared mean of the melt speed at the melted region width. One can obtain it, considering melt speed be linear at the coordinate (Veiko, 1983): $\sqrt{\overline{U_l^2}} = \sqrt{\frac{1}{l}\int_0^l U_l^2 dx} = \frac{U}{\sqrt{3}}$.

The following differential equation is obtained from (23) and (24):

$$\ell\frac{dU}{dt} = \frac{3\sigma_T}{\rho_f}. \tag{25}$$

The half–width of the melt region ℓ decreases due to melt transfer and increases due to melting of solid material, adjoining the melted region:

$$\frac{d\ell}{dt} = V_m - U. \tag{26}$$

The initial conditions to equations (25), (26): $\ell = h_f$, $U = 0$ by $t = 0$. $V_m = \frac{L}{2n(\tau - t_H)}\beta(\tau)$ is speed of the melt boundary transfer, L is the whole width of the irradiated zone of the film, $t_H = \frac{\varepsilon_H}{\varepsilon}\tau$ is the time of melting beginning, n is the number of "hot points" at the distance L, $\beta(t) = \beta(\tau)\frac{t}{\tau - t_H} = \frac{\varepsilon}{\varepsilon_m - \varepsilon_H}\frac{t}{\tau}$ is ratio of mass of the melted part of the film to whole mass of the film in the irradiated zone at the moment t, which is calculated from melting beginning, $\varepsilon_H = \rho_f c_f h_f T_{ml}$, $\varepsilon_m = \rho_f c_f h_f T_{ml} + \rho_f h_f L_{ml}$, L_{ml} is specific melting heat.

The differential equation of change of the half–width of the melted region due to compression action $\Delta = \int_0^t U dt$ one can obtain, integrating the equation (26) and using (25). Value of Δ is connected with compression tension by correlation:

$$\sigma_T = E\left(\alpha_f T_{ml} + \delta_L \beta(t) - \frac{\Delta}{L}\right), \tag{27}$$

where δ_L is specific linear enlargement of the material by its melting.

So, the differential equation for change of the compression tension σ_T is:

$$(\sigma_T)''_{tt}\left[\frac{\sigma_T}{E} - \left(\alpha_f T_{ml} - \frac{h_f}{L}\right) - \frac{\varepsilon}{\varepsilon_m - \varepsilon_H}\left(\bar{\delta}_L - \frac{1}{2n}\right)\frac{t}{\tau}\right] = -\frac{3E}{\rho_f L^2}\sigma_T . \tag{28}$$

The initial conditions: $t = 0$, $\sigma_T = E\alpha_f T_{ml}$, $(\sigma_T)'_t = E\dfrac{\delta_L}{\tau}\dfrac{\varepsilon}{\varepsilon_m - \varepsilon_H}$. As film melt stops after pulse finishing, $\bar{\delta}_L = \begin{cases}\delta_L & by\ t \le \tau - t_H \\ 0 & by\ t > \tau - t_H\end{cases}$. Velocity of film fragment movement is defined from the equation (28) as $V = \sigma_T(t = t_*)/\sqrt{E\rho_f}$.

The film fragment velocity was defined from equation (28) by using a numerical method for 10^{-7} m W film irradiated by 10^{-8} s laser pulse. It was shown, that the velocity of the film fragment depends considerably on the number of melting regions, i.e. on spatial distribution of laser fluence. Velocity of the film fragment V on laser fluence ε, calculated for W film ($h = 10^{-7}$ m) in all considered regimes is given in Fig. 7. The obtained results are in a good agreement with known experimental data (Tóth, 1995).

3.1.4. Film degradation

The considered conception of film buckling can be used for explanation of the phenomenon of film degradation, i.e. film destruction under the action

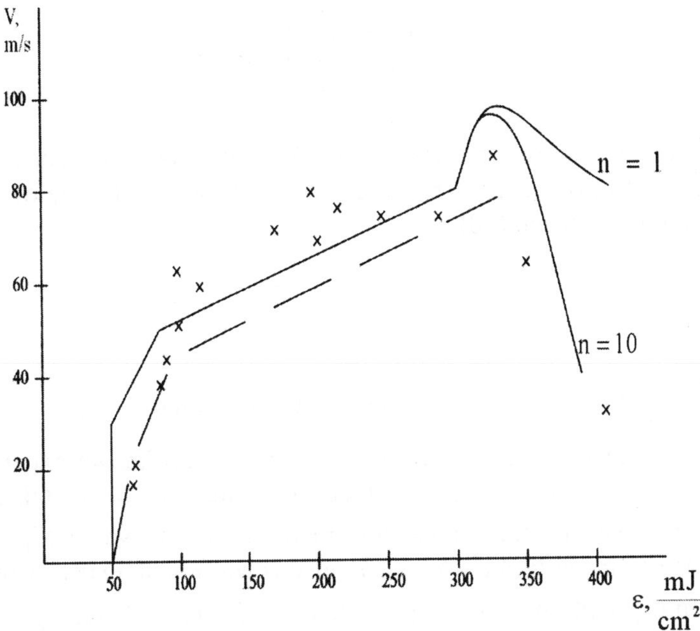

Fig. 7. The dependence of velocity of fragment of W film ($h_f = 10^{-7}$ m, $\tau = 10^{-8}$ s) on laser fluence; dash line – elimination of adhesion occurs due to compression energy, solid line - elimination of adhesion occurs due to other factors, x - experimental data (Tóth, 1995).

of many laser pulses, each of them possesses fluence many times less than ablation threshold fluence. The density of compression energy to volume is given by (21). Threshold fluence value ε_0 defines the density of compression energy, by which adhesion eliminated in the whole irradiated zone:

$$Q_0 = \frac{E}{2}\left(\frac{\alpha_f A_f \varepsilon_0}{\rho_f c_f h_f}\right)^2 . \qquad (29)$$

By the action of a series of N pulses, with fluence each of them $\varepsilon < \varepsilon_0$, the partial adhesion change occurs, it is proportional to $Q(\varepsilon)$. In the intervals between the pulses its value decreases due to relaxation. So, adhesion decreases from pulse to pulse, until film ablates after some N–th pulse:

$$Q_0 = N\frac{E}{2}\left(\frac{\alpha_f A_f \varepsilon}{\rho_f c_f h_f}\right)^2 - \frac{\omega}{h_f}(N-1), \tag{30}$$

ω is relaxation energy per surface unit. One can obtain the dependence of the number of pulses, that is necessary for ablation, on fluence of a single pulse, equating (29) and (30):

$$N = \frac{\varepsilon_0^2 - \varepsilon_*^2}{\varepsilon^2 - \varepsilon_*^2}, \tag{31}$$

$\varepsilon_* = \sqrt{\frac{2qh}{E}\frac{\rho c}{\alpha}}$ is some value, connected with relaxation.

The dependence (31) is in a good agreement with known experimental data. The calculated curve, in which the value of ε_* is taken from experimental data (Tóth, Private information), are shown in Fig. 8. The dependence of ε_* on film thickness h_f also is confirmed: experimental data for Cr films, with thickness $h_{f1} = 210\,\text{nm}$ and $h_{f2} = 105\,\text{nm}$, give value $\varepsilon_*(h_{f1})/\varepsilon_*(h_{f2}) = 1.6$. The calculated value is $\varepsilon_*(h_{f1})/\varepsilon_*(h_{f2}) = 1.4$.

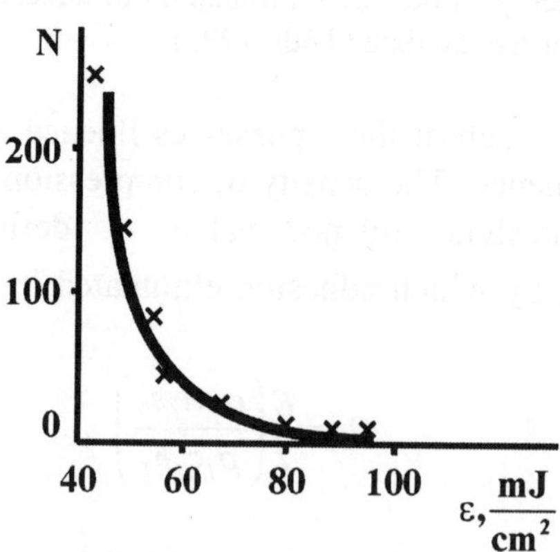

Fig. 8. The dependence of the number of pulses, that is necessary for film ablation in multiple–pulse regime, on fluence of a single pulse, x - experimental data (Tóth, Private information).

3.2. Laser cleaning by film shaking–off

Film shaking-off is the most probable physical mechanism of before–evaporation ablation of unelastic films, such as wax.

Similar to dry laser cleaning of the surface from particles, laser heating of the substrate and film causes their thermal expansion and correspondingly displacement of the film mass center.

Quantitative evaluation of thresholds of laser cleaning surfaces from film by its shaking–off can be obtained by using formulae (4), (9), (10), where adhesion force per mass unit $F_a(z_0)/m$ can be represented as $P_a/\rho_f h_f$ (P_a is adhesion force per surface unit, h_f is film thickness, ρ_f is its density), α_p, A_p, ρ_p, c_p is to be replaced to α_f, A_p, ρ_p, c_f (corresponding parameters of the film). Threshold parameters of laser cleaning of the surface from film in the multipulse regime can be also considered analogously as for particles (see (11), when particle size d is to be replaced to film thickness h_f).

3.3. The conditions of action of shaking–off and buckling mechanisms of film ablation

The physical mechanism of film removal from a solid surface is mainly defined by elastic properties of the film. The most probable mechanism of removal of unelastic films as wax is only film shaking–off. Removal of elastic films generally is possible both by film buckling and by its shaking–off. Contribution of each of the mechanisms is defined by correlation between values of kinetic energies (i.e. velocities) of film fragment movement from the substrate surface under the action of buckling and shaking–off.

The maximum velocity of the film fragment movement by film buckling is (see Table 1):

$$V_1 = \sqrt{\frac{E}{\rho_f}}\frac{\alpha_f A_f \varepsilon}{\rho_f c_f h_f}. \qquad (32)$$

Estimation of the maximum velocity of absorbing film fragment movement by its shaking–off can be obtained from formulae (1), (2):

$$V_2 = q_m k \frac{\alpha_f A_f}{\rho_f c_f}. \tag{33}$$

Ratio V_1/V_2 for triangular temporal form of the pulse, $\varepsilon = q_m \tau/2$, and for uniform film heating on its thickness ($k = 1/2$) is:

$$\frac{V_1}{V_2} = \sqrt{\frac{E}{\rho_f}} \frac{\tau}{h_f}. \tag{34}$$

So, the action of the buckling mechanism of film ablation prevails by film thickness $h_f < h_*$, where $h_* = \tau\sqrt{E/\rho_f}$ is some critical value. The action of the shaking–off mechanism prevails by $h_f > h_*$. Particularly, value of h_* for metal films by nanosecond laser pulse duration $h_* \sim 10$ μm. So, for 0.1 μm metal films the action of buckling mechanism 100 times exceeds the action of shaking–off mechanism.

3.4. Other mechanisms of laser cleaning of solid surfaces from films

Besides aforesaid thermomechanical mechanisms of laser cleaning of solid surfaces from films, blasting mechanisms are possible. They have been considered before in our paper on mechanisms of before–evaporation film ablation by local laser transfer of films (Veiko, 1997). Blasting mechanisms can be the main mechanisms of laser cleaning in some particular cases, for instance cleaning of metal surfaces from corrosion (especially in case of porous oxide films, as mixture of FeO–Fe_2O_3–Fe_3O_4) or by sufficiently high fluence of laser radiation. Film ablation by laser cleaning under the action of blasting mechanisms is caused by one of the following phenomena:

1. Gasification of the ground material into the cavities of film ground material interface (if ground material is thermally unstable). By this, film destruction occurs under the action of pressure of products of ground material gasification. Film removal is possible if gasification pressure is larger than adhesion value. Gasification pressure can be estimated as

$$P_g = P_{0g} \exp\left[L_g\left(1-\frac{T_{0g}}{T}\right)\right], \tag{35}$$

where T is ground material interface temperature, P_{0g} is gasification pressure by the temperature T_{0g}, L_g is specific gasification heat. Estimation of threshold gasification pressure value, above which solid film removal occurs, is $P_g^{th} \sim 200-300$ Pa.

2. Thermodesorption of gas adsorbed at the defects of the interface (if ground material is thermally stable). The desorbed gas pressure value can be found as

$$P_d = \frac{k_B T}{S_0 r_d}, \tag{36}$$

where k_B is Boltzmann constant, S_0 is the section of the adsorbed gas molecule, r_d is defect size. If $r_d \sim 1$ μm and $T \sim 10^3$ K, the threshold pressure of film ablation is $P_d^{th} \sim 10^3$ Pa.

3. Film evaporation into the closed cavities of film/ground material interface, or in the films cavities, if cavity sizes are small enough. The pressure P_1 in the cavity is about twice larger than pressure at the free film surface ($P_1 = P_{t1} + P_{t2}$, P_{t1} is evaporated particles reactive pressure, P_{t2} is pressure of evaporated particles back flow, in the closed cavity $P_{t1} \approx P_{t2}$ at the free surface $P_{t2} \approx 0$). This can cause internal explosion in the cavities the sizes of which are larger than critical value r_{cr}, in before–evaporation regimes when laser fluence is insufficient for film surface evaporation.

By larger values of laser fluence, laser cleaning of a solid surface from films or particles can be the result of their evaporation.

4. Conclusion

The worked out physical mechanisms of laser cleaning of solid surfaces from particles and films let estimate the threshold parameters of the action and to prognosticate effectiveness of the cleaning process.

Acknowledgments

We are grateful to Prof. B. S. Luk'yanchuk for critical comments and discussions. We would also like to thanks Dr. Yong Feng Lu and Dr. Johannes Boneberg for reprints of their papers. The work was supported by grant TOO-6.5 of Ministry of Education of Russian Federation.

References

Boneberg J., Mosbacher M., Dobler V., Leiderer P., Chaoui N., Siegel J., Solis J., Afonso S.N., *The EOS/SPIE International Symposia on Industrial Lasers and Inspection*, Munich, Germany, 1999.

Kikoin I. K., *Tables of physical properties. Reference book,* Moscow, Atomizdat, 1976 (in Russian).

Lu Y.F., Song W.D., Zhang Y., Low T.S., Proc. SPIE, vol. **3550**, 7 (1998)

Luk'yanchuk B. S., Zheng Y. W., Lu Y. F., *A new mechanism of laser cleaning*, Proc. SPIE, vol. **4423**, 115 (2001)

Mosbacher M., Chaoui N., Siegel J. et al., *A comparison of ns and ps steam laser cleaning of Si surfaces*, *Appl. Phys.*, **A 69**, S331 (1999)

Tam A. C., Leung W. P., Zapka W., Ziemlich W. J., Appl. Phys., **71**, 3515 (1992)

Tóth Z., Hopp B., Kántor Z., Ignácz F., Szörényi T., Bor Z., *Dynamics of excimer laser ablation of thin tungsten films as followed by ultrafast photography*, J. Appl. Phys., **A 60**, 431 (1995)

Tóth Z., *Private information.*

Veiko V. P., Kaidanov A. I., Tutchkova E. A., Jakolev E. B., *Melt transfer by laser evaporation of metal films*, Electronic treatment of materials (Russia), No. **3**, 18 (1983)

Veiko V. P., Shakhno E. A., *Laser induced ablation and condensation on to the close–to–target substrate*, Proc. SPIE, vol. **3093**, 276 (1997)

Yavas O., Leiderer P., Park H. K. et al., *Optical and acoustic study of nucleation and growth of bubbles at liquid–solid interface induced by nanosecond pulsed – laser heating*, J. Appl. Phys., **A 58**, 407 (1994)

Part 4. Laser Cleaning of Artworks

Chapter 8

LASER ABLATION IN CLEANING OF ARTWORKS

V. Zafiropulos

Laser cleaning of artworks and antiquities is reaching a mature enough stage both in terms of the procedures employed and the understanding of the interaction processes involved. To this effect, a series of studies have been pursued aiming to process optimization. Appropriate model systems have been studied for elucidating the interaction mechanisms both in polymerized phases as well as inorganic encrustation. An overview of these issues presented and their implications for future directions outlined.

Keywords: Laser ablation, artworks cleaning, conservation by laser, polymerized phase, resins, stone, encrustation, composites

PACS: 42.62.Cf, 52.38.Mf, 52.77.Bn, 68.35.Np, 81.05.Lg, 81.05.Qk, 81.05.Rm, 81.40.Tv, 81.65.-b, 81.65.Cf, 82.80.Ch, 89.90.+n

1. Introduction

Laser-based techniques play an important role in the preservation of cultural heritage. Amongst the modern applications are spectroscopic techniques for composition analysis (Anglos, 2001; Clark, 1995), holographic interferometry for damage assessment (Asmus, 1973; Amadesi, 1974; Bertani, 1982; Tornari, 1998, 2000 a, b, c) and diagnosis of environmental influences (Boon & Van Nieuwenburgh, 1993). Laser is also used for removal of surface encrustations from marble sculptures and stonework (Asmus, 1986; Watkins, 1995; Cooper, 1998; Maravelaki, 1997, 1999; Gobernado-Mitre, 1997; Klein, 2000) or aged varnish from paintings (Zergioti, 1997; Fotakis, 1997; Georgiou, 1998; Zafiropulos, 1998, 2000, 2001; de Cruz, 2000). The successful implementation of these techniques relies on a close collaboration of conservators with laser scientists. Current trends consider the need for scientific approaches that may complete or even replace some of the traditional empirical methods applied in art conservation. Extensive literature on laser-based conservation techniques exists in LACONA Conference proceedings (König & Kautek, 1997; Salimbeni & Bonsanti, 2000).

Cleaning of artworks and antiquities by lasers provides the advantage for selective removal of undesired surface layers. In principle, it is possible to leave the original delicate surface or substrate entirely unaffected. For the on-line monitoring and in situ control at all stages of laser cleaning the LIBS (Laser Induced Breakdown Spectroscopy) technique is valuable (Gobernado-Mitre,1997; Klein, 1999; Maravelaki, 1997, 1999; Zafiropulos, 1998, 2001c).

This paper consists of two sections based on the nature of the layer to be removed. First, we consider complex polymeric substrates where UV lasers of nanosecond or picosecond pulse's duration are necessary. In the second section, we consider the divestment applications concerning inorganic encrustation on marble or stone surfaces. In this case the self-limiting processes are important.

2. Laser ablation of complex polymerized materials

2.1. Optimization of laser parameters

The ablation of polymers by short UV laser pulses has been discussed due to the plethora of technological applications (Fogarassy & Lazare, 1992; Dyer, 1992; Miller, 1994). Comprehensive review of the subject can be found in the book of Bäuerle, 2000.

In UV-laser ablation of polymers the ablation products typically consist small fragments (atoms and small molecules) and medium sized fragments, such as monomers and their parts (Srinivasan, 1989, 1993, 1994). The breaking of covalent bonds creates these fragments by direct photodissociation or thermal decomposition. Photothermal and photochemical ablation could be considered as limiting cases of the photophysical mechanism, where the removal of electronically excited species from the surface is taken into account (Luk'yanchuk, 1993a, b, 1994, 1996).

The photochemical modifications induced by UV laser ablation of doped PMMA films were investigated by Georgiou, 1998 and Athanassiou,

2000, in simulating the ablation processes in natural resins. However, we should mention that polymerized materials found in paintings are rather complex. For example the aged polymerized resins consist of hundreds of different organic molecules irregularly cross-linked in a 3D network. There are also different salts in the form of molecules, clusters or particulates embedded into the polymeric network.

Here we attempt to draw the basic guidelines for choosing the proper laser parameters for *safe* laser ablation of unwanted surface layers without affecting the underlying usually valuable substrate. First we make the assumption that the adequate laser wavelength for the ablation of these materials is the UV up to the limiting wavelength 248 nm or in some cases 266 nm. At these UV laser wavelengths the optical penetration depth $\ell_\alpha = \alpha^{-1} << d$ (α is the linear absorption coefficient and d is the total thickness of the unwanted surface layer), assuring the safety of the underlying substrate. For highly polymerized natural resins l_α is of the order of 0.1 μm for λ_L = 193 nm and 1 μm for λ_L = 248 nm, while d is usually 10-100 μm. For the pulse duration, we assume the use of the easily accessible nanosecond lasers without ignoring the ultrashort UV pulse lasers, which are not yet accessible for mass applications. Another point that must be clearly noted here is the usual choice of the KrF laser over the ArF Excimer laser. Although the latter offers a better guarantee than the former owing to the much lower l_α, its overall efficiency (including the commercially available power levels) is one order of magnitude lower. Therefore, from the application point of view we concentrate to λ_L = 248 nm presuming that for most delicate interventions the choice will be a shorter wavelength.

Apart from the wavelength and the pulse duration of the laser, the most important parameter is the laser energy density or fluence. According to the special requirements of the specific applications, it is of major interest to quantify the transmission of photons through the resin layer as a function of incident fluence. Aged varnish consists of unidentified complex polymerized material with random inorganic inclusions. An experimental methodology has been developed for deriving the optimal fluence for each particular application. Three different approaches have been considered: Ablation Efficiency studies, Light transmission studies and Chemical alteration studies.

2.1.1. Ablation efficiency studies

Common problems encountered in painting conservation have to do with aging associated to polymer phase formation, cross-linking and photochemical degradation of the varnish layer, accumulation of various kinds of pollutants on the surface that can be further polymerized and cross-linked to the resin, or even overpaints on the original painting that must be removed. Removal of these surface layers must be accomplished in such a way that the integrity of the original work is guaranteed.

Laser cleaning is based on the removal of well-defined surface layer under fully controlled conditions. In the case of homogeneous resin layers, e.g. aged resin, the ablation rate is characteristic for the entire surface to be processed. Therefore, a step-by-step scanning of the laser beam over the surface results in a layer removal of controlled thickness.

Characteristic ablation rate curves as a function of laser fluence obtained from model and real samples using KrF excimer laser (emitting at 248 nm, pulse duration of 25 ns) are shown in Fig. 1a. The four representative polymerized materials presented here are two types of artificially aged varnish (dammar and gum lac), a multi-layer polymerized linseed oil-based overpainting and a black polyurethane paint film. The way such measurements are made, as well as the dependence of ablation rate on laser wavelength and pulse duration are described elsewhere (e.g. see Zafiropulos *et al*, 1995). The data show the mean removed depth per laser pulse as a function of laser energy fluence in units of J/cm^2. It can be seen that the resolution attained by the KrF excimer laser can be as low as 0.1 μm per pulse, which is simply impossible when using other mechanical or chemical techniques. Fig. 1a also demonstrates the attainable definite control of varnish removal, that is, it is possible to completely adjust the depth of each layer removal by choosing the laser fluence and or the number of pulses. The horizontal axis has intentionally a logarithmic scale for demonstrating the linear response of ablation rate to the logarithm of the laser energy fluence, at low fluence values excluding the Arrhenius tail (Bäuerle, 2000). For every polymerized material this linearity is always strictly localized within a narrow range of laser fluence.

The ablation efficiency curves see in Fig. 1b, provide the optimum laser fluence for processing the material. Here the word "optimum" refers to the fluence where the ablation becomes most efficient, and it should not be confused with the fluence corresponding to the highest step-resolution. In

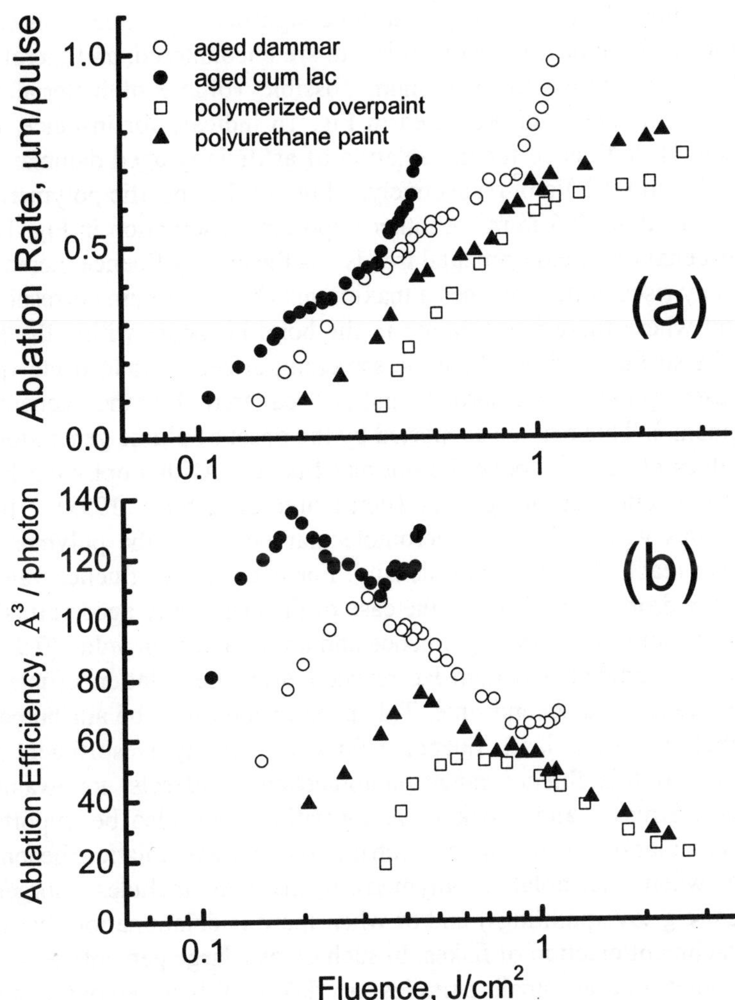

Fig. 1. Ablation rate (a) and ablation efficiency data (b) for four different complex polymerized materials using KrF excimer laser. The error is less than 10%.

the next section it is shown that at optimum fluence the transmission becomes minimal. The efficiency is a quite significant parameter in cases where large surfaces or thick varnish layers are encountered, and therefore, the goal is to remove the maximum possible volume of material per available photon. The data presented in Fig. 1b indicate for instance, that the optimum laser fluence for the ablation of artificially aged dammar and gum lac is 0.3 and 0.2 J/cm^2 respectively, while for the specific polymerized over paint it is about 0.6 J/cm^2. Another important observation in Fig. 1b is the similar behavior of experimental trends. As the energy fluence increases, the efficiency rises until it reaches a maximum. This maximum corresponds to the point where there is saturation in the bond breakage within the first layers of the surface and/or where the screening effect starts to occur (part of the laser pulse is absorbed and/or scattered by the generated photofragments). This may be supported by the onset of plume generation at fluence values at or just above the optimal fluence. At this optimum laser fluence, the excited chromophores (Pettit and Sauerbrey, 1993) - light absorbing sites within the 3D macromolecular lattice of the polymerized phase - have reached a maximum density. For even higher fluence values, the efficiency drops. Especially in the case of the artificially aged varnishes as we further increase the energy fluence and above a certain value (0.9 and 0.35 J/cm^2 for dammar and gum lac respectively), the ablation efficiency starts to increase for a second time. This phenomenon may be attributed to the fact that, for such laser fluence values the energy is sufficient for vaporization. At this fluence range photomechanical effects, for example thermoelastic stresses and shock wave formation, may also be important during laser ablation. Their direct contribution in laser ablation becomes significant when the ablated polymerized material includes inorganic particulates (e.g. overpaintings) and/or when there is delamination that can lead to detachment/ejection of flakes. In such cases a large percentage of the material comes out as small particles or flakes. When removing aged varnish on the other hand, the shock wave produced seems to have a negligible effect especially when working near the optimum laser fluence. Of course, this applies to the contribution of the photomechanical effects in the removal process itself. The possible long-term consequence of the shock wave is a different issue that is currently under investigation (e.g. see Tornari, 1998).

In general, incubation phenomena are very important especially when fluence values near the ablation threshold are used (Bäuerle, 2000).

Although a detailed analysis is essential to study the effects occurring in this low energy regime, it finally turns out that in the applications under consideration the optimal fluence is well above the ablation threshold, where incubation was found to be negligible.

Here it is worthwhile to present some results obtained using ultrashort laser pulses. Fig. 2 shows the ablation rate and efficiency data of artificially aged dammar resin (same as in Fig. 1) using laser pulses of 500 fs duration and $\lambda_L = 248$ nm. The nanosecond data are also plotted in the same graph for comparison. The maximum ablation efficiency for both pulse durations is reached at about the same laser fluence (~ 0.3 J/cm^2). For this particular resin and for the encountered low degree of polymer formation (5-7 monomers) it seems that the ablation rate depends on laser fluence and not flux. One of the many different reasons for such a behavior can be the long relaxation time in the particular material.

Finally, for a highly polymerized naturally aged varnish the situation is expected to be different. In such a case the sub-nanosecond data have been found to be clearly distinctive compared to the nanosecond data. Fig. 3 shows the results for three different pulse durations (500 fs, 5 ps and 25 ns) at $\lambda_L = 248$ nm obtained from an unknown highly polymerized varnish undergone a natural aging of more than 60 years. Here the relaxation times must be of the order of picoseconds as in polymers (Bäuerle, 2000). The expected non-thermal operative mechanisms with sub-picosecond UV laser pulses on highly polymerized materials may be also presumed by considering the SEM images of the same resin after laser ablation using the three different pulse durations. Fig. 4a shows the SEM picture of the reference surface, while Figs. 4 b-d present the resin after removal of ~ 1 μm film thickness with pulses of different durations (500 fs, 5 ps, and 25 ns) at $\lambda_L = 248$ nm. The laser ablation was carried out under scanning mode of operation with overlapping laser spots (see description in section 2.2). The white particles in the images are salts mainly containing Ca and Na, always present in the natural resins (also see section 2.2). The surface left after using nanosecond pulses has a very smooth texture, alike a surface undergone melting and freezing, see in Fig. 4d. Additionally, the inorganic particles have also undergone a melting-freezing transformation. On the contrary, the surface of the resin and the morphology of the inorganic particles remain as in the reference after ablation by sub-picosecond or picosecond pulses; see in Figs. 4b and 4c, respectively.

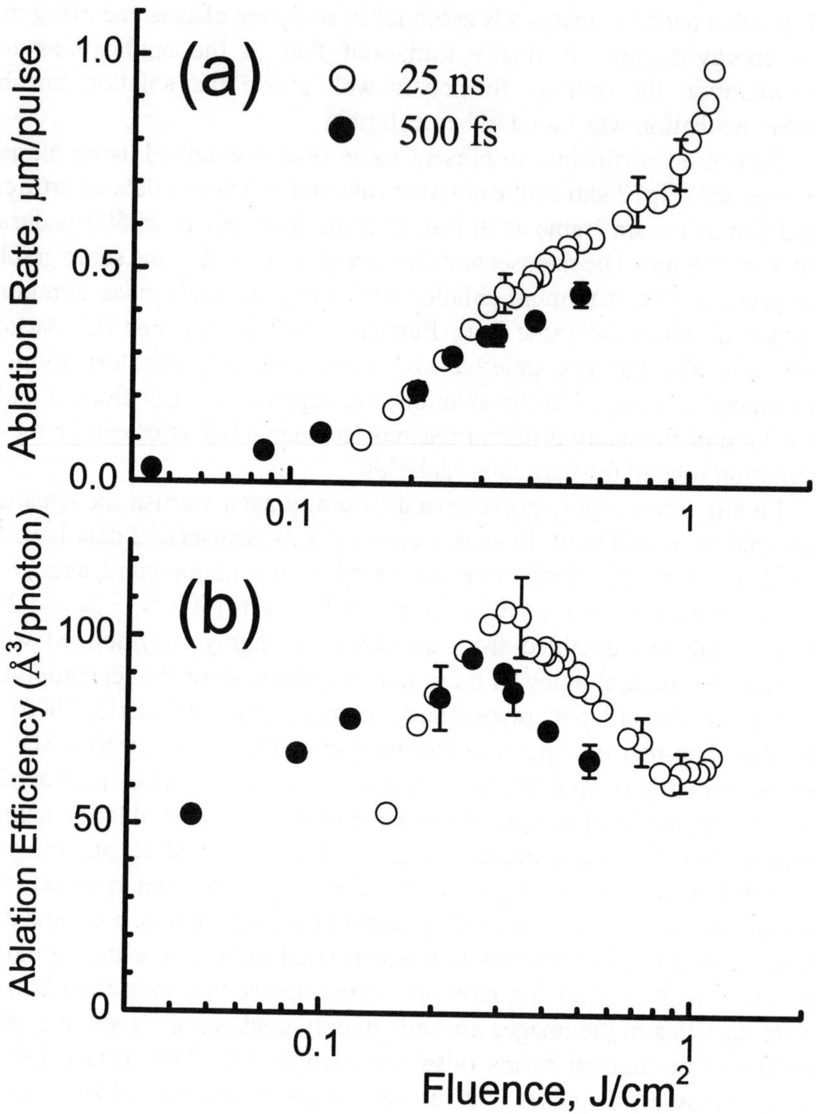

Fig. 2. Ablation rate (a) and ablation efficiency data (b) of artificially aged dammar resin using laser pulses of 25 ns and 500 fs duration (λ_L = 248 nm).

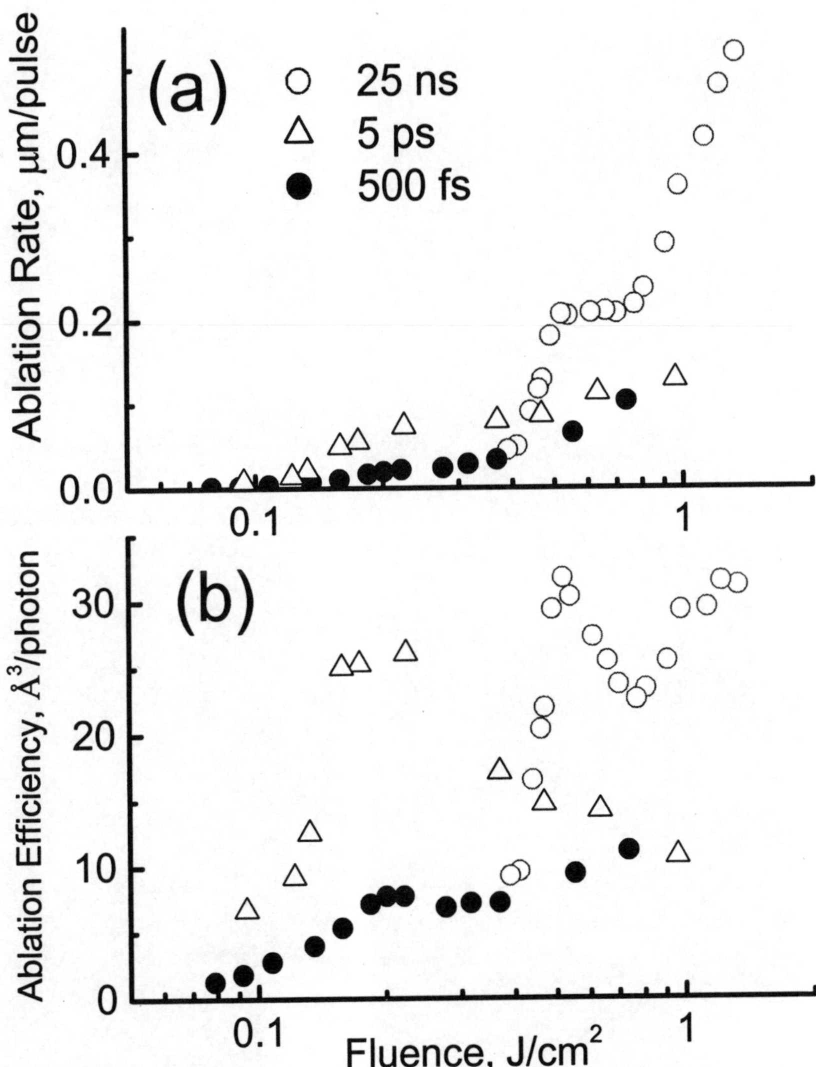

Fig. 3. Ablation rate (a) and ablation efficiency data (b) of a naturally aged unknown resin using laser pulses of 25 ns, 5 ps and 500 fs duration (λ_L = 248 nm). The error is less than 10%.

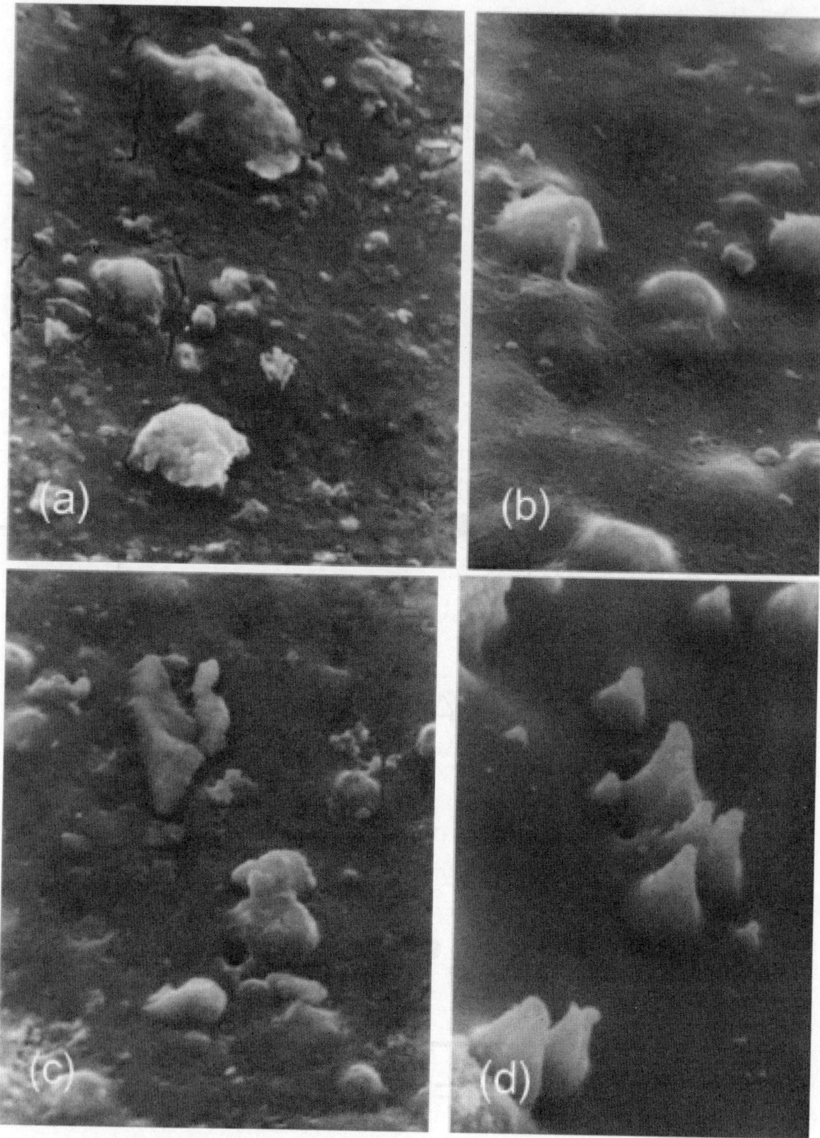

Fig. 4. SEM pictures (x3500, 40° angle of observation) of reference surface (a) and after the removal of 1 μm film using laser pulses of 500 fs (b), 5 ps (c) and 25 ns (d). See text for details.

2.1.2. Light transmission studies

During the laser ablation of polymerized resins a certain portion of the laser light is transmitted through the resin, finally reaching the underlying substrate (e.g. pigment layer). Therefore, quantification of the optical penetration depth is significant. In the experiments presented here, the goal was to quantify the transmitted photon energy and to find the optimal laser parameters for minimizing it.

The experiments were designed in such a way that the transmission of laser photons through model resin samples on quartz is measured. A similar experiment on polyimide has been reported by Pettit (1994). During laser ablation, the energy of laser light penetrating the sample is determined for each successive laser shot. This provides the effective absorption coefficient and the energy of UV transmitted light as a function of depth. For this, the laser beam is appropriately shaped and directed on the sample. A sensitive energy meter measured the energy of transmitted light for every consecutive laser pulse. The signal is triggered by a fast photodiode and is displaced on the oscilloscope. For each laser fluence value used, a series of laser pulses were fired until full removal of the resin. In Fig. 5 plots of $log\ (I_{trans}/I_0)$ versus d are presented for different fluence values, where I_0 is the intensity of the incident on the film laser beam, I_{trans} is the intensity of the transmitted light, and d is the film thickness left after the laser pulse. In this case the sample was an artificially aged mastic film prepared on a quartz plate. The light reflected by the latter is taken into account by normalizing the I_{trans} measurements, comparing them with the transmission after the total removal of the film. The slope of the graphs represents the effective absorption coefficient for the laser light. This graph provides the following information:

(a) The total energy density of transmitted light. This can be calculated if we know the initial film thickness, the fluence and ablation rate, and the number of laser pulses fired. The total transmitted energy density is crucial when the substrate is sensitive to the laser light.

(b) The optimal laser fluence can be identified, where "optimal" in this case is defined in terms of least transmitted light.

Here it is interesting to compare the transmission data with the ablation efficiency data obtained from the same samples. Figs. 6a and 6b show the results of ablation rate and ablation efficiency studies, respectively, of the

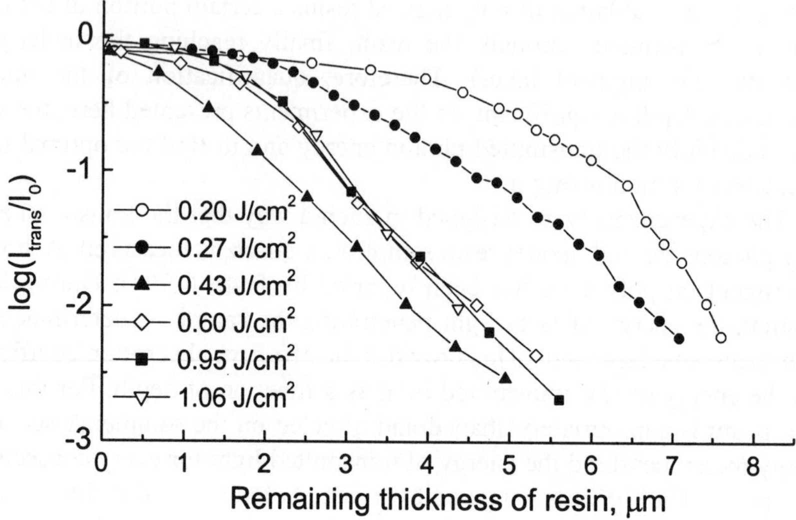

Fig. 5. Light transmission measurements of aged mastic during laser ablation (λ_L = 248 nm, 25 ns pulse duration), presented as log (I_{trans}/I_0) versus d, for six different laser fluence values. All points represent consequent laser pulses, except at 0.20 J/cm^2. The corresponding errors for log (I_{trans}/I_0) and d are less than 5% and 10% respectively.

same aged mastic sample as in Fig. 5. The six different fluence values that were used in obtaining the transmission data presented in Fig. 5 are marked in Fig. 6b for comparison. At the low fluence of 0.2 J/cm^2 just above the ablation threshold the transmission of laser light is rather high. For slightly higher fluence values we observe a sharp drop of the transmission, reaching a minimum at a fluence, which corresponds to the highest ablation efficiency. For higher fluence values the transmission curves tend to coincide, possibly owing to plume absorption. Fig. 6c shows the transmission as a function of fluence when the remaining mastic resin is 3 μm. A comparison between Figs. 6b and 6c shows that the optimum fluence defined in terms of transmission coincides with the optimum fluence defined in terms of efficiency. This has been found to be a general rule for a

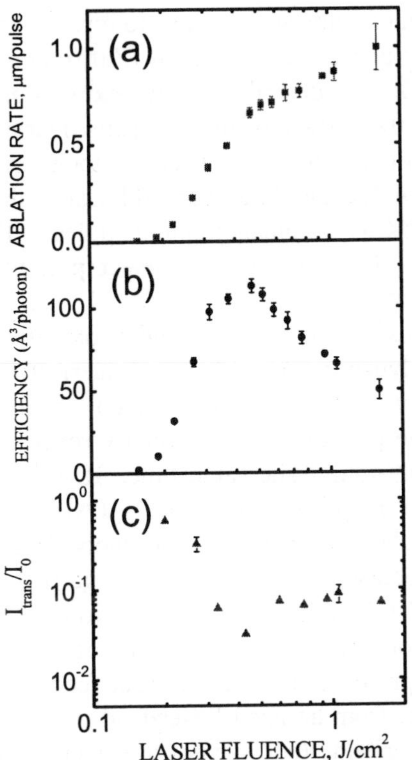

Fig. 6. Ablation rate (a) and ablation efficiency data (b) of artificially aged mastic resin (same as in Fig. 5) using KrF excimer laser. Plot (c) shows the transmission as a function of fluence for a hypothetical remaining mastic film of 3μm thickness (data extracted from Fig. 5).

series of different polymerized resins. The results are in general in accordance with the three-level chromospheres model of Pettit, 1994.

The implications of these observations are essential for the safe use of UV pulsed lasers in the removal of superficial resin layers from paintings or from other delicate substrates. It is also important to note that in such applications the approach cannot be empirical and certainly cannot be the same as in the removal of inorganic encrustation. In the latter case, the end user starts from low fluence values progressively increasing the energy density until a "cleaning threshold" is reached. In the case of polymerized resins, however, such an approach could be catastrophic.

2.1.3. Chemical alteration of substrate

The most of known pigments used in paintings consist of inorganic salts mixed in a binding medium, usually linseed oil. On top of the pigments a resin layer has been usually applied, mainly for protecting the pigments from dirt accumulation. It has been found (De la Rie, 1987) that resins also protect pigments from the ambient UV light that induces free radical formation and subsequent oxidation reactions both in the resin and in the binding medium. At the same time the most salts are sensitive to direct laser irradiation undergoing chemical and/or phase transformation (Zafiropulos, 2001a; Luk'yanchuk & Zafiropulos, 2002). As described in the next section, the aged resins must be removed from the surface of the pigments. During laser-assisted divestment, the transmitted light may lead to long-term deterioration of the original surface of the painting. For addressing the issue, laser ablation was combined with Gas Chromatography coupled with Mass Spectroscopy (GC-MS) analysis. The goal was to detect possible oxidation products induced by laser radiation.

The process of degradation of linseed oil medium under oxidative conditions involves a simultaneous polymerization and de-polymerization reaction. In general, unsaturated fatty carbon chains undergo various oxidation reactions resulting to unstable peroxides. These finally result in conjugated products, up to nine carbon atoms. The same time, 3D cross-linking occurs as a basic consequence of the presence of radicals. Oxidative cleavage and production of low molecular weight volatile products as the end-result of the possible oxidative processes, is routinely investigated by gas chromatography (Mills and White, 1987). After transesterification of all possible glyceryl esters with hydrochloric methanol, a mixture of lower molecular weight methyl esters of all present fatty acids is produced and analyzed. Such a gas chromatogram e.g. from an aged linseed oil medium reveals that: (a) Long unsaturated fatty chains (e.g. oleate) are decreased due to oxidation. (b) Saturated fatty chains are basically unchanged. (c) A number of oxidation products, such as dimethyl esters of azelaic (nonadioic) acid are formed. The quantities of esterified fatty acid residues such as methyl palmitate (C16:0), stearate (C18:0), oleate (C18:1), as well as monomethyl and dimethyl azelate (C9) have been previously used as a measure of various oxidative routes in oils (Mills and White, 1987). Here the two numbers correspond to the number of C atoms and the number of double bonds, respectively. Oleate is slowly oxidized (decrease in concentration), while azelate is produced as an oxidation product, thus enabling an evaluation of the aging process. Therefore, low C18:1/C18:0 or C18:1/C16:0 and high C9/C16:0 or C9/C18:0 ratios suggest a strongly oxidized medium.

Based on the above, model samples of cinnabar pigment in linseed oil covered with dammar varnish (~ 20 μm thickness) were made and artificially aged using UV lamp and elevated temperature. Then for each sample a predetermined thickness of varnish was removed by choosing the laser fluence (KrF excimer) based on the ablation curves of dammar (see Fig. 1). Note that the same dammar samples were used to obtain the ablation rate data and the GC-MS data. Finally, the remaining pigment/medium/ remaining varnish of each sample was analyzed by GC-MS (for details see Zafiropulos *et al.*, 2000).

In general, the results show no increase of oxidation products when at least a thin varnish layer is left over the pigment and the optimum laser fluence is used. On the contrary, when high laser fluence is used the oxidation products increase. In more detail, Fig. 7 presents the C18:1/C16:0 and C9/C18:0 ratios of the GC-MS peaks for different values of removed resin layer. Based on the results described in the previous sections, the optimal fluence for the ablation of dammar is 0.3 J/cm^2, while a fluence of 0.8 J/cm^2 is considered rather high, see Fig. 1(b). Figs. 7a and 7b correspond to the results using these two values of energy fluence, respectively. Fig. 7a shows that there is no evidence of any oxidation change compared to the reference sample, even when the remaining varnish layer is only 2 μm. On the contrary, when using a laser fluence of 0.8 J/cm^2 the concentration of the sensitive to oxidation oleic component (C18:1) is decreased as the thickness of the removed layer increases, see Fig. 7b. At the same time the concentration of the C9 oxidation product increases - the exception of the last point may be due to the further oxidation of C9 to lower molecular weight products. Therefore, it is concluded that the chemical alteration of the binding medium can be avoided when using the optimal laser fluence that is different for each resin and polymerization parameters. At the same time a portion of the resin must be left for "filtering out" the transmitted light.

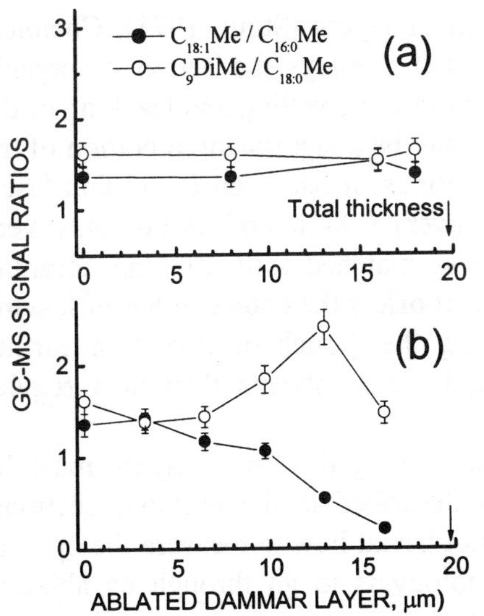

Fig. 7. GC-MS signal ratios as a function of removed layer thickness using KrF excimer laser at fluence 0.3 J/cm^2 (a) and 0.8 J/cm^2 (b).

2.2. Laser-assisted removal of aged varnish from paintings

The application of varnish layers has been shown to play a significant role in protecting painted surfaces (De la Rie, 1987). During aging, the varnish degrades through oxidation, auto-oxidation, and cross-linking processes, catalyzed by the absorption of light (De la Rie, 1988, 1989). In particular, the absorption of UV light leads to formation of products playing a significant role in the polymer phase formation (cross-linking). For aged natural resins it has been recently found (Zafiropulos, 2001b) that certain optical properties related to polymerization and/or oxidation (e.g. UV or IR absorption) are scalar the vector being the depth from the surface. This is owing to the exponential decrease of ambient light (catalyzes the formation of radicals and therefore the polymerization processes) as it propagates into the resin. The gradient of oxidation across the film thickness is directly related to an equivalent gradient of the solubility throughout the thickness of varnish layer (Theodorakopoulos and Zafiropulos, 2002). Consequently, the deeper one goes into the varnish layer the more dilute solution (of the appropriate chemical agent) is required to remove the remaining layer. As a general rule, the required concentration drops by a factor of two when going from the surface to a depth of about 8-10 μm.

Owing to preservation reasons the artwork must be relieved from the degraded varnish that can affect the inner layers (Stout, 1975). Chemical cleaning is a complicated procedure that employs the action of organic solvents potentially able to initiate a process of swelling and leaching of the film (Sutherland, 2001). The outcome may be a significant alteration of the constitution of the image layers of paintings. It has been found that laser-assisted removal of the top varnish layers (~8-10 μm) is not only very helpful but also unique since it cannot be matched with any other cleaning method (Doulgeridis, 1997). Moreover, it offers the choice either of leaving the rest of the resin intact (preserving the "touch of time") or further removing it, using less aggressive and toxic solvents than in a regular chemical treatment.

The laser-assisted removal of the top aged varnish layers must be performed at the optimal fluence, as described in the previous sections. When working on real paintings, it is not trivial how to measure the optimal fluence. The only established methodology is to go through an ablation rate/efficiency study using a small part of the aged varnish (typically at the edge of the painting) for generating the ablation spots. An area of 5 x 50

mm^2 is enough, where a number (usually 7-10) of ablation spots are created, for each using a different fluence value. The total ablation depth of each spot must never exceed the 80% of the varnish thickness for obvious reasons. It must be noted that the ablation efficiency is calculated from the mean depth of the ablated material rather than the maximum depth, found by profilometry measurements and subsequent data analysis. A description of such measurements is given elsewhere (Zafiropulos, 1995). A second methodology, which is though still under investigation, is based on the onset of plume generation. Plume starts to appear at fluence values near the optimal fluence.

The plume observation has been also associated to the well known monitoring technique LIBS or Laser Induced Breakdown Spectroscopy (e.g. see Anglos, 2001), being a micro-destructive technique. Here, however, the observed emission during laser-assisted divestment is a useful byproduct used for on-line controlling the exact depth advancement. (Gobernado-Mitre, 1997; Klein, 1999; Maravelaki, 1997, 1999; Zafiropulos, 1998, 2001c). During laser ablation the produced emission is characteristic of the composition of the ablated material. In most cases, the layer synthesis changes as the material removal progresses towards the final step, and it is exactly this change that can be used to guide and control the laser ablation process.

For an automatic control during the laser cleaning process, the light emission spectra are continuously recorded in a pre-selected spectral region. The data acquisition processor is programmed to calculate certain peak intensity ratios that are then used as inputs in an algorithm, which produces an output value after each laser shot. When this value is not within certain boundaries, the laser stops firing and the motorized X-Y stage moves the painting or the laser beam to a new position where the cleaning process starts again. At this new position, laser pulses are delivered until the output value moves again out of the chosen confines. The distance that the sample is moved in every step is determined by an overlapping protocol for the sequential sample movement in both X and Y directions.

At this point it is worthwhile to present another evidence for the existing scalar properties in the resin, which have been used for the on-line control during the laser-assisted removal of aged varnish. Except polymer phase formation, there is also natural oxidation (De la Rie, 1988) taken place especially at the outer layers. In the case of a single resin layer, the elemental composition throughout a cross-section of the layer does not change with aging. This fact could be prohibitive in using LIBS as a monitoring technique, since LIBS is in general an atomic analytical method. However, in LIBS experiments that were performed on naturally aged resin samples we found out that there is a change in the spectrum of the produced plume - as the varnish removal progresses towards the lower varnish layers - originating from produced excited dimers. Fig. 8 shows a typical series of LIBS spectra observed when successive laser pulses ablate aged varnish. The four spectra correspond to the 2nd, 3rd, 4th, and 5th pulse respectively. A comparison of the four spectra reveals the gradual decrease of the CO $B^1\Sigma$-$A^1\Pi$ bands (Angstrom bands) and the simultaneous increase and then stabilization of the two most intense C_2 $A^3\Pi_g$-$X'^3\Pi_u$ (Swan bands). The CO bands originate from the carbonyl groups (products of oxidation, e.g. see De la Rie, 1988) whose density is expected to be higher at the outer varnish layers than at the inner ones. This is verified by observing the micro-FTIR spectra (reflection mode) of the surface after each consecutive laser pulse. Fig. 9 shows a representative FTIR spectrum (Chryssoulakis, 1996). The ratio of the peak attributed to carbonyl groups (1650-1730 cm^{-1}, marked as A) versus the characteristic C-H bending of methyl and methylene groups (1400-1470 cm^{-1}, marked as B) was found to decrease after each laser shot (see small graph in Fig. 9). This is in accordance with the LIBS observations shown in Fig. 8. In this particular varnish system, the limiting value of A/B was found to be 1.4 ± 0.1..

Here we will present a few test case studies just for demonstrating the proper application of laser-assisted top-layer removal in real paintings. The research has been carried out in collaboration with Michael Doulgeridis, director of the Conservation department of the National Gallery of Athens. The first example was an 18th century Flemish tempera painting on a wooden panel of size 27x37 cm^2. It suffered from an unsuccessful restoration some decades ago, and the varnish that was used to protect the painting had undergone severe polymerization (hard as glass). The optimal fluence was found to be 0.38 J/cm^2 with an ablation rate of 0.25 μm/pulse (λ_L = 248 nm, 25 ns pulse duration). The final fluence used was 0.42 J/cm^2 while LIBS on-line control was employed as described above with the mean thickness of removed varnish being 9 μm. Fig. 10a shows an initial stage of processing where a small area (left, towards the bottom) has been laser-cleaned, while in Fig. 10b the two thirds of the painting (top) have been

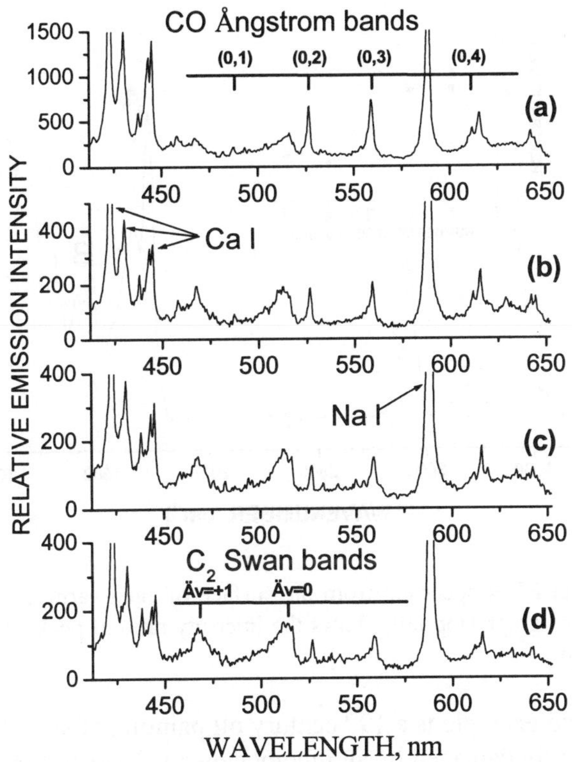

Fig. 8. LIBS spectra of unknown naturally aged varnish using KrF excimer laser at a fluence value of 0.5 J/cm^2. The four spectra correspond to the 2nd (a), 3rd (b), 4th (c) and 5th (d) laser pulse.

additionally processed. At the very bottom of Fig. 10b and just below the small initially laser-cleaned area, we can also see two little vertical zones that have been cleaned using a mixture of solvents for comparison. The complete exposure of the pigments in the latter case can be compared with the delicate top-layer laser-assisted removal. Fig. 10c presents the interface between the original surface of oxidized resin (top) and the uncovered layer of the less oxidized resin (bottom). Fig. 10d shows the final result after completion of the laser processing and the subsequent removal of the exposed soft varnish using a mild solvent (Doulgeridis, 1997).

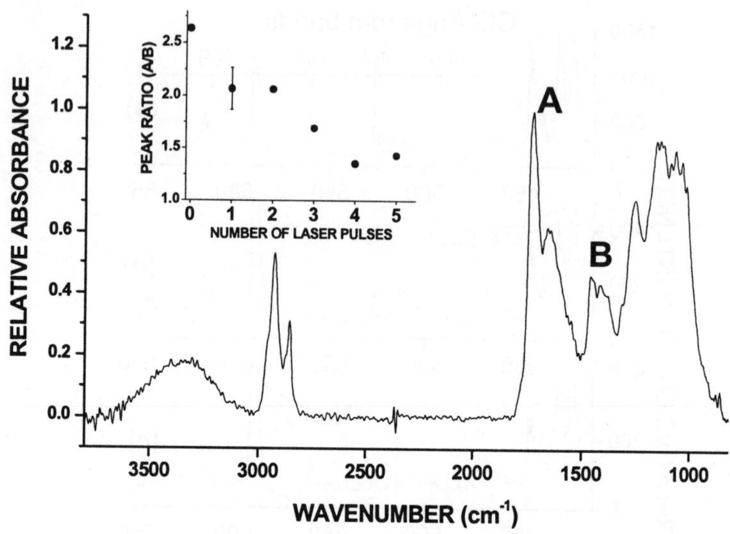

Fig. 9. Typical FTIR spectrum from the surface of aged varnish (same as in Fig. 8). The enclosed graph (top-left) shows the intensity ratio of peaks A versus B after each laser pulse.

The second example is a 19th century oil painting of size 49x77 cm^2 that has been severely damaged by an inhomogeneous layer of lime mortar over the surface of aged varnish, see Fig. 11a. The latter was only a few microns thick (~7 μm), evenly spread over the original pigment. Any attempt to remove the lime particles and/or the aged varnish using the traditional technique of a cotton swab wet in the appropriate solvent would have been catastrophic, owing to the transformation of the cotton/particles system into a form of sandpaper. The strategy of KrF excimer laser-assisted divestment in this case was to ablate the top aged resin layers resulting to the ejection of lime particles together with the resin photofragments. At this particular case, a fluence value of 0.30 J/cm^2 (slightly below the optimal fluence of 0.33 J/cm^2) was used in order to minimize the etching step. The thickness of removed varnish was ~2 μm with the synchronous removal of the over sitting lime particles. The final result of the restoration (Doulgeridis, 1997) is presented in Fig. 11b.

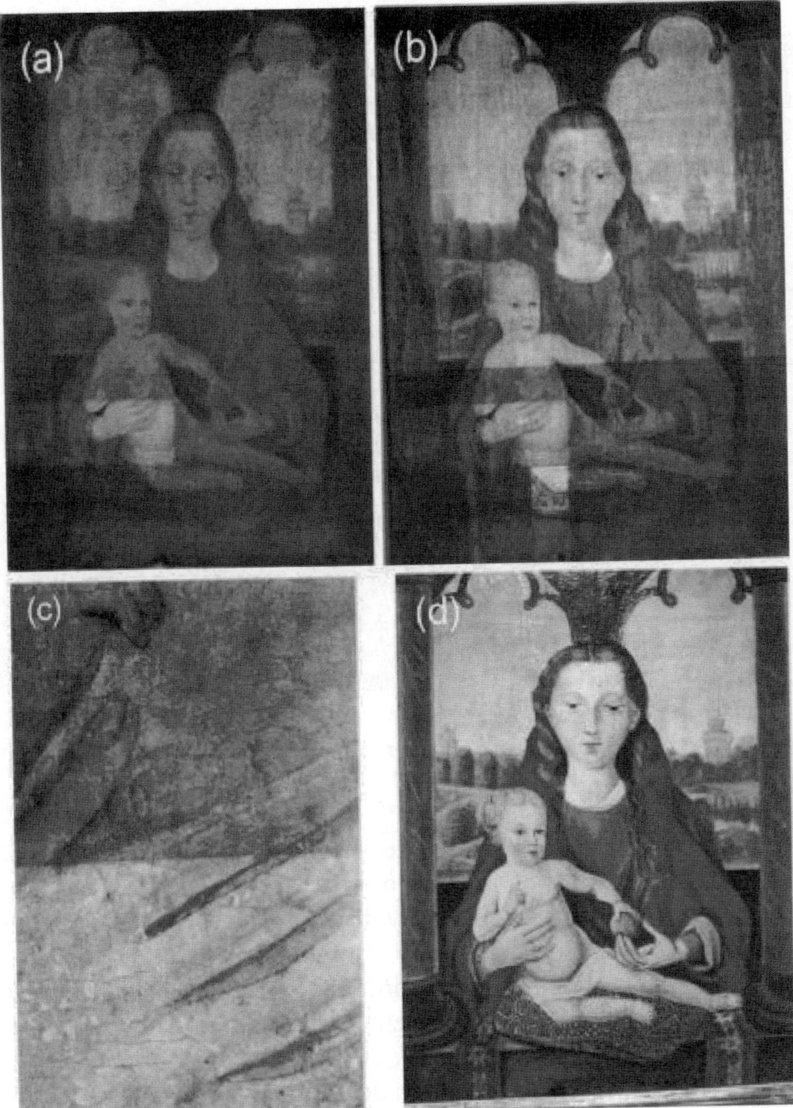

Fig. 10. (a) Original surface except a small laser-cleaned area in lower left part. (b) Additionally, the top part has been laser-cleaned. Two small zones (bottom) have been chemically cleaned for comparison. (c) A detail at the interface between original resin (top) and laser-cleaned surface (bottom). (d) After laser intervention followed by traditional restoration.

Fig. 11. (a) Original surface and (b) after laser intervention followed by traditional restoration.

In the case of evenly applied resins, we can detect evidence of the gradient polymer formation in terms of crack formation. Fig. 12 shows part of a tempera icon painting covered with an unknown varnish aged over a period of more than 100 years. A KrF excimer laser (fluence 0.54 J/cm^2) has been used to remove the resin at ten different zones. The zones are indicated by increasing numbers that represent an increased removal of varnish (Theodorakopoulos, 2002). At zones 1-5 the increase of depth-step is 3.3 μm, i.e. zone 1 corresponds to 3.3 μm depth from the original surface, while zone 5 corresponds to 16.5 μm. At zones 6-10 the depth-step intervals are larger, with a maximum depth of 82 μm at zone 10, where the remaining varnish is ~20 μm. The observed cracks in the resin are due to the stiffness of the aged resin near the surface. The gradual decrease of cracks going from zone 1 to 10 is owing to the polymer phase gradient as a function of depth from the surface. The more material is removed the fresher the

Fig. 12. Part of painting covered with naturally aged varnish. Ten laser cleaned areas span a thickness of varnish removal from 3.3 μm (area 1) to 82 μm (area 10).

remaining resin is and therefore the less cracks are revealed. In this case, the end user can choose the endpoint according to the desired appearance and/or solubility of the remaining varnish.

2.3. Laser-assisted removal of paint from composite materials

In a recent study (Zafiropulos, 1995) on the photoablation of polyurethane (PU) films used as paint systems on delicate composite materials, short-pulse ultraviolet laser ablation has been studied in terms of ablation rate/efficiency behavior and time of flight mass spectroscopy of the positively charged photofragments. The results of the influence of energy fluence on the ablation rate were tested against different existing photoablation models. The upper limit of the mean activation energy for desorption has been found to be considerably lower (~ 3 to 15 times) than the average energy required to break single covalent bonds, indicating the importance of photophysical types of processes (Luk'yanchuk, 1993a, 1993b, 1994, 1996). At least in one case (λ_L = 193 nm) it was found that photochemical processes (Pettit, 1993, 1994) predominate in the operative mechanism. It was also found that coherent and incoherent multiphoton absorption must significantly contribute to the ablation in the case of femtosecond and picosecond laser pulses. Apart from these general conclusions it is extremely unsafe to get into details of the operative ablation mechanisms.

A particular application of Laser-assisted paint removal from composite materials is the so-called "paint stripping" from composite surfaces, e.g. aeronautical parts (Manz and Zafiropulos, 1997). As in varnished paintings where the underlying pigment layers are sensitive to photons transmitted through the resin, the paint layers on composite structures can be removed only if the integrity of the composite's resin is guaranteed. Therefore, the use of on-line monitoring techniques is also essential for this type of application. From fatigue testing measurements it has been found (Pantelakis, 1995) that there is always a measurable fatigue of the composite when laser removes the total paint system. Such an example is shown in the SEM pictures of Fig. 13, where an epoxy/carbon fiber composite structure painted with a Si-based polymeric paint system has been divested using a XeCl excimer laser at 1.4 J/cm^2. Fig. 13a shows the reference painted surface, Fig. 13b corresponds to a processing where the paint has been partially removed, while in Fig. 13c we see the breakage of carbon fibers when the paint coating is totally removed. Here, the epoxy primer – a 3-μm layer between the composite substrate and the paint system – had been applied as a continuation of the epoxy resin where the carbon fibers were embedded in. Therefore, if the ablation process is terminated after all the paint is removed, the laser affects the epoxy resin of the composite substrate. It is possible though, to choose the endpoint in such a way that there is no fiber damage succeeding an almost complete removal of the top coating. Fig. 14 shows the results of paint stripping on such a composite component, with the end-point of the final processing corresponding to the state shown in Fig. 13b.

In cases where the primer is a well-defined layer, removing the top coating and leaving part of the primer paint on top of the composite's resin has been found to be safe for the substrate. Fig. 15 shows the SEM image of a PU based paint system applied on Kevlar composite. In the laser-divested area (KrF excimer laser, 0.5 J/cm^2) the PU primer has been uncovered. The

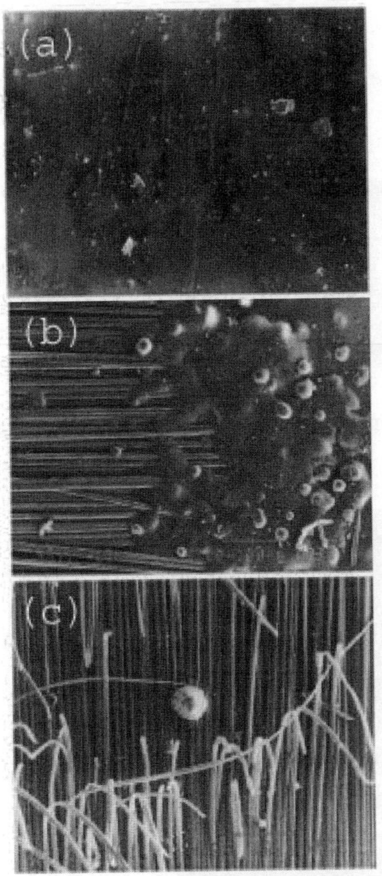

Fig. 13. SEM pictures of painted epoxy /carbon fiber composite (x 200): (a) reference surface, (b) after laser-assisted partial paint removal and (c) after total paint removal.

optimal laser parameters for this particular system have been investigated by Zafiropulos, 1995. Fig. 16 shows a large area of the composite where eight different KrF excimer laser processes at different depths have been performed. The area at the top-right has been divested until the primer has been uncovered, while in the remainder areas the moralization of the composite has been uncovered in varying levels. At the three sections in the middle row the resin of the metallisation layer has been partially ablated, while at the three areas of the lower line the ablation has stopped just after the removal of the primer. Finally, Fig. 17 shows the results of paint stripping of a non-moralized Kevlar composite – painted with the same PU

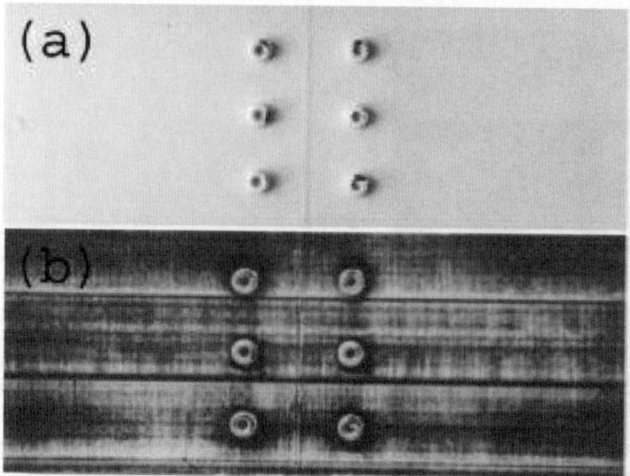

Fig. 14. Epoxy/carbon fiber composite component: (a) Reference surface and (b) after laser-assisted partial paint removal.

Fig. 15. SEM picture of painted (PU) metallized Kevlar composite (x 350, 35° angle of observation). In the center the material has been ablated down to the anticorrosive primer PU.

Fig. 16. Painted metallized Kevlar composite paint stripped down to eight different depths (see text).

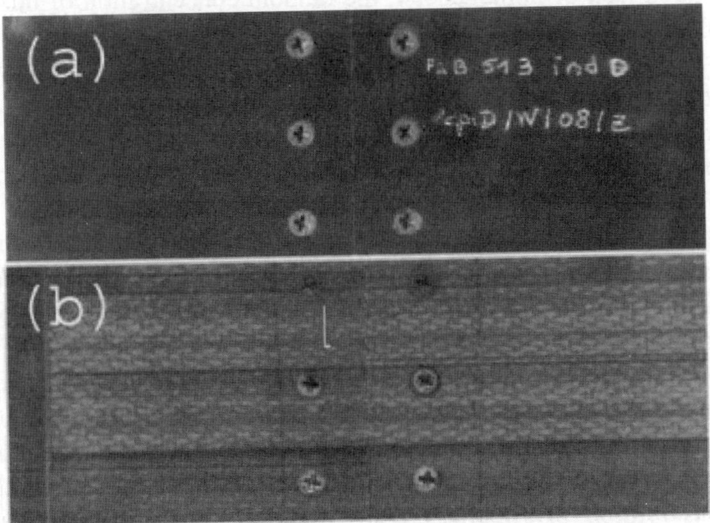

Fig. 17. Painted non-metallized Kevlar composite before (a) and after paint stripping (b).

paint system – down to the outer epoxy resin layer of the composite. In this particular example, this outer resin layer had a thickness of about 10 μm giving the possibility of uncovering it without affecting the fibers of the composite.

3. Laser divestment of encrustation

3.1. Major operative mechanisms and associated optical phenomena

In contrast to UV laser ablation of complex layers of various polymeric phases where the dominant mechanism is the bond rapture within the macromolecules and the subsequent ejection of mostly small molecular fragments, in nanosecond laser divestment of inorganic encrustation the prevailing processes are the selective explosive vaporization and the spallation induced by shock wave. A mathematical description of the former has been given by Asmus *et al.* (1976) and Watkins (1997), while the results of shock wave action cannot be safely quantified. This is owing to the encountered lack of homogeneity, the random concentration of mixtures of minerals, and the irregular sequence of different encrustation layers that makes any modeling a difficult task. In the limited extent of this work we will not attempt any detailed analysis of the operation of the ablation mechanisms on these complex substrates. Rather, we will try to elucidate the problem by presenting evidence for the operation of these mechanisms, as well as the accompanying optical phenomena such as yellowing.

Black encrustation divestment on marble/stone artifacts using a Q-Switched Nd: YAG laser (λ_L = 1064 nm) has been fluently implemented in conservation during the last years (e.g. see Asmus, 1976, 1986, 1987; Cooper, 1998; Vergès-Belmin, 1997; Maravelaki, 1997). Other pulse durations (e.g. see Asmus, 1976; Siano, 1997, 1999, 2000) and wavelengths (Skoulikidis, 1995; Maravelaki, 1999; Klein, 2000; Marakis, 2000, 2002) have been used as well. Actually, laser-assisted cleaning of sculptures and façades is by far the most frequently met application in the field of *Lasers in the Conservation of Artworks (LACONA)*. This is owing to the "self-limiting" divestment process, that is, the spallation threshold for the encrustation is usually much lower than for the substrate marble or stone. Therefore, by choosing proper laser fluence values between the two

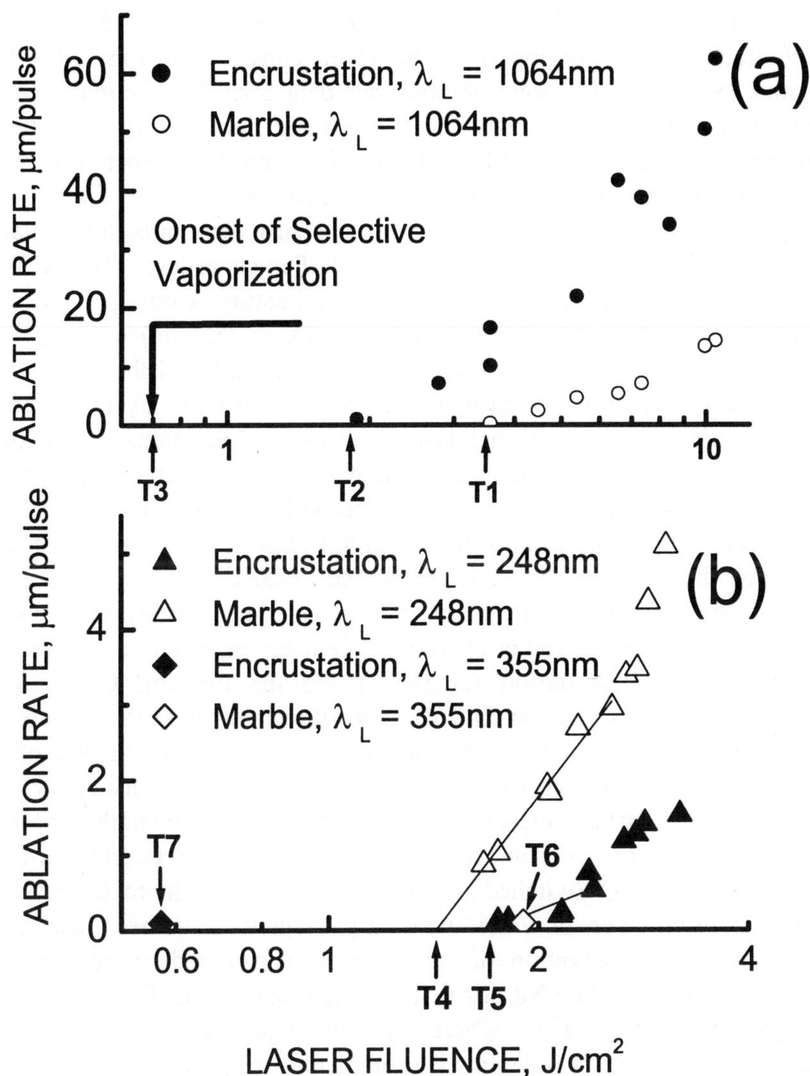

Fig. 18. Ablation rate data for Pentelic marble and a particular dendritic encrustation, a cross-section of which is shown in Fig. 19, obtained using a Q-Switched Nd: YAG laser (a) and a KrF excimer laser and the 3rd harmonic of a Q-Switched Nd: YAG laser (b). Arrows designate the various thresholds described in text.

thresholds one can remove the encrustation without affecting the substrate. Shock wave formation, however, may induce some structural damage (e.g. micro cracks) when working at fluence levels considerably above threshold (Asmus, 1976). Fig. 18 shows ablation rate data from Pentelic marble and a thick encrustation. In Fig. 18a the two sets of data clearly indicate the spallation thresholds of marble and crust (T1 and T2 respectively) when using the fundamental of a Q-Switched Nd: YAG laser. T1 (3.5 J/cm^2) is the absolute maximum fluence for processing the particular marble/encrustation system. The corresponding thresholds for a KrF excimer laser [T4 and T5 in Fig. 18b] are not in favor of such an application. Contrary, the third harmonic of a Q-Switched Nd: YAG laser (λ_L = 355 nm) offers a huge threshold gap in this particular case [see T6 and T7 in Fig. 18b].

The usual superposition of more than one encrustation layers prevents a more analytical approach to such laser divestment applications. At this point a generalization for the structure of the encrustation layer is very crucial. From previous studies (e.g. see Skoulikidis, 1981; Cooper, 1998; Maravelaki, 2001), one can notice that in most cases there is always a layer that must be at least partially preserved located just on top of the marble/stone substrate. Although numerous types of encrustation – including biological, mixed or even multiple-layer encrustation – have been considered, the encrustation removed using the fundamental of a Q-Switched Nd: YAG laser at fluences near threshold presents a couple of unique characteristics. The first characteristic is a yellowish hue of the resulted surface compared with the non-cleaned reference surface, or even with the same surface cleaned using other laser wavelengths or other techniques (e.g. micro sandblasting). The second characteristic of the substrates after the Q-Switched Nd: YAG laser action is the maintenance of the gypsum-rich layer, in which the dark particulates have been selectively vaporized. The preservation of this layer is considered to be the great advantage of Q-Switched Nd: YAG laser cleaning of stone. Even in cases of multiple encrustation layers where the outer black layer (rich in carbon particles and other pollutants) is completely removed, the remaining gypsum-rich encrustation layer is left clean of black particles. Therefore, we could consider such a layer as being the most common encrustation encountered at the last stage of laser cleaning, below other existing layers. A typical cross-section of a 200-μm single-layer encrustation on marble is shown in Fig. 19a. Fig. 19b shows a cross-section of the same sample corresponding to a laser processed area using the fundamental of a Q-switched Nd:YAG laser at a fluence (0.8 J/cm^2) just above the threshold of selective vaporization [shown in Fig. 18a at 0.7 J/cm^2] of the particles. In Fig. 19b it is shown that the gypsum-based layer is preserved, while the dark carbon/iron/aluminum/silicate particles (Maravelaki, 1999, 2001) are selectively vaporized.

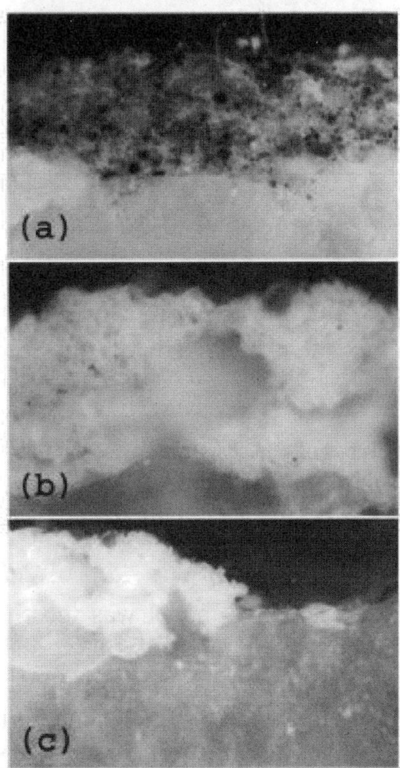

Fig. 19. Cross-section (x 200) of dendritic type of encrustation on Pentelic marble: (a) reference encrustation, (b) after selective vaporization of particles using 50 pulses at 0.8 J/cm^2 and (c) after partial removal of the gypsum rich layer using 50 pulses at 1.2 J/cm^2 of a Q-Switched Nd: YAG laser. The width of each picture corresponds to 500 μm.

In general, there are different mechanisms of laser-assisted removal of encrustation, even operating at the same time on a particular substrate (Watkins, 1997). For example in the case of multiple encrustation layers, the top black layer may be removed by thermal vaporization and simultaneous spallation (shock wave action). The above observations, however, point to "selective explosive vaporization" as the main mechanism responsible for the last step of stone cleaning – a selective vaporization of the dark particulates within the body of the preserved layer. This selectivity is due to the absorptivity of 1064 nm laser light, which is 3-4 times higher for the black particles than for the gypsum layer or the substrate (Asmus, 1976). The onset of this process occurs at much lower fluence values than the spallation threshold [T3 and T2 in Fig. 18a, respectively].

Asmus *et al.* (1976) have applied a one-dimensional surface-heating model to predict the required laser fluence levels for selective vaporization. In this simplified model the temperature increase at the surface is found to depend on the surface absorbance, the incident laser

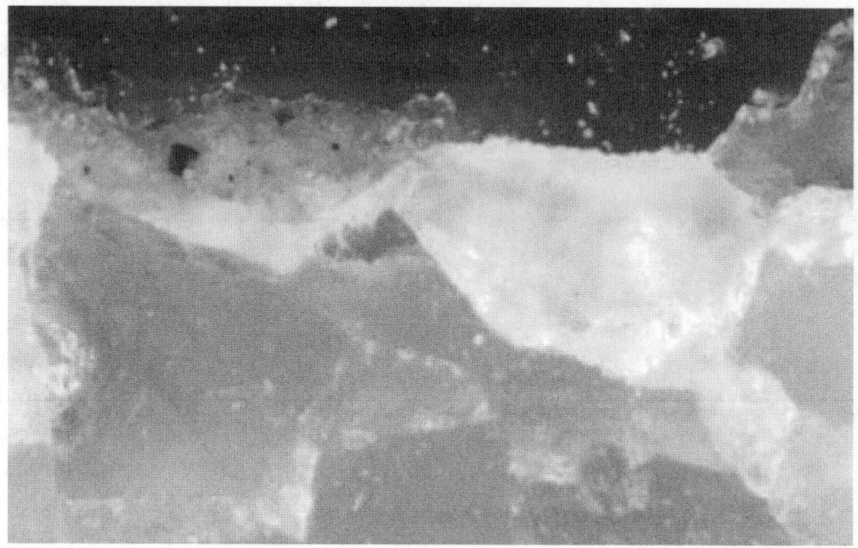

Fig. 20. Cross-section (x 200) of dendritic type of encrustation on Pentelic marble: Reference encrustation (left), after removal of encrustation (right) using 50 pulses of the third harmonic of a Q-Switched Nd: YAG laser at 0.7 J/cm^2. The width of the picture corresponds to 500 μm.

flux, and the thermal parameters of the material. This model was mostly applied for a normal-pulse laser (pulse duration of the order of 10^{-3} s) and the results agreed very well with the experimental fluxes. In principle, it can be also applied to nanosecond pulses for fluxes below the spallation threshold of the surface layer, where selective vaporization of particles embedded in the gypsum-rich layer occurs. Finally, it can be corrected for non - planar laser/matter interac-tion, which is the case of particle vaporization. For the highly variable, inhomogeneous and poorly characterized encrustations, the model is satisfactory in projecting laser effects and fluence requirements for particular laser-divestment projects.

The experimental fluence value for the onset of particle vaporization in the encrustation of Fig. 19 using the fundamental of a Q-Switched Nd: YAG laser is 0.7 J/cm^2 [T3 in Fig. 18a], which is close to calculated values using different sets of parameters. For laser fluence values between 1.0 J/cm^2 and 1.8 J/cm^2, it was found that the generated shock wave had an additive effect on the delamination and finally the fragmentation of the encrustation when many pulses were used. Fig. 19c shows a cross-section of the same sample after irradiation of 50 pulses at 1.2 J/cm^2, where all particles are removed as well as part of the gypsum-rich layer. For fluence values > 1.8 J/cm^2 spallation of the encrustation started right at the first laser pulse. Fig. 20 shows a cross-section on the interface between the reference encrustation (left) and the laser-ablated area (right) using the third harmonic of a Q-Switched Nd: YAG laser (50 pulses at 0.7 J/cm^2). The self-limiting process at this laser wavelength (like also in the case of λ_L = 1064 nm), as well as, the different type of operative mechanism [comparison with Fig. 19b] can be easily identified.

When the fundamental of a Q-Switched Nd: YAG laser is employed to selectively vaporize the dark particles, the color of the surface turns into yellowish while use of the third harmonic leaves the surface color unchanged. This is shown in Fig. 21 where both 1064 nm and 355 nm laser wavelengths have been used, the numbers representing the laser fluence used in units J/cm^2. In this particular example the encrustation had a thickness of 25 ± 5 μm. Here, the threshold of selective vaporization for λ_L = 1064 nm was also 0.7 J/cm^2, while the threshold of spallation using the third harmonic was 0.3 J/cm^2 (compared to 0.57 J/cm^2 value in the previous example). The lower value is owing to the thinner and looser character of this encrustation. On the right side of the picture we see a zone processed with λ_L = 1064 nm and on the left sine there are small areas processed with different laser fluence values of λ_L = 355 nm. In the latter case, although there is a varying cleaning effect going from 0.3 to 0.5 J/cm^2 the yellow hue appears nowhere on the processed areas. Therefore it becomes obvious that yellowing is an optical phenomenon that must be further elucidated. Zafiropulos, 2002 a, b, have carried out initial studies and the first results have shown that yellowing is a light scattering problem. A summary of the up to now results is presented below.

As a result of the selective vaporization of particles, the preserved gypsum-rich surface layer becomes more porous than its previous state not only because of the created voids, but also because of the decrease of the specific volume of the remaining layer - partial conversion from gypsum to anhydrites and hemihydrates (Maravelaki, 1999). Moreover, when the removed particles are of the order of 10 - 20 μm, the voids in the gypsum-rich layer are even more apparent [see Fig. 19b]. In such a situation we expect the elements that the particles are composed from to remain in the

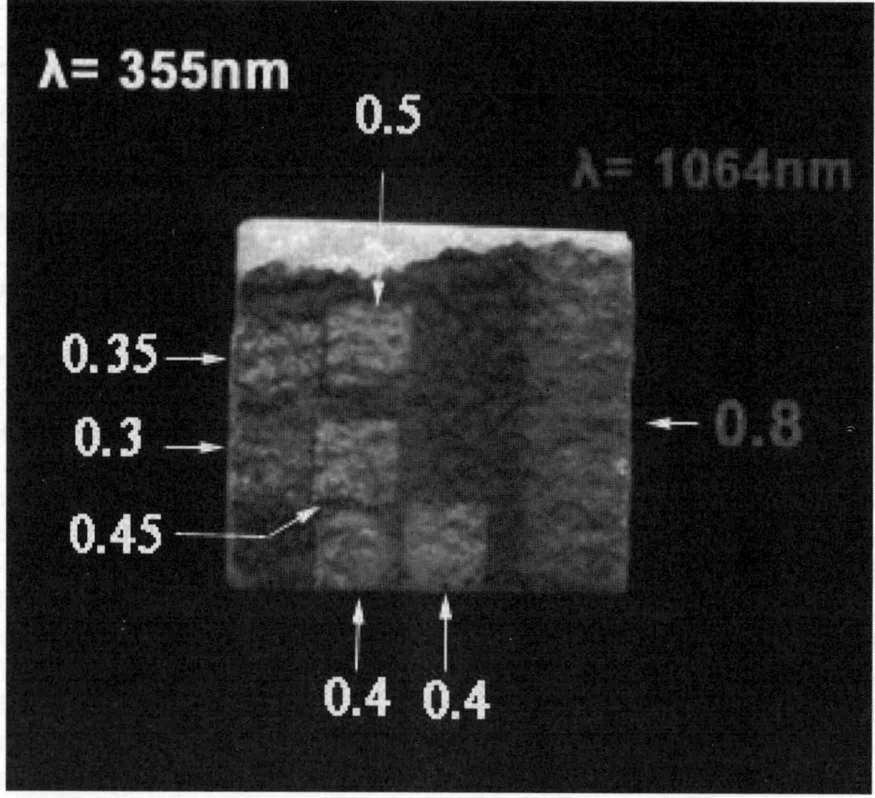

Fig. 21. Thin encrustation on Pentelic marble divested using λ_L = 355 nm (left) and λ_L = 1064 nm just above the onset of selective vaporization (right). The numbers represent different fluence values used in units of J/cm^2.

preserved layer. This has been verified using LIBS and SEM-EDX analysis on the cleaned surface (Maravelaki, 1997). This is also in accordance to the results of a recent study (Klein, 2001), where it has been found that the surface left after 1064 nm laser cleaning of an artificially soiled sample contains Fe as in the original layer. All these observations lead to conception of a simple light scattering model that consists of a thin absorbent layer (corresponding to the cleaned gypsum-rich layer) and a virtual top layer considered being a single scattering layer. The light transport is diffusive while the angular distribution of the reflectance is influenced by the last scattering event, process that depends on the

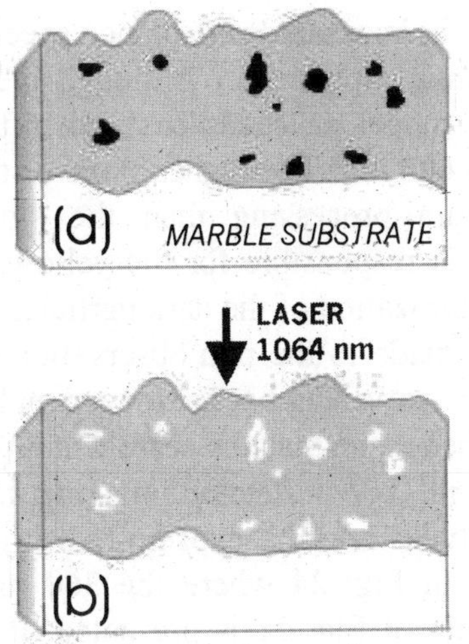

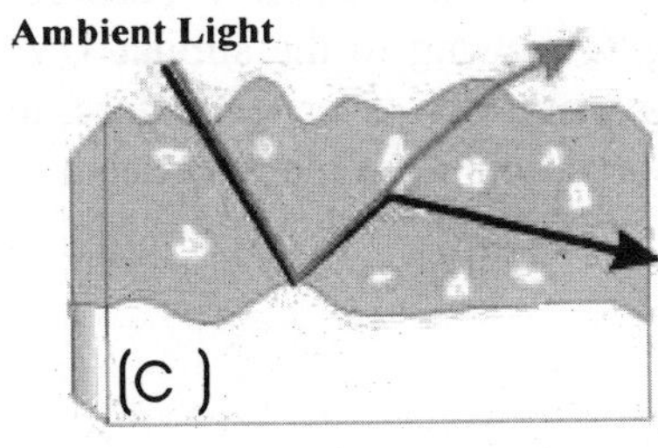

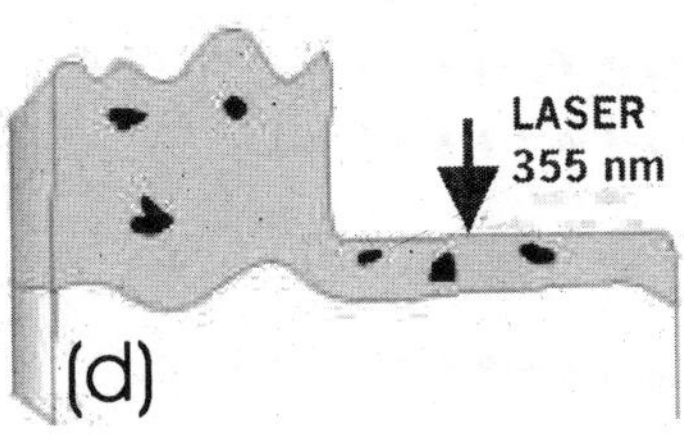

Fig. 22. Schematic representation of the encrustation (a) and the two different ablative mechanisms: the selective vaporization using low fluence at λ_L = 1064 nm (b), resulting to the optical phenomenon of yellowing (c); and the spallation using λ_L = 355 nm (d).

scattering anisotropy of these last scattering centers (voids). Although multiple scattering usually occurs, the final path of a ray is governed by the last scattering event. Fig. 22 schematically represents the conception of this simple model in relation to the two different operative mechanisms, the selective vaporization Fig. 22b and spallation Fig. 22d. Fig. 22c simplifies the possible path of two collinear rays of different color, where the higher λ ray finally escapes while the lower λ ray scatters back into the material.

Following the approach by Popescu (2000), the angular distribution of light scattered by the scattering centers (created micro-voids) is related to the integrand of the product of the probability to scatter the light within the elementary solid angle times the angular transmission probability for the diffusive medium. Since the former is a function of λ and refractive index, the spectral distribution of light emerging from the sample can be calculated. Using this simplified scattering model, we can calculate the reflectance spectra in the backward direction for different sizes *d* of the scattering centers ($d \gg \lambda$). The result (monotonic to λ) can be further corrected for the spectral absorption of the diffusive medium by multiplying the scattering spectrum with the squared transmission of the layer (Zafiropulos, 2002a,b). The result for *d*

= 5 μm is plotted in Fig. 23 together with the measured reflectance spectra using a homemade hyper-spectral imaging apparatus. Although very simplified to a first approximation, this model succeeds to predict the measured relative reflectance spectra up to 600 nm. The higher slope of the reflectance spectrum after λ_L = 1064 nm processing than the slope corresponding to λ_L = 355 nm processing is responsible for the yellowish appearance of the surface after selective vaporization of the dark particles.

At this point a cross-reference can be made to the non-observation of the yellowing effect when using shorter wavelengths or longer pulse durations on a wet surface. In these cases the reason must be searched out to the different operative ablation mechanism. UV reflection imaging has shown (Marakis, 2002) that the black particles and the gypsum-rich layer absorb 355 nm equally. This is revealed in Fig. 24 where the 360 nm-centered UV reflection image of the sample of Fig. 21 is presented. It is observed that the zone corresponding to selective vaporization of particles (right side in Fig. 21) is not visible in Fig. 24, owing to the similar UV-absorption characteristics of particles and gypsum-rich layer. Therefore, at λ_L = 355 nm the mechanism of selective vaporization is not likely to occur.

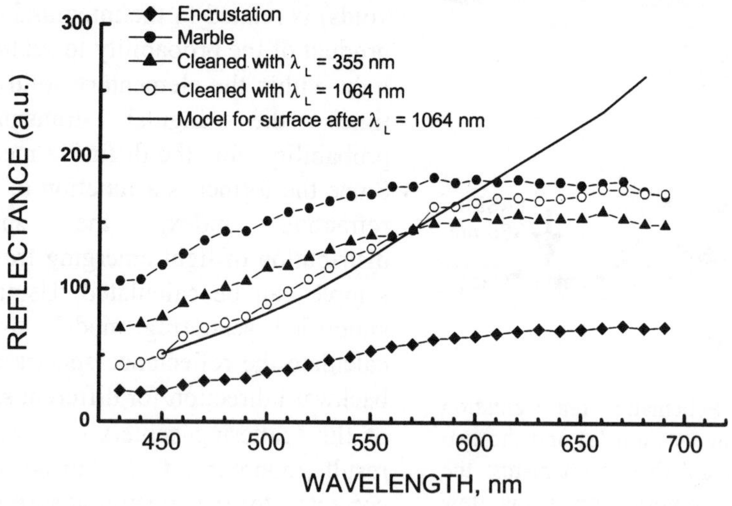

Fig. 23. Reflectance spectra taken from the surface of the sample shown in Fig. 21, recorded using a hyper-spectral imaging apparatus. The error is < 10%. The solid line represents the result of calculations for d = 5μm multiplied with the squared transmission of the surface layer.

In such cases, cross-sections have revealed the simultaneous and with no discrimination removal of black particles and gypsum-rich layer down to the substrate, as previously presented in Fig. 20. In practice, there is always a gypsum-interface left on top of the environmentally altered marble surface (Marakis, 2000). Consequently, at λ_L = 355 nm we would expect the observed reflectance to resemble the reflectance characteristics of the reference surface, which is actually the case (see Fig. 23). As with λ_L = 355 nm nanosecond pulses, in the case of water-assisted microsecond pulse Nd: YAG laser cleaning, extensive studies have revealed that the superficial layers can be removed down to a desired depth (Sabatini, 2000; Siano, 2000; Pini, 2000).

Fig. 24. UV-reflectance image of the sample also presented in Figure 21. The window of observation is 320-400 nm centered at 360 nm.

In this case, stratigraphic observations have also shown that controlled "chopping" of the superficial layer occurs leaving the remaining layer without any visual alteration or voids, the final result being also the absence of yellowing.

3.2. Removal of encrustation - Test case studies

Cooper (1998) has presented many test case studies on sculpture laser cleaning applications using the fundamental of a Q-Switched Nd: YAG laser. Here a few examples of laser-assisted encrustation-divestment are presented just for demonstrating the particular application. In contrast to the fundamental, the third harmonic of a Q-Switched Nd: YAG laser has been also used with very good results. The beam delivery is usually through a flexible optical articulated arm, although transmission through fibers has

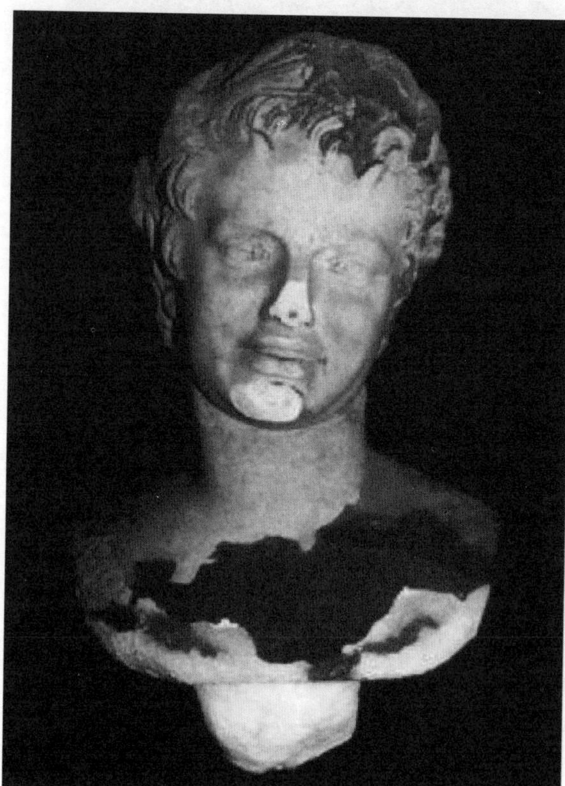

Fig. 25. Sculpture after partial cleaning using the fundamental of a Q-Switched Nd: YAG laser. By kind permission of Martin Cooper and the National Museums and Galleries on Merseyside.

been also reported for λ_L = 1064 nm (Boquillon, 1997). The marble sculpture in Fig. 25 is from the Roman Imperial period, which forms part of the Ince-Blundell collection of Classical Greek and Roman sculpture. Years of exposure to a polluted environment have led to formation of an unsightly and potentially damaging black crust, which obscures the surface of the sculpture. In some areas, weathering has led to the loss of the surface and severe degradation of the underlying crystalline structure. Fig. 25 shows the sculpture after partial cleaning: a Q-Switched Nd: YAG laser (1064 nm, 5-10 ns) has been used to carefully remove the hard black pollution encrustation (Cooper, 2001), whilst preserving the underlying patina. Beam delivery was via articulated arm and an approximate fluence in the range 0.5 - 1.0 J/cm^2 was used. The repetition rate varied from 1-10 Hz, depending on the condition of the surface being worked on. Alternative cleaning methods could not be used in this case due to the hardness of the black crust and the fragile nature of the marble.

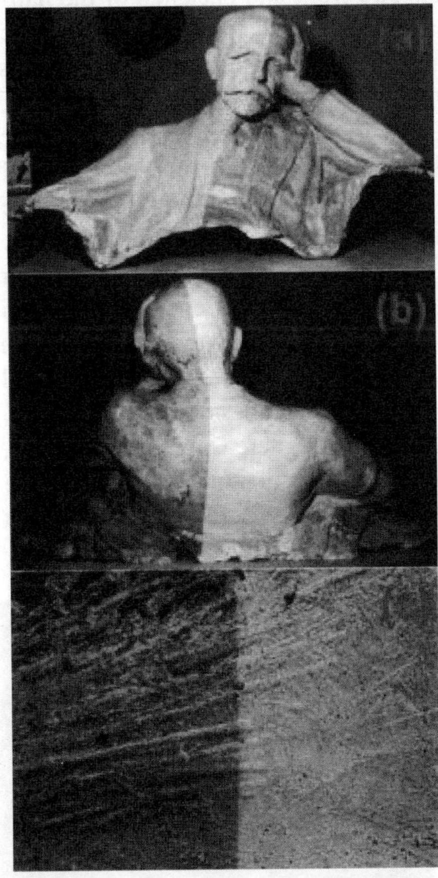

Fig. 26. Front (a), backside (b) and a detail (c) of half-cleaned plaster sculpture using the 3rd harmonic of a Q-Switched Nd: YAG laser at 0.55 J/cm^2. Taken from Marakis (2000).

Another example, using this time the 3rd harmonic of a Q-Switched Nd: YAG laser, is presented in Fig. 26. The original surface (plaster, made 1937) was kept in a good condition with the main problem being the soot-type pollution on the surface. At some points there were also spots of organic nature, most probably fungi. Traditional techniques of cleaning would have been catastrophic, e.g. water-based poultices would have partially dissolve the surface gypsum layer, while micro-sandblasting would have destroyed the sculpture's details owing to the softness of the material (Marakis, 2000). The spallation threshold for the plaster substrate using λ_L = 355 nm was found to be 0.62 J/cm^2, while the onset of soot removal was at 0.3 J/cm^2 with the efficiency increasing with fluence. Based on stereomicroscopic observations, the optimal fluence was found at 0.55 J/cm^2. The repetition rate was at 10 Hz. The front and backside of the half-cleaned sculpture is shown in Figs. 26a and 26b respectively, while Fig. 26c presents a detail of the surface showing the preservation of the sculpture's details.

The photomechanical character of soot removal using the 3rd harmonic of a Q-Switched Nd: YAG laser is demonstrated by the successful use of very low fluence values when the soot particles are loosely attached on the surface. Fig. 27 shows a modern plaster relief artificially covered with soot, which has been half-divested using a fluence of 0.2 J/cm^2. The same laser has been also successfully used to remove compact black encrustation from stone façades (Marakis, 2000). Fig. 28 presents such an example of a façade from the outer wall of a church undergone decades of environmental pollution. In such cases of uneven encrustation thickness and morphology, the divestment is carried out in 2-3 steps with increasing laser fluence, always being below the spallation threshold of the substrate. In the example of Fig. 28a first scan was made at 0.4 J/cm^2 while a second one followed at selective points using 0.5 J/cm^2.

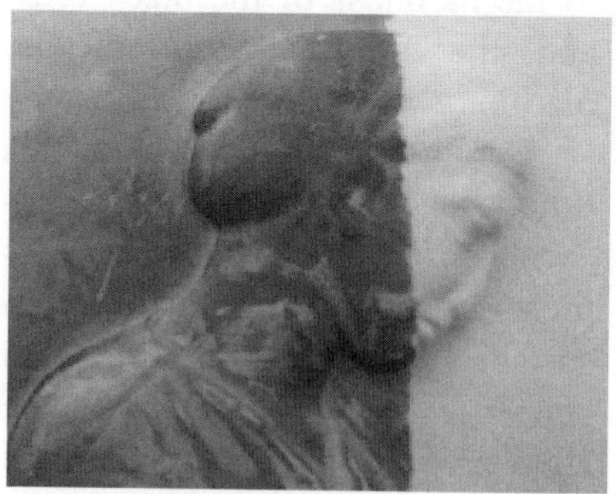

Fig. 27. Soot-divestment from plaster relief using the 3rd harmonic of a Q-Switched Nd: YAG laser at 0.2 J/cm^2. Taken from Marakis (2000).

Fig. 28. Encrustation-divestment from stone façade using the 3rd harmonic of a Q-Switched Nd: YAG laser at 0.4 and 0.5 J/cm^2. Taken from Marakis (2000).

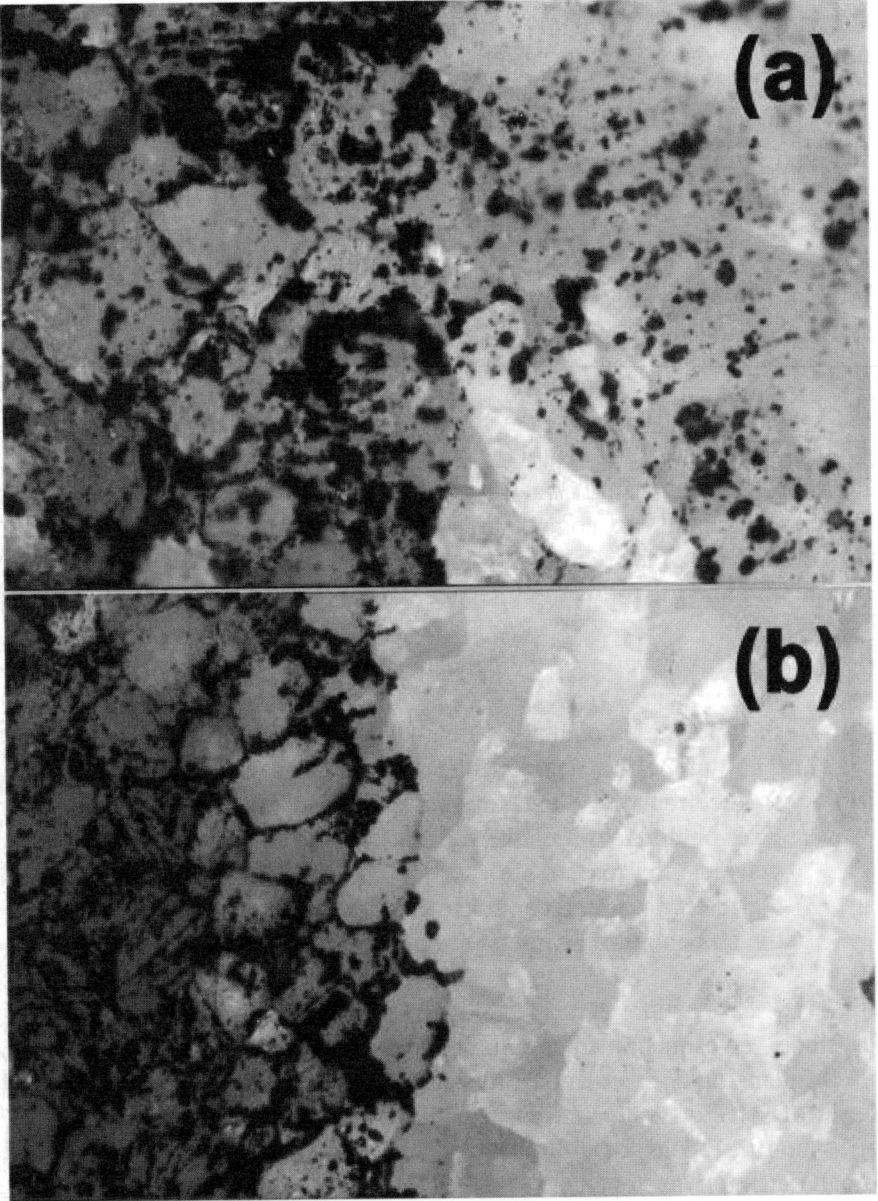

Fig. 29. Divestment of biological encrustation using λ_L = 1064 nm (a) and λ_L = 355 nm (b), respectively. Magnification is x 50.

Finally, the 3rd harmonic of a Q-Switched Nd: YAG laser is the only alternative when the encrustation is of biogenic character including combination with inorganic encrustation. Owing to the high absorption of UV photons from biological molecules, λ_L = 355 nm is very successful in removing fungi, lichen and other biological encrustation. Fig. 29 shows the results of λ_L = 1064 and 355 nm on biological encrustation using 1.85 and 0.84 J/cm^2, respectively. In this particular case, the marble spallation thresholds for the two laser wavelengths were 3.1 and 2.6 J/cm^2, respectively (Marakis, 2000). It is noticeable that λ_L = 355 nm gives superior results compared to λ_L = 1064 nm.

4. Conclusions

It has been shown that in principle, it is possible to divest delicate substrates such as paintings or composite structures from polymeric layers (e.g. network of polymerized resin or polymeric paints, respectively) when bearing in mind the following fundamental canons:

(1) The first layer over the delicate substrate – usually a layer of a few micrometers – must be always left intact, acting as a light filter for the transmitted photons.
(2) The optimal laser parameters such as laser wavelength and energy fluence must be found for succeeding the minimal transmission of laser photons towards the substrate. The shortest available UV-laser wavelength is preferable.
(3) On-line monitoring/control techniques must be used for identifying the end point of the intervention.

In the case of inorganic or biological encrustation on surfaces of marble or stone, divestment is possible via selective vaporization or spallation mechanisms. In both situations, it is advisable to utilize the gap between the onset of selective vaporization or encrustation removal in general and the spallation threshold for the substrate. Such a gap usually exists covering a broad range of wavelengths.

The operation of selective vaporization is accompanied with a certain optical phenomenon, the yellowing. Preliminary studies revealed that yellowing is associated to: (a) the light scattering occurring at the voids created by selective vaporization and (b) to the modified spectral absorption of the gypsum-rich remaining layer. Different research groups have started systematic studies on the subject.

Acknowledgements

This work has been supported in part by EPET II Programme (General Secretariat for research and Technology, Greece), Project No 640; by the E.U.-DGXII BRITE/EURAM project BRE2-CT92-0151 and by the Ultraviolet Laser Facility operating at *FO.R.T.H. – IESL* under the large Installations Plan of E.U.-DGXII program ERBCHGE-CT92-0007. We would like to acknowledge Giorgos Marakis, Alexandra Manousaki, John Petrakis, Charis Theodorakopoulos and Anastassia Galyfianaki in carrying out the most of the experiments presented in this work, as well as Costas Fotakis, Costas Balas, Pagona Maravelaki-Kalaitzaki, Demetrios Anglos, Savas Georgiou, Apostolis Englezis, Vivi Tornari and Vassilios Kilikoglou for their time in stimulating discussions. I would also like to greatly acknowledge Aristide Dogariu from CREOL – University of Central Florida for his precious advise, Yannis Chryssoulakis and Sofia Sotiropoulou from Ormylia Art Diagnosis Centre for providing the micro-FTIR analysis, Martin Cooper from the National Museums and Galleries on Merseyside for making available his test case study and Stamatis Boyatzis from The University of Patras for the GC-MS analysis. Finally, we would like to express our great gratitude to Michael Doulgeridis, director of the Dept. of Conservation in the National Gallery of Athens for his continuous support throughout the work on painting-divestment applications.

References

Amadesi S., Gori F., Grella R., Guattari G., *Holographic methods for painting diagnostics*, Appl. Opt. **13**, 2009 (1974)

Anglos D., *Laser-Induced Breakdown Spectroscopy in Art and Archaeology*, Appl. Spectroscopy, **55**, 186A, (2001)

Asmus J. F., Guattari G., Lazzarini L., Musumeci G., Wuerker R. F., *Holography in the Conservation of Statuary*, Studies in Conservation **18**, 49 (1973)

Asmus J. F., Seracini M., Zetler M. J., *Surface morphology of laser cleaned stone*, Lithoclastia **1**, 23 (1976)

Asmus J. F., *More light for art conservation*, IEEE Circuits and Devices Magazine (March), p. 6 (1986)

Asmus J. F., *Light for art conservation*, Interdisciplinary Science Reviews **12**, 171 (1987)

Athanassiou A., Lassithiotaki M., Anglos D., Georgiou S., Fotakis C., *A comparative study of the photochemical modifications effected in the UV laser ablation of doped polymer substrates*, Appl. Surface Sci. **154-155**, 89 (2000)

Bäuerle D., *Laser Processing and Chemistry*, (Springer, Berlin, Heidelberg. 2000)

Bertani D., Cetica M., Molesini G., *Holographic tests on the Ghiberti panel: The life of Joseph*, Studies in Conservation **27**, 61 (1982)

Boone P., Van Nieuwenburgh D., *Optical analysis of environmental influences on artifacts*, In: Optics for Protection of Man and Environment against Natural and Technological Disasters, p. 183 (Elsevier, Amsterdam, 1993)

Boquillon J. P., *Laser light transmission through optical fibers for laser cleaning applications*, In: Restauratorenblätter, Sonderband – Lacona I, Ed. by E. König and W. Kautek, p. 103 (Verlag Mayer & Comp., Vienna, 1997)

Chryssoulakis Y, Sotiropoulou S., Zafiropulos V., *unpublished results* (1996)

Clark R. J. H., *Raman microscopy: application to the identification of pigments on medieval manuscripts*, Chem. Soc. Rev. **24**, 187 (1995)

Cooper M., *Cleaning in Conservation: an Introduction* (Butterworth Heinemann, Oxford, 1998)

Cooper M., *kind permission of the National Museums and Galleries on Merseyside* (2001)

De Cruz A., Wolbarsht M. L., Hauger S. A., *Laser removal of contaminants from painted surfaces*, J. Cult. Heritage **1**, S173 (2000)

De la Rie, E. R., *The Influence of Varnishes on the Appearance of Paintings*, Studies in Conservation **32**, 1 (1987)

De la Rie E. R., *Photochemical and thermal degradation of films of dammar resin*, Studies in Conservation **33**, 53 (1988)

De la Rie, E. R., *Old Master Paintings: A Study of the Varnish Problem*, Analytical Chemistry **61,** 1228 (1989)

Doulgeridis M., *private communication* (1997)

Doulgeridis M., *The Word of Artwork and the space-time*, in Greek, (Bastas publications, Athens 2000)

Dyer P. E., *Photochemical Processing of Electronic Materials* (Academic, London 1992)

Fogarassy E., Lazare S. (Eds.), *Laser ablation of electronic materials. - Basic mechanisms and applications* (North-Holland Elsevier, Amsterdam 1992)

Fotakis C., Zafiropulos V., Anglos D., Georgiou S., Maravelaki P. V., Fostiridou A., Doulgeridis M., *Lasers in art conservation*, In: *The Interface between Science and Conservation,* Ed. S. Bradley, p. 83 (The Trustees of the British Museum, 1997)

Georgiou S., Zafiropulos V., Anglos D., Balas C., Tornari V., Fotakis C., *Excimer laser restoration of painted artworks: Procedures, Mechanisms and Effects*, Appl. Surface Sci. **127-129**, 738 (1998)

Gobernado-Mitre I., Prieto A.C., Zafiropulos V., Spetsidou Y., Fotakis C., *On-line monitoring of laser cleaning of limestone by laser induced breakdown spectroscopy*, Appl. Spectrosc. **51**, 1125 (1997)

Klein S., Stratoudaki T., Zafiropulos V., Hildenhagen J., Dickmann K., Lehmkuhl T., Laser-induced breakdown spectroscopy for on-line control of laser cleaning of sandstone and stained glass, *Appl. Phys. A* **69**, 441 (1999)

Klein S., Stratoudaki T., Marakis Y., Zafiropulos V., Dickmann K., *Comparative study of different wavelengths from IR to UV applied to clean sandstone*, Appl. Surface Sci. **157**, 1 (2000)

Klein S., Ferksanati F., Hildenhagen J., Dickmann K., Uphoff H., Marakis Y., Zafiropulos V., *Discoloration of marble during laser cleaning by Nd: YAG laser wavelengths*, Appl. Surface Sci. **171**, 242 (2001)

König E., Kautek W. (Eds.), *Proceedings of the First International Conference LACONA I - Lasers in the Conservation of Artworks*, In: Restauratorenblätter, Sonderband – Lacona I (Verlag Mayer & Comp., Vienna 1997)

Luk'yanchuk B., Bityurin N., Anisimov S., Bäuerle D., *The role of excited species in UV-laser materials ablation. Part I: Photophysical ablation of organic polymers*, Appl. Phys. A **57**, 367 (1993a)

Luk'yanchuk B., Bityurin N., Anisimov S., Bäuerle D., *The role of excited species in UV-laser materials ablation; Part II: The stability of the ablation front*, Appl. Phys. A **57**, 449 (1993b)

Luk'yanchuk B., Bityurin N., Anisimov S., Bäuerle D., *Photophysical ablation of organic polymers*, In: Excimer Lasers, Ed. by L.D. Laude, p. 59 (Kluwer Academic Publishers, Dordrecht, 1994)

Luk'yanchuk B., Bityurin N., Anisimov S., Arnold N., Bäuerle D., *The role of excited species in ultraviolet-laser materials ablation. III. Non-stationary ablation of organic polymers*, Appl. Phys. A **62**, 397 (1996)

Luk'yanchuk B. S., Zafiropulos V., *The model for discoloration effect in pigments at cleaning of artworks by laser ablation*, Proc. SPIE, vol. **4426** (2002)

Manz C., Zafiropulos V., *Laser ablation strips thin layers of paint*, Opto & Laser Europe (OLE) **45**, 27 (1997)

Marakis G., *Comparative study on the removal of encrustation from marble using the fundamental and the third harmonic of Q-Switched Nd: YAG laser*, In Greek, Diploma thesis (Technical Educational Institute & FORTH-IESL, Athens, Heraklion, 2000)

Marakis G., Pouli P., Zafiropulos V., Maravelaki-Kalaitzaki P., *Comparative study on the application of the 1st and the 3rd harmonic of a Nd: YAG laser system to clean black encrustation on marble*, J. Cult. Heritage, (2002)

Maravelaki P.V., Zafiropulos V., Kylikoglou V., Kalaitzaki M., Fotakis C., *Laser Induced Breakdown Spectroscopy as a Diagnostic Technique for the Laser Cleaning of Marble*, Spectrochimica Acta B **52**, 41 (1997)

Maravelaki-Kalaitzaki P., Zafiropulos V., Fotakis C., *Excimer laser cleaning of encrustation on Pentelic Marble: procedure and evaluation of the effects*, Appl. Surface Sci. **148**, 92 (1999)

Maravelaki-Kalaitzaki P., Anglos D., Kilikoglou V., Zafiropulos V., *Compositional characterization of encrustation on marble with laser induced breakdown spectroscopy*, Spectrochimica Acta B **56**, 887 (2001)

Miller J. C. (Ed.), *Laser Ablation - Principles and Applications* (Springer-Verlag, Berlin 1994)

Mills J. S., White R., *The Organic Chemistry of Museum Objects*, (Butterworths, London 1987), and references therein.

Pantelakis S., Despotopoulos A., Lentzos G., Kermanidis T., *Influence of the laser paint stripping process on the mechanical behaviour of fiber reinforced composites*, Proc. of the 4th European Conference on

Advanced Materials and Processes, p. 345 (Padua/Venice, Italy, 1995)

Pettit G. H., Sauerbrey R., *Pulsed Ultraviolet laser ablation*, Appl. Phys. A **56**, 51 (1993)

Pettit G. H., Ediger M. N., Hahn D.W., Brinson B. E., Sauerbrey R., *Transmission of polyimide during pulsed ultraviolet laser irradiation*, Appl. Phys. A **58**, 573 (1994)

Pini R., Siano S., Salimbeni R., Piazza V., Giamello M., Sabatini G., Bevilacqua F., *Application of a new laser cleaning procedure to the mausoleum of Theodoric*, J. Cult. Heritage **1**, S93 (2000)

Popescu G., Mujat C., Dogariu A., *Evidence of scattering anisotropy effects on boundary conditions of the diffusion equation*, Phys. Rev. E **61**, 4523 (2000)

Sabatini G., Giamello M., Pini R., Siano S., Salimbeni R., *Laser cleaning methodologies for stone facades and monuments: laboratory analyses on lithotypes of Siena architecture*, J. Cult. Heritage **1**, S9 (2000)

Salimbeni R., Bonsanti G. (Eds.), *Proc. of the International Conference LACONA III - Lasers in the Conservation of Artworks*, J. Cult. Heritage **1**, Suppl. 1. (2000)

Siano S., Margheri F., Mazzinghi P., Pini R., Salimbeni R., *Cleaning processes of encrusted marbles by Nd: YAG lasers operating in free running and Q-Switching regimes*, Appl. Opt. **36**, 7073 (1997)

Siano S., Pini R., Salimbeni R., *Variable energy blast modeling of the stress generation associated with laser ablation*, Appl. Phys. Lett. **74**, 1233 (1999)

Siano S., Fabiani F., Pini R., Salimbeni R., Giamello M., Sabatini G., *Determination of damage thresholds to prevent side effects in laser cleaning of Pliocene sandstone of Siena*, J. Cult. Heritage **1**, S47 (2000)

Skoulikidis Th., Papakonstantinou-Ziotis P., *The mechanism of sulfation by atmospheric SO2 of limestones and marbles of the ancient monuments and statues. I. Observations in situ and measurements in the laboratory; activation energy*, Br. Corros. J. **16**, 63 (1981)

Skoulikidis Th., Vassiliou P., Papakonstantinou P., Moraitou A., Zafiropulos V., Kalaitzaki M., Spetsidou I., Perdikatsis V., Maravelaki P., *Some remarks on Nd: YAG and Excimer UV Lasers for cleaning soiled sulfated monument surfaces*, In: LACONA –

Workshop on "Lasers in the Conservation of Artworks", (Heraklion, Crete, Book of Abstracts, 1995)

Srinivasan R., Braren B., *Ultraviolet Laser ablation of organic polymers*, Chem. Rev. **89,** 1303 (1989)

Srinivasan R., *Ablation of polyimide (kapton) films by pulsed (ns) ultraviolet and infrared (9.17 μm) lasers; a comparative study*, Appl. Phys. A **56**, 417 (1993)

Srinivasan R., *Interaction of laser radiation with organic polymers*, In: Laser Ablation, Ed. by J. C. Miller, p. 107 (Springer, Berlin, Heidelberg, 1994)

Stout G. L., *The Care of Pictures* (Dover Publications, INC, New York 1975)

Sutherland K., *Solvent extractable components of oil paint films*, Ph.D. thesis, (University of Amsterdam 2001)

Theodorakopoulos C., Zafiropulos V., *Uncovering of scalar oxidation within a naturally aged varnish layer*, J. Cult. Heritage, (2002)

Tornari V., Fantidou D., Zafiropulos V., Vainos N. A., Fotakis C., *Photomechanical effects of laser cleaning: A long-term non-destructive holographic interferometric investigation on painted artworks*, Proc. SPIE, vol. 3411, 420 (1998)

Tornari V., Zafiropulos V., Bonarou A., Vainos N. A., Fotakis C., *Modern technology in artwork conservation: a laser based approach for process control and evaluation*, Optics and Lasers in Engineering **34**, 309 (2000a)

Tornari V., Bonarou A., Zafiropulos V., Fotakis C., Doulgeridis M., *Holographic applications in evaluation of defect and cleaning procedures*, J. Cult. Heritage **1**, S325 (2000b)

Tornari V., Zafiropulos V., Fantidou D., Vainos N. A., Fotakis C., *Discrimination of photomechanical effects after laser cleaning of artworks by means of holographic interferometry*, In: Optics and Lasers in Biomedicine and Culture, Ed. G. von Bally, p. 208 (Springer-Verlag, Berlin, 2000c)

Vergès-Belmin V., *Comparison of three cleaning methods – micro-sandblasting, chemical pads and Q-Switched Nd: YAG laser – on a portal of the cathedral Notre-Dame in Paris, France*, In: Restauratorenblätter, Sonderband – Lacona I, ed. by E. König and W. Kautek, p. 17 (Verlag Mayer & Comp., Vienna, 1997)

Vergès-Belmin V., Dignard C., *Laser yellowing, myth or reality?*, J. Cult. Heritage (2002), and references therein.

Watkins K. G., Larson J. H., Emmony D. C., Steen W. M., *Laser cleaning in art restoration: a review*, In: Proceedings of the NATO Advanced Study Institute on Laser Processing: Surface Treatment and Film Deposition, p. 907 (Kluwer Academic Publishers, 1995)

Watkins K. G., *A review of materials interaction during laser cleaning in art restoration*, In: Restauratorenblätter, Sonderband – Lacona I, Ed. by E. König and W. Kautek, p. 7 (Verlag Mayer & Comp., Vienna, 1997)

Zafiropulos V., Petrakis J., Fotakis C., *Photoablation of polyurethane films using UV laser pulses*, Optical and Quantum Electronics **27**, 1359 (1995)

Zafiropulos V., Fotakis C., *Lasers in the Conservation of painted Artworks*, In: Laser in Conservation: an Introduction, Ed. by M. Cooper, p. 79 (Butterworth Heineman, Oxford, 1998)

Zafiropulos V., Galyfianali A., Boyatzis S., Fostiridou A., Ioakimoglou E., *UV-laser Ablation of Polymerized Resin Layers and Possible Oxidation Process in Oil-based painting media*, In: Optics and Lasers in Biomedicine and Culture, Ed. G. von Bally, p. 115 (Springer-Verlag, Berlin 2000)

Zafiropulos V., Stratoudaki T., Manousaki A., Melesanaki K., Orial G. *Discoloration of pigments induced by laser irradiation*, Surface Engineering **17**, 249 (2001a)

Zafiropulos, V., Manousaki A., Kaminari A., Boyatzis S., *Laser Ablation of aged resin layers: A means of uncovering the scalar degree of aging*, Proc. SPIE, vol. **4430**, 181 (2001b)

Zafiropulos V., Maravelaki P., Fotakis C., *Ablation Rate studies and LIBS as an on-line control technique in the removal of unwanted selected layers*, In: Restauratorenblätter, Sonderband – Lacona II, Ed. by E. König and W. Kautek (Verlag Mayer & Comp., Vienna, 2001c)

Zafiropulos V., Balas C., Manousaki A., Marakis G., Maravelaki-Kalaitzaki P., Melesanaki K., Pouli P., Stratoudaki T., Klein S., Hildenhagen J., Dickmann K., Luk'yanchuk B.S., Mujat C., Dogariu A., *Yellowing effect and discoloration of pigments: Experimental and Theoretical studies*, J. Cult. Heritage, (2002a)

Zafiropulos V., Marakis G., Balas C., Mujat C., Dogariu A., *Optical phenomenon associated to laser-assisted divestment of encrustation*, (2002b) to be published.

Zergioti I., Petrakis A., Zafiropulos V., Fotakis C., Fostiridou A., Doulgeridis M., *Laser Applications in Painting Conservation*, In: Restauratorenblätter, Sonderband – Lacona I, Ed. by E. König and W. Kautek, p. 57. (Verlag Mayer & Comp., Vienna, 1997)

Chapter 9

ON THE THEORY OF DISCOLORATION EFFECT IN PIGMENTS AT LASER CLEANING

B. S. Luk'yanchuk, V. Zafiropulos

The physical mechanism and the theoretical model for pigment discoloration at laser ablation are discussed. The model includes two subsequent steps: 1) examination of ablation in subthreshold (Arrhenius tail) region and 2) examination of the nucleation process and structural modification within the volume of material close to ablation front. The results of calculations show that in the wide range of parameters exists the strong sharpening effect, when the thickness of modified layer (a few nanometers) is significantly smaller than the heat penetration depth or depth of optical absorption. These results are in good agreement with experimental data on discoloration effect.

Keywords: Laser Ablation, Pigments, Discoloration effect, Modelling, Cinnabar, Metacinnabar

PACS: 42.62.Cf, 81.65.Cf, 68.35.Np, 81.65.-b, 85.40.-e, 81.07.Wx

1. Introduction

Laser ablation by short laser pulse is successfully used for cleaning in art conservation (Zafiropulos, 1995, 1998; Georgiou, 1998). Artworks are usually made of quite complex organic and inorganic materials in contrast to comparably simple materials in microelectronics (technical surfaces). Thus, in art conservation a significant part of the restoration work is related to diagnostics of the degraded superficial varnish layers that cover the painting's surface (Zafiropulos, 1998), as well as to diagnostics of "origin" layer and pigment identification (Klein, 1999; Burgio, 2000; Anglos, 1997; Stratoudaki, 2000).

Although the removal of unwanted surface layers can be solved with help of pulsed laser ablation, the problem to preserve the original artwork needs examination of effects, which arise within the "origin" layer due to thermal, mechanical, thermochemical and photochemical effects of laser action (Bäuerle, 2000; Karlov, 2000; Lassithiotaki, 1999).

Amongst the many interesting problems to be tasked, there are two issues of increased interest, the so-called yellowing and discoloration effects (Zafiropulos, 2000 a). The first, yellowing effect, was discovered in laser cleaning of marble (Skoulikidis, 1995). This effect coexists with the so-called "patina" effect (Cooper, 1998). While the patina effect is attributed to the chemical alteration of the surface, the yellowing effect seems to be caused by physical modification of the material. One of the recent explanations (Zafiropulos, 2002 a, b) consider the yellowing effect originated by scattering of light by microvoids. These voids are produced during evaporation of absorbing inclusions immersed in gypsum reach layer.

Another "nonchemical" effect is the discoloration effect occurring on pigments. The discoloration of a pigment may appear when the encrustation is directly on top of it. Then by using a Q-Switched Nd: YAG laser to remove the encrustation, after the removal of the last crust layer, the pigment underneath becomes black or gray. The experimental examination of this effect shows *the change of crystalline phase* at the very top layer (1-5 μm) of the pigment (Zafiropulos, 2001). The XRD measurements before and after laser pulse action show the small although very distinct change. The example we present here is a simple pigment, the cinnabar red (HgS). In XRD we see change of cinnabar to metacinnabar (black), but still the cinnabar peaks dominate the spectrum. From very rough calculations we found that the X-ray penetration is only ~1 μm for a few degrees of incidence. But a cross section of the changed phase shows a thickness of a full monolayer of crystals. So the domination of cinnabar peaks after the laser irradiation is not explained. A possible explanation may be that the changed phase is only superficial (on the surface of the crystals). In any case it refers to high sharpening effect in the distribution of metacinnabar. In the present paper we suggest the possible origin of this sharpening effect.

It is well known that the cinnabar transfers to metacinnabar at temperature above 350 ^{0}C (Lide, 2002). Thus one can explain this effect by residual heating of non-ablated surface material. The problem is however that the phenomenon is practically independent on the parameters, which influence laser heating of the material: number of pulses, fluence, wavelength and pulse duration. Even with 100 fs laser pulses discoloration occurs.

Thus, we examine theoretically the main peculiarities of the ablation – heating – phase transformation problem. Because some parameters of the material are not known with sufficient accuracy, we were mainly interested in qualitative features of the discussed discoloration effect. The results of calculations show that in the wide range of parameters exists the strong sharpening effect, when the thickness of metacinnabar modified layer is significantly smaller than the heat penetration depth. These results are in good agreement with experimental data on discoloration effect.

Here we present our theoretical model which includes two steps: 1) analysis of the ablation-heating process in subthreshold regime and calculation of the overheating in nonablated material near the surface; 2) dynamics of overheating is used further for examination of the nucleation process and metacinnabar cluster growth.

2. The thermal ablation model

At typical experiments the size of laser beam, r_0, (by order of 1 cm) was significantly bigger than the heat penetration depth, thus the problem can be treated as 1D problem. First, we want to estimate the thickness of the non-ablated surface material, which was heated above the 350 ^{0}C.

We will consider a *purely thermal surface* evaporation and will use the non-stationary averaging technique to find an approximate solution to the *nonlinear* heat equation. This method was developed in (Arnold, 1998 a, b, c), where arbitrary temperature dependencies of the optical and thermo-physical parameters were taken into account.

For the analysis of thermal ablation one should solve *nonstationary* heat equation with ablation velocity $v = v\,(t)$ changing during the pulse. We write the thermal problem in terms of volumetric enthalpy H:

$$H(T) = \rho \int_{T_\infty}^{T} c(T_1)\, dT_1 \qquad (1)$$

where c is the specific heat (per unit mass) of the condensed phase, and $T_\infty = 300\ K$ is the ambient temperature. The density of condensed phase ρ = *const*. One-dimensional heat equation in the moving reference frame fixed with the ablation front reads (Anisimov, 1971):

$$\frac{\partial H}{\partial t} = v\,\frac{\partial H}{\partial z} + \frac{\partial}{\partial z}\left(\kappa\frac{\partial T}{\partial z}\right) + I_s\,\alpha\,exp\left(-\alpha z\right) \equiv B\left(z,t\right), \tag{2}$$

where we introduced the notation B for the r.h.s. The heat conductivity κ and the source term may depend on temperature T. Latent heat of phase transformations (e.g., melting, structure phase transition) can be included into (1). The intensity within the solid should obey Bouguer equation. I_s is the intensity absorbed at the surface. Henceforth index "s" will refer to the quantities at the surface z = 0. We adopt the following approximations. Following to Anisimov, 1971; Mao, 1997, we relate I_s to laser pulse intensity $I(t)$ by

$$I_s = I\;A\;exp\left\lfloor -\alpha_g\;h\right\rfloor, \tag{3}$$

where A is the absorptivity and α_g the vapor absorption coefficient recalculated to the density of solid. Both may depend on T_s. The case when one takes into account the volume variation in the absorption coefficient is more complex, because for this case one should explicitly solve the Bouguer equation to find the heat source term

$$\frac{\partial I}{\partial z} = -\alpha(T)I, \quad I\big|_{z=0} = I_s, \tag{4}$$

where the absorption coefficient $\alpha(T)$ depends on the local value of the temperature, $T = T(z,t)$. This dependence is the most difficult to treat by moment method. Nevertheless we can easily solve equation (2) with α = $\alpha(T_s)$. With strong absorption, when parameter $p = \alpha\sqrt{\chi t_\ell} \gg 1$ (χ is the heat diffusivity and t_ℓ is the duration of laser pulse) situation is close to surface absorption and the volume temperature variations in the absorption coefficient does not lead to a great variation in the temperature.

The temporal profile of the laser pulse (pulse shape) is given by $I = I(t)$. For the typical excimer laser pulse it can be approximated by the smooth function (Bäuerle, 2000):

$$I(t) = I_0 \frac{t}{t_\ell} exp\left[-\frac{t}{t_\ell}\right]. \quad (5)$$

The laser fluence is given by $\Phi = I_0 t_\ell$. The pulse duration, t_P (at the full widths at half-maximum), is related to parameter t_ℓ by $t_P = 2.446 t_\ell$.

The surface evaporation (ablation) rate v should be given by (Anisimov, 1971; Bäuerle, 2000):

$$v = v_0 \exp(-T_a / T_s), \quad (6)$$

where T_a is the activation temperature (in Kelvin), and v_0 is of the order of sound velocity. The boundary condition for the heat equation (2) at the surface z = 0 assumes negative thermal flux J_S (Anisimov, 1971; Luk`yanchuk, 1997):

$$-\kappa \frac{\partial T}{\partial z}\Big|_{z=0} \equiv -\chi_s \frac{\partial H}{\partial z}\Big|_{z=0} = -v\left[L + H_g(T_s) - H(T_s)\right] \equiv J_s, \quad (7)$$

where $\chi_s \equiv \kappa_s / \rho c_s$ is the thermal diffusivity at T_S.

The expression in square brackets is the enthalpy difference between the vapor and condensed phase at $T = T_S$. L is the latent heat of vaporization per unit volume at $T_S = T_\infty$, and $H_g(T) = \rho \int_{T_\infty}^{T} c_g(T_1) dT_1$ is the enthalpy of vapor per unit volume of the condensed phase. c_g is the specific heat per unit mass of the vapor phase. The boundary conditions at infinity $T\big|_{z\to\infty} = T_\infty$ (or $H\big|_{z\to\infty} = 0$) and initial condition $T\big|_{t=0} = T_\infty$ are obvious.

Though somewhat simplified this model is acceptable for the *quantitative* description of purely thermal surface laser ablation. It contains nonlinearities and nonstationary effects, which can be analyzed numerically. This requires big computational time (Peterlongo, 1994). With the Moments technique (Arnold, 1998 a, b, c) that we use in the further analysis this problem can be reduced to three coupled nonlinear ordinary differential equations (ODE) with a small loss in accuracy. A fast algorithm of ODE solving included in many computational packages can routinely solve these equations.

3. Method of moments

The idea of the "moments method" or "nonstationary averaging method" is simple (see for details in Samarsky, 1987; Zwillinger, 1989). The exact solution of the boundary value problem (1)-(7) fulfills (2) identically. If one uses trial solution $H = H_p(z, t)$, the identity (2) will be violated. Nevertheless one can use $H_p(z, t)$ as an *approximate* solution, if it obeys "conservation laws" for the moments M_n:

$$\dot{M}_n = \int_0^\infty z^n B\left[H_p(z,t)\right]dz \text{ , where } M_n = \int_0^\infty z^n H_p(z,t)\,dz \text{ .} \tag{8}$$

Here the equation (2) was multiplied by z^n and integrated over z, and dot stands for differentiation with respect to time. The obtained set of equations has a clear physical sense. For example, the equation for M_0 reflects the *time-dependent* energy balance. The number of differential equations in (7) should be equal to the number of time-dependent parameters characterizing $H_p(z, t)$. We consider two such parameters -- surface temperature $T_s(t)$ and characteristic "thermal length" $\ell(t)$ for the enthalpy distribution. Therefore, we must use *two* first moments of *enthalpy* distribution. Equations (8) yield:

$$\dot{M}_0 = -vH_s + J_s + I_s \equiv -v(L + H_{gs}) + I_s, \tag{9}$$

$$\dot{M}_1 = -vM_0 + \int_{T_\infty}^{T_s} \kappa(T)\,dT + \frac{I_s}{\alpha}. \tag{10}$$

Here $H_s \equiv H(T_s)$, $H_{gs} \equiv H_g(T_s)$. We set the trial solution $H_p(z, t)$, in the form:

$$H_p = \frac{1}{1-\alpha\ell}\left[(H_s - \frac{\ell J_s}{\chi_s})e^{-\alpha z} - (\alpha\ell H_s - \frac{\ell J_s}{\chi_s})e^{-\frac{z}{\ell}}\right]. \tag{11}$$

This form satisfies the obvious requirement $H_p(z = 0, t) \equiv H_s(t)$ and boundary condition (7). The first term in (11) is directly related to absorption of radiation, while the second term describes the changes in enthalpy distribution due to heat conduction. With $\ell = 0$ equation (11) gives calorimetric solution. With v = *const* equation (11) coincides with the stationary ablation wave solution (Anisimov, 1971). For the first two moments defined in (9), (10), equation (11) yields:

$$M_0 = \frac{1}{\alpha}\left[(1+\alpha\ell)H_s - \frac{\ell J_s}{\chi_s}\right], \tag{12}$$

$$M_1 = \frac{1}{\alpha^2}\left[(1+\alpha\ell+\alpha^2\ell^2)H_s - (1+\alpha\ell)\frac{\ell J_s}{\chi_s}\right], \tag{13}$$

When (12) and (13) are substituted into (9) and (10), one obtains two ordinary differential equations for $T_s(t)$ and $\ell(t)$. All quantities should be finally written in terms of the surface temperature T_s. Namely, one should insert J_s and χ_s from (7), I_s and v from (3), (5) and (6), and H_s from (1). Depending on particular temperature dependencies of $c(T)$ and $\kappa(T)$ the resulting equations can be sufficiently complex, but it is not necessary to resolve the resulting equations with respect to dT_s/dt and $d\ell/dt$ for the numerical computations that were done with the "*Mathematica*" software package (Wolfram, 1999).

The equation for the thickness of ablated material (*coupled* to (9) and (10) via screening (3)) is given by:

$$\dot{h} = v = v_0 \exp\left[-T_a/T_s\right]. \tag{14}$$

Thus, the initial problem (2)-(6) is reduced to three *ordinary* differential equations for T_s, ℓ, and h that should be solved with the initial conditions:

$$T_s(t=0) = T_\infty, \quad \ell(t=0) = 0, \quad h(t=0) = 0. \tag{15}$$

The study (Arnold, 1998 a, b, c) shows that their solution coincides with known analytical and numerical solutions of the initial problem (2)-(6) typically, within 10-15% accuracy. We want to emphasize that a wide range of materials were tested in (Arnold, 1998 a, b, c) from metals to polymers with different temperature dependencies of parameters.

We can simplify the solution for the particular case of constant parameters. We also neglect the difference in the enthalpies of vapor and condensed phases, considering $H_g(T_s) = H(T_s)$ in the boundary condition (7). The system of resulting equations has a relatively simple form:

$$\frac{dT_s}{dt} = F(T_s, \ell), \qquad \frac{d\ell}{dt} = \Psi(T_s, \ell), \tag{16}$$

where F and Ψ functions are given by (Anisimov & Luk`yanchuk, 2002)

Table 1. Parameters used in the present study.

	Value	Units
Density of solid, ρ	8.1	g/cm^3
Heat capacity, c	0.208	J/g K
Heat conductivity, κ	0.01	W/cm K
Latent heat of vaporization, L	1500	J/g
Absorptivity for $\lambda = 1.06$ μm	0.5	Dimensionless
Absorption coefficient, α for $\lambda = 1.06$ μm	$3.3\ 10^5$	cm^{-1}
Activation energy, T_a	18000	K
Preexponent, v_0	10^6	cm/s
Absorption coefficient in gas phase, α_g	$3.3\ 10^5$	cm^{-1}
Duration of laser pulse, t_P FWHM	6	ns
Laser fluence, Φ	0.3	J/cm^2
Temperature of phase transition, T_c	623	K
Initial temperature, T_∞	300	K

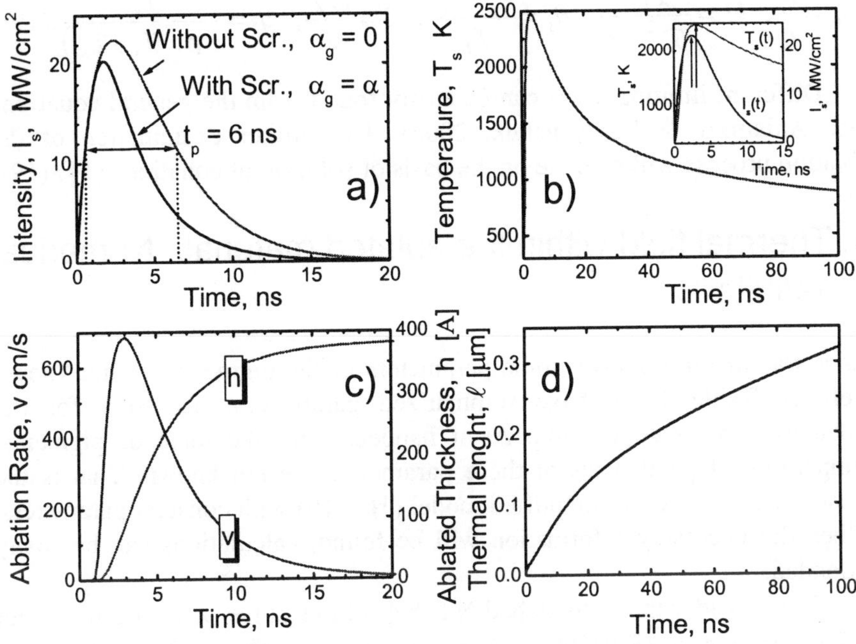

Fig. 1. a) Intensity $I_s(t)$ with and without screening effect. b) Surface temperature $T_s(t)$ (insertion shows the initial stage of the process, max temperature delayed the max intensity. Arrows indicate corresponding moments). c) Ablation Rate, $v(t)$ and thickness of ablated material, $h(t)$. d) Variation in thermal length $\ell(t)$.

$$F = \alpha v\, T_L \,\frac{(2q+1)\dfrac{1}{\xi} - \left(1 + \dfrac{\xi}{\alpha\,\ell}\right)\theta_s - 2}{2 + \alpha\,\ell + \Pi}, \tag{17}$$

$$\Psi = \frac{\chi}{\ell}\,\frac{\alpha\ell\left[q - \theta_s(1-\xi) + (1 + \alpha\ell + \Pi)\left(1 - q - \dfrac{1}{\xi}\right)\right] + (1 + \theta_s\xi)(1 + \Pi)}{(1 + \theta_s\xi)(2 + \alpha\ell + \Pi)}, \tag{18}$$

and different dimensionless parameters are presented by

$$T_L = \frac{L}{c},\ \Pi = \frac{T_a T_L}{T_s^2}\frac{v\ell}{\chi},\ \theta_s = \frac{T_s - T_\infty}{T_L},\ \xi = \frac{\alpha\chi}{v},\ q = \frac{I_s}{c\,\rho v\,(T_s - T_\infty)}. \quad (19)$$

Different limiting cases can be easily found from the general equations (see Anisimov & Luk`yanchuk, 2002). The further examination of the ablation process will be done on the basis of solution of equations (14)-(19).

4. Thermal field within the ablated material. Numerical results

As it was mention above, some parameters of the pigments are not known well. In Table 1 we have summarized parameters, necessary for the calculation of laser ablation. As it happens with the most of pigments, temperature dependencies of these parameters are not known. That is the reason why we use the simplified model (16)-(19) with constant parameters. When the necessary information will be found, calculations can be easily extended.

We consider the Q-Switched Nd: YAG laser pulse (λ = 1.06 μm) with pulse duration t_P (FWHM) = 6 ns and fluence Φ = 0.3 J/cm^2, which corresponds to conditions where the discoloration effect is well seen experimentally. So, for parameters λ, t_P, Φ, T_∞, and T_c one can consider as "strictly defined". Among the other parameters, given in Table 1, parameters ρ, c, κ, α and A are "well-defined". The remnant parameters, T_a, v_0, L, and α_g are not known and we use for estimations of these parameters characteristic values, typical for some organic materials and polymers (see e.g. Arnold, 1998 a, b, c). We call this set of parameters as "generic". In reality, even some "well defined" parameters, e.g. heat conductivity, κ, are known with accuracy by factor 2.

Using "genetic" set of parameters we solve numerically the equations (14)-(16). Results of the solution are presented in Fig. 1. The qualitative behavior of the ablation process was discussed previously in (Arnold, 1998 a, b, c; Anisimov, 2002), and we cannot see any new features of different quantities in Fig. 1.

Variation of different parameters in Table 1 in the "reasonable limits" shows that the maximal surface temperature can vary in the range of 1800-2800 K, and the total thickness of ablated material is somewhere in the range of 50-500 Å per pulse. In any case, this is below the ablation threshold, the so-called "Arrhenius tail" region (Arnold, 1998 a, b, c). This Arrhenius tail was found experimentally for many materials, see e.g. (Küper, 1993). The physical reason for this tail has been already discussed in the papers (Luk`yanchuk, 1997; Arnold, 1998 a; Anisimov, 2002)

The heat penetration depth, $\ell(t)$, quite soon reaches optical penetration depth, α^{-1}. After this moment it varies closely to usual heat conduction law, $\ell \propto \sqrt{\chi t}$. We should emphasize that the cooling stage is sufficiently long due to 1D character in the heat conductivity. The typical time, until the surface is heated above the temperature of phase transformation T_s (t) >T_c, consists a few hundred nanoseconds and even a few µs, within the region of "reasonable parameters". The 3D effects in the heat conductivity, which leads to faster cooling, take place just from the moment $t \approx r_0^2 / \chi$. This time in the discussed experiments consists tens of seconds.

Using the formula (11) we can estimate the thickness of overheated layer. Namely, we can examine the overheating, $\Delta T = T(z,t) - T_c$, inside the material at different distances z from the ablation front. This overheating for the genetic case is shown in Fig. 2a.

Thus, points, which are closer to ablation front are heated to the higher temperature, and duration of the overheated stage is longer. At some distance $z > z_m$ points do not reach temperature of phase transition at all. We shall call this distance as thickness of the transformed material. For the genetic case it is equal z_m = 2300 Å. Varying the parameters in the "reasonable limits" one can find that z_m is typically in the range of 0.2 - 2 µm. One important point that should be emphasized is that the mentioned coordinate z is geometric coordinate (distance from the ablation front). If one is interested in the overheating of the material point (i.e. fixed small volume of the material) he should take into account the motion of ablation front. Due to this motion, material point situated at the distance z from the material surface at initial moment t = 0, will be situated at the distance $z - h(t)$ at the moment t, i.e. material moves into the hot region. For the points situated far from the surface this motion leads to relatively small increase of the temperature, but near the distance, comparable with the optical penetration depth, α^{-1}, this increase in the maximal temperature is significant (see in Fig. 2 b). Naturally, we are not interesting in the thermal dynamics of material points, which will be deleted during the ablation, i.e.

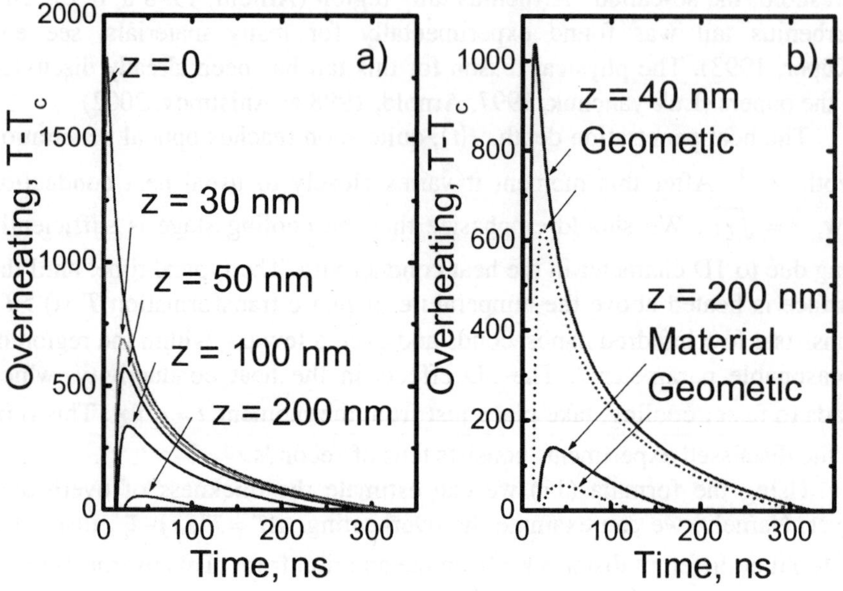

Fig. 2. a) Overheating, T - T_c, inside the ablated material at different position from the ablation front. b) Overheating, T - T_c, inside the ablated material for geometric point with coordinate z (dot), and for material point with running coordinate $z - h(t)$ (solid).

points with initial position $z < h(\infty) \approx 400$ Å for genetic case. To characterize thermal dynamic of points, which remain in the material when ablation process is completed we plot in Fig. 3 maximal overheating versus coordinate $z_1 = z - h(\infty)$. Here z_1 is the true distance of material point from the surface after ablation. One can see from Fig. 3 that near the surface maximal temperature varies very sharp, it decreases ≈ 250 K within the material layer with thickness 100 Å (gradient in the overheating near the surface consists $\approx 2.5\,10^8$ K/cm).

It is clear that this big gradient on overheating may lead to drastically change material properties near the surface. To estimate this effect we should examine more detailed kinetics of phase transition.

5. Kinetics of phase transition and surface modification

We shall consider that the transition of cinnabar to metacinnabar arises as *homogeneous* phase transition. It is different from the heterogeneous transition on the surface of moving front (say, process, similar to motion of melting on crystallization front). Thus, we consider that the process occurs within the *volume* of material. As it is known from the kinetic theory of the first order phase transitions (see, e.g. Lifshitz & Pitaevsky, 1987), the phase transition has two well-separated stages: 1) nucleation, and 2) forthcoming growth of overcritical nucleus. We consider that transition of cinnabar to metacinnabar starts when the temperature of cinnabar exceeds the equilibrium temperature T_c. The kinetics of phase transition is governed by superheating

$$\theta = \frac{T - T_c}{T_c}. \qquad (20)$$

It is similar to volume boiling process. We consider also, that nuclei are moving together with material. Naturally, we should include the latent heat of phase transition into the effective heat capacity in (1). Nevertheless we neglect this heat for the first estimation, considering that it does not change a lot of the heat balance, which is defined by interplay of absorbing laser energy, ablation and heat diffusion.

To describe the phase transformation we introduce the quantity x, which denotes the degree of transformation (i.e. $x = 0$ corresponds to pure cinnabar and $x = 1$ to pure metacinnabar). We define ν as the number of the cluster (per molecule of cinnabar). We suppose that each cluster consists of g molecules. Then the degree of transformation is given by:

$$x = \nu g . \qquad (21)$$

Here we consider that all clusters at the given Lagrangian point are of the same size (narrow distribution function). This assumption is self-consistent for fast transformation process, but it can be violated for sufficiently slow transformation process, where the stage of coalescence (see Lifshitz-Slyozov theory: Lifshitz, 1958, 1984) can be important.

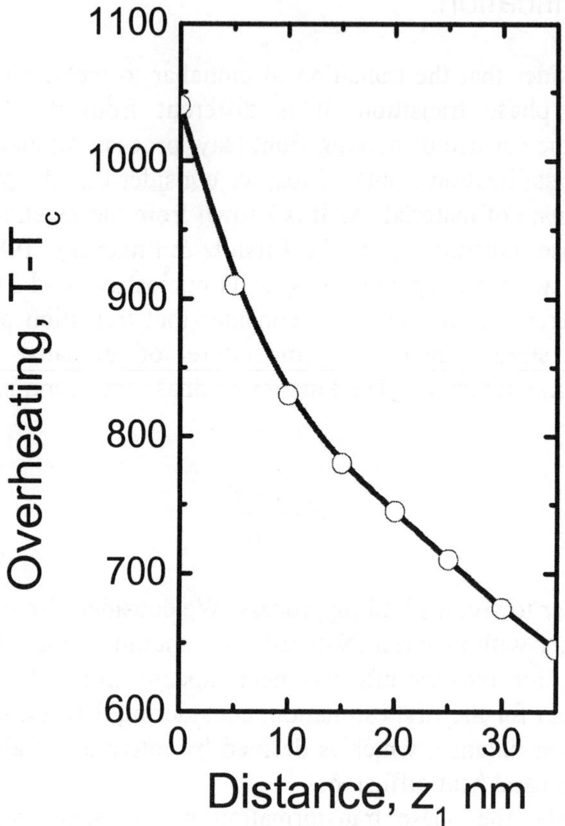

Fig. 3. Variation of the maximal overheating reached within the material near the surface.

Under the assumption of narrow distribution function for clusters at the given Lagrangian point, the differential equation for x and ν follows from (21):

$$\frac{dx}{dt} = g\frac{d\nu}{dt} + \nu\frac{dg}{dt}. \tag{22}$$

The first term in (22) describes the change in x due to formation of new nucleus (crystallites), while the second one describes variations due to the crystalline growth.

The rate of formation of overcritical nucleus can be found from the homogeneous nucleation theory, developed in classical papers of Becker & Dőring, 1935, Kramers, 1940 and Zeldovich, 1942. The contribution of Frenkel, 1955 should be mentioned also. The classical nucleation theory is based on the kinetic equation of Fokker-Planck type, which is written in "space of sizes":

$$\frac{\partial f}{\partial t} + B_s f = \frac{\partial}{\partial g}\left(D_s \frac{\partial f}{\partial g} \right), \tag{23}$$

where f is the size distribution function and D_s is the "diffusion coefficient" in the space of sizes g. Information related to coefficient B_s can be found, e.g. in (Lifshitz & Pitaevsky, 1984; Zeldovich, 1942).

The quasi-stationary solution of this equation (see Zeldovich, 1942 for details) describes the production rate (nucleation rate) of subcritical nucleus:

$$\frac{d\nu}{dt} = k_\nu (1-x) \exp\left[-\frac{T_\nu}{T\theta^2} \right], \quad \nu\big|_{t=t_c} = 0. \tag{24}$$

This production rate has a clear physical sense; it describes a thermally activated process similar to tunneling of a Brownian particle across a potential barrier (Kramers, 1940). Moment t_c in the initial condition corresponds to the moment of time, when the temperature for the first time (heating stage) reaches value $T = T_c$. Correspondingly, when the temperature cross temperature T_c for the second time (during the cooling stage) the phase transition is terminated (quenching in crystallization).

The constant k_ν is related to characteristic vibration frequency. The constant T_ν is the activation energy (in Kelvin). By the order of magnitude it is the same activation energy, which characterizes the solid-state self-diffusion. We do not know these constants for cinnabar, thus we shall use different "reasonable estimations".

The equation for the cluster (crystalline) growth can be written under the assumption that the growth of nuclei occurs in kinetically controlled regime. It leads to the equation of the type (see details in [33]):

$$\frac{dg}{dt} = k_g g^{2/3}(1-x)\left\{1 - exp\left[\frac{q}{T}\left(\theta - \beta g^{-1/3}\right)\right]\right\}, \quad g\,|_{t=t_c} = g_0, \qquad (25)$$

where the constant k_g once again is related to characteristic vibration frequency, q is latent heat of vaporization, and value β describes the critical nuclei (it contains $(\beta/\theta)^3$ pigment molecules).

A similar nonstationary model of phase transition has been suggested by Raizer, 1960; Zeldovich & Raizer, 1966. We call this model as ZR-model. We applied ZR model to discuss formation of nanoclusters within the expanding vapor, produced by laser ablation (Luk'yanchuk, 1998, 1999, 2000; Kuwata, 2001).

Because of the very sharp dependence of the nucleation rate on superheating θ in (24), practically all nuclei are ejected at the moment $t = t_e$, when the superheating reaches maximum. It is convenient to recalculate all initial conditions to this moment. For example, the number of molecules within the smallest critical nuclei at ejection time is given by (Luk'yanchuk, 1998)

$$g\big|_{t=t_e} = g_{min} = \left(\frac{\beta}{\theta_{max}}\right)^3 >> 1. \qquad (26)$$

Now we can choose the set of reasonable parameters (see in Table 2) and integrate equations (22), (24), (25), using the superheating (20), where temperature $T = T(z - h(t), t)$ is found from the solution of the ablation problem. Typical result of this examination is shown in Fig. 4.

One can see in Fig. 4, that superheating rapidly decreases inside the material. As a result number of nucleation centers decreases as well.

Although the averaged size of cluster increases, it does not influence practically the fast decrease in total concentration. The final spatial distributions of clusters in the surface layer of material (after ablation) are shown in Fig. 5.

Table 2. Parameters, which have been used in calculations of nucleation and crystal growth.

	Value	Units
k_v	2 10^7	s^{-1}
T_v	5000	Kelvin
k_g	2 10^7	s^{-1}
q	560	Kelvin
β	2	Dimensionless

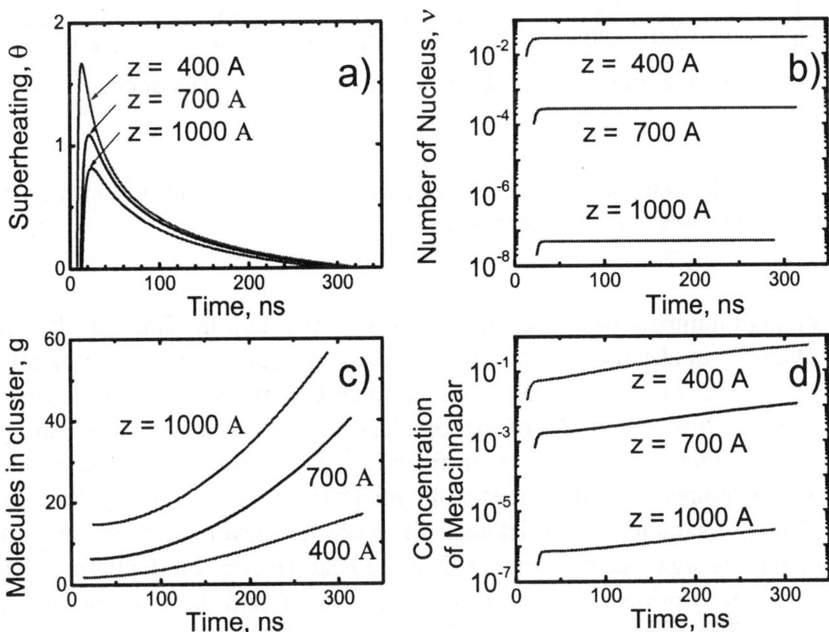

Fig. 4. Dynamics of the nucleation process at three material points (z = 400, 700 and 1000 Å). a) Superheating, b) Number of nucleus, c) Number of molecules in cluster, d) Concentration of metacinnabar.

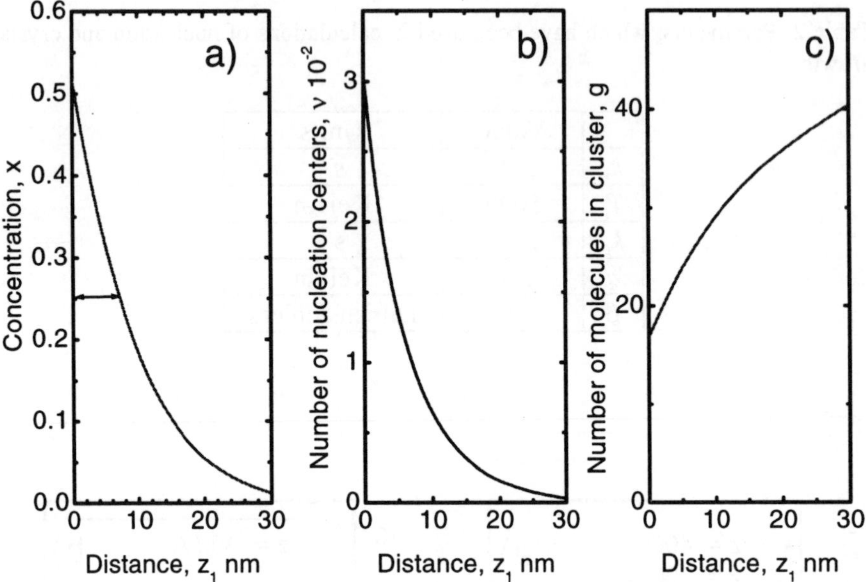

Fig. 5. Spatial distributions of metacinnabar concentration (a), Number of nucleation centers (b) and cluster size (c) after ablation.

Although numbers of concentration can be corrected, the final result will not change qualitatively. It shows that metacinnabar is typically concentrated in *thin surface layer*. Its characteristic thickness consists from few tens to few hundreds of nanometers. The reason for this strong sharpening effect can be easily seen from Fig. 5 b: it occurs because of the very sharp temperature dependence in nucleation rate, given by formula (24). Thus, characteristic thickness of modified layer (a few tens of nanometers) is significantly smaller than the heat penetration depth or depth of optical absorption. Namely this effect was found in laser-produced discoloration in pigments at laser cleaning. Because of its general reason this sharpening effect may play some role in the discoloration of other pigments as well. Finally, it may be also used for superficial phase/chemical modification of various particles. In fact sharpening effect is a general property of any first order phase transition stimulated by laser radiation in the volume of solid.

Acknowledgements

We wish to thank Prof. S. I. Anisimov, Prof. C. Fotakis, Dr. N. Arnold and Dr. N. Bityurin for discussions. B. L. is thankful to Russian Basic Research Foundation (grants 01-02-16136 and 01-02-16189). This work is also partially supported by the "Ultraviolet Laser Facility" which operates at FO.R.T.H within the European Union TMR project of DGXII, ERBFMGECT 950021.

References

Anisimov S. I., Imas Ya. A., Romanov G. S., Khodyko Yu. V., *Action of High -Power Radiation on Metals*, (National Technical Information Service, Springfield, Virginia, 1971

Anisimov S. I., Luk'yanchuk B. S., *Selected Problem of Laser Ablation,* Uspekhi Fizicheskih Nauk, vol. **172**, issue 3, pp. 301-333 (2002)

Arnold N., Luk'yanchuk B., Bityurin N., *A fast quantitative modeling of ns laser ablation based on nonstationary averaging technique,* Appl. Surf. Sci. **127-129**, 184 (1998 a)

Arnold N., Luk'yanchuk B., Bityurin N., Bäuerle D., *Nonstationary effects in laser ablation of Indium: Calculations based on spatial moments technique,* Laser Physics, **8**, 47 (1998 b)

Arnold N., Luk'yanchuk B., Bityurin N., Bäuerle D., *A fast quantitative modeling of ns laser ablation based on nonstationary averaging technique (spatial moments technique),* Proc. SPIE, **3343**, 484 (1998 c)

Anglos D., Couris S., Mavromanolakis A., Zergioti I., Solomidou M., Liu W.-Q., Papazoglou T.G., Zafiropulos V., Fotakis C., Doulgeridis M., Fostiridou A., *Artwork Diagnostics. Laser Induced Breakdown Spectroscopy (LIBS) and Laser Induced Fluorescence (LIF)*, in "Restauratorenblätter, Sonderband – Lacona I, Lasers in the Conservation of Artworks", Eds. E. König, W. Kautek (Verlag Mayer & Comp., Vienna, 1997) pp. 113-118

Backer von R., Dőring W., *Kinetiche Behandlung der Keimbildung in übersattigten Dämpfen*, Annalen der Physik, **24**, 719 (1935)

Bäuerle D., *Laser Processing and Chemistry*, 3rd Ed. (Springer-Verlag, Berlin 2000)

Burgio L., Clark R. J. H., Stratoudaki T., Anglos D., Doulgeridis M.: *Pigment Identification. A Dual Analytical Approach employing Laser*

Induced Breakdown Spectroscopy (LIBS) and Raman Microscopy, Appl. Spectrosc. **54**, (2000)

Cooper M., *Laser Cleaning in Conservation: an Introduction*, (Butterworth Heinemann, Oxford, 1998)

Frenkel Ya. I., *Kinetic Theory of Liquid*, (Dover, New York, 1955)

Georgiou S., Zafiropulos V., Anglos D., Balas C., Tornari V., Fotakis C., *Excimer laser restoration of painted artworks: Procedures, Mechanisms and Effects*, Appl. Surf. Sci. **127-129**, 738 (1998)

Karlov N. V., Kirichenko N. A., Luk'yanchuk B. S., *Laser Thermochemistry. Fundamentals and Applications,* (Cambridge International Science Publishing, Cambridge, UK, 2000)

Klein S., Stratoudaki T., Zafiropulos V., Hildenhagen J., Dickmann K., Lehmkuhl T., *Laser-induced breakdown spectroscopy for on-line control of laser cleaning of sandstone and stained glass*, Appl. Phys. **A** **69**, 441 (1999)

Kramers H. A., *Brownian Motion in a Field of Force and the Diffusional Model of Chemical Reactions*, Physica, **7**, 284 (1940)

Küper S., Brannon J., Brannon K., *Threshold behavior in polyimide photoablation single-shot rate measurement and surface-temperature modeling*, Appl. Phys. **A. 56**, 43 (1993)

Kuwata M., Luk'yanchuk B., Yabe T., *Nanoclusters formation within the vapor plume, produced by ns-laser ablation: Effect of the initial density and pressure distributions*, Jpn. J. Appl. Phys., vol. **40**, Part 1, No. 6A, pp. 4262-4268 (2001)

Lassithiotaki M., Athanassiou A., Anglos D., Georgiou S., Fotakis C., *Photochemical Effects in the UV Laser Ablation of Polymers: Implications for Laser Restoration of Artwork*, Appl. Phys. A **69**, R363, (1999)

Lide D. R. (Ed.), *CRC Handbook of Chemistry and Physics*, 83rd Edition, 2002

Lifshitz I. M., Slyozov V. V., *On the kinetics of diffusional decay of supersaturated solid solutions*, JETP, **35**, 478 (1958)

Lifshitz E. M., Pitaevsky L. P., *Physical Kinetics* (Pergamon, Oxford 1984)

Luk'yanchuk B., Bityurin N., Himmelbauer M., Arnold N., *UV-laser ablation of polyimide: From long to ultrashort laser pulses,* Nucl. Instr. and Methods in Phys. Research **B 122**, 347 (1997)

Luk'yanchuk B., Marine W., Anisimov S. I., *Condensation of vapor and nanoclusters formation within the vapor plume, produced by ns- laser ablation of Si*, Laser Physics, **8**, 291 (1998)

Luk'yanchuk B. S., Marine W., Anisimov S. I., Simakina G. A., *Condensation of vapor and nanoclusters formation within the vapor plume, produced by ns-laser ablation of Si, Ge and C*, Proc. SPIE, **3618**, 434 (1999)

Luk'yanchuk B. S., Luches A., Blanco A., Orofino V., *Physical Modelling of the Interstellar Dust* Proc. SPIE, **4070**, 154 (2000b)

Mao X., Russo R. E., *Observation of plasma shielding by measuring transmitted and reflected laser pulse temporal profiles*, Appl. Phys. **A 64**, 1 (1997)

Peterlongo A., Miotello A., Kelly R., *Laser-pulse sputtering of aluminum: Vaporization, boiling, superheating, and gas-dynamic effects*, Phys. Rev. **E 50**, 4716 (1994)

Raizer Yu. P., *Condensation of a Cloud of Vaporized Matter Expanding in Vacuum*, JETP, **37**, 1229 (1960)

Samarsky A. A., Kurdyumov S. P., et al: *Regimes with Sharpening in Problems of Quasi-Linear Parabolic Equations*, (Nauka, Moscow, 1987)

Skoulikidis Th., Vassiliou P., Papakonstantinou P., Moraitou A., Zafiropulos V., Kalaitzaki M., Spetsidou I., Perdikatsis V., Maravelaki P., *Some remarks on Nd: YAG and Excimer UV Lasers for cleaning soiled sulfated monument surfaces*, In: LACONA – Workshop on "Lasers in the Conservation of Artworks", Book of Abstracts (Heraklion, Crete, 1995)

Stratoudaki T., Xenakis D., Zafiropulos V., Anglos D., *Laser Induced Breakdown Spectroscopy in the Analysis of Pigments in Painted Artworks. A database of pigments and spectra*, In: "Optics and Lasers in Biomedicine and Culture", – Series of the International Society on Optics Within Life Sciences, Vol. V, Series Ed. G. von Bally (Springer-Verlag, Berlin, 2000), pp. 163-168

Wolfram S., *Mathematica*, Fourth Edition, (Cambridge University Press, 1999)

Zafiropulos V., Petrakis J., Fotakis C., *Photoablation of polyurethane films using UV Laser pulses*, Opt. and Quantum Electronics **27**, 1359 (1995)

Zafiropulos V., Fotakis C., *Lasers in the Conservation of Painted Artworks*, Chapter 6 In: "Laser Cleaning in Conservation: An Introduction", Ed. M. Cooper (Butterworth Heinemann, Oxford, 1998)

Zafiropulos V., Stratoudaki T., Manousaki A., Melesanaki K., Oral G., *Discoloration of pigments induced by laser irradiation*, Surface Engineering **17** (2001)

Zafiropulos V., Balasa C., Manousaki A., Marakis Y., Maravelaki-Kalaitzaki P., Melesanaki K., Pouli P., Stratoudaki T., Klein S., Hildenhagen J, Dickmann K., Luk'yanchuk B. S., Mujat C, Dogariu A., *Yellowing effect and discoloration of pigments: Experimental and Theoretical studies,* Paper presented at LACONA IV. To be published in Journal of Cultural Heritage (Elsevier, 2002 a)

Zafiropulos V., *Laser Ablation in cleaning of Artworks*, Chapter 8 in this volume (2002 b)

Zeldovich Ya. B., *Theory of new-phase formation. Cavitation*, JETP, **12**, 525 (1942)

Zeldovich Ya. B., Raizer Yu. P., *Physics of Shock Waves and High Temperature Hydrodynamics Phenomena*, (Academic Press, N.Y., 1966)

Zwillinger D., *Handbook of Differential Equations*, (Academic Press, Boston, 1989)

Part 5. Applications of Laser Cleaning

Chapter 10

LASER CLEANING OF SILICON FIELD EMITTER ARRAY

M. Takai, N. Suzuki, O. Yavas

Laser cleaning of FEA (field emitter array) surfaces with four different wavelengths has been performed to improve the emission characteristics of FEAs. Wavelength dependent (i.e., energy dependent) cleaning effect was found to take place for Si FEAs, when the FEAs were irradiated by a Nd: YLF laser.

Keywords: Laser cleaning, field emitter array

PACS: 42.62.Cf, 81.65.Cf, 68.35.Np, 81.65.-b, 85.40.-e, 81.07.Wx

1. Introduction

A tiny electron source and its arrays embedded in a solid, i.e., a field emitter (FE) and its array (field emitter array: FEA), have drawn much attention because of their features such as high brightness, low energy spread, and possible applications to vacuum microelectronics such as high power radio frequency amplifiers, high-voltage switches, and field emission displays (FEDs) with high brightness, quick response, and wide viewing angle (Schwoebel, 1995; Temple, 1998, 1999). Various types of emitters have been investigated to date, for example, Spindt-type emitters with Mo tips, gated Si emitters, and various carbon emitters such as diamond like carbon, or carbon nano tube (CNT) emitters (Temple, 1999). Spindt-type and gated Si emitters have widely been investigated as FEAs among various emitters because of the process compatibility with semiconductor processing.

The issues with such FEAs are stable and homogeneous emission with high current, requiring homogeneous tip surface processing over an array for high emission performance. Considerable efforts are therefore being made to improve the emission performance of the FEAs (Takai, 1995, 1996, 1998a, 1998b, 1998c). It has been shown that performance improvement could be achieved by various surface modifications, such as anodization (Takai, 1995), gas ambient operation (Takai, 1998c), silicidation (Takai, 1998b), and deposition of thin films such as diamond and diamond-like carbon films (Yavas, 1999).

One of the most important issues on Si field emitters has been instability of the tip surface, which was easily oxidized or contaminated with carbon or H_2O molecules during processing. The cleaning of contaminated emitter surfaces after processing is an important issue, since contaminants increase the work function and, hence, decrease the emission current, and lead to irregular emission behavior due to uncontrolled desorption during operation, resulting in flicker noise, step-like or spike noise during electron emission.

Various tip-surface cleaning methods such as plasma cleaning or gas-ambient emission (Takai, 1998c) for FEAs have been tried to improve or stabilize emitter-tip surfaces. Among various cleaning methods, ultra violet (UV) laser light irradiation seems a gentle process which can only desorbs surface contaminants through light enhanced desorption processes. Laser surface cleaning has extensively been investigated using excimer laser lights (Tam, 1992; Lu, 1994a, 1994b) in which the laser photon energy was close to or slightly higher than the binding energies of contaminants. Therefore the laser photons can directly dissociate the surface contaminants as shown in Fig. 1.

Conventional cleaning techniques for FEAs, based on high temperature treatment or high-energy ion bombardment, are not desirable due to the limited substrate temperature such as 500 °C for glass substrates and the likelihood of fatal damage to the emitter arrays. A prolonged emission, i.e., aging, requires a plenty of time before vacuum sealing. On the contrary, laser cleaning can be used as an efficient but 'soft' cleaning technique for decontamination of FEA surfaces, yielding a significant increase in emission current (Takai, 1998a; Yavas, 1998, 2000) and it shortens aging time.

In this study, the harmonics of a diode-pumped Nd: YLF laser were used to excite the top surface layers of Si FEAs in order to investigate the effect of UV light irradiation on electron emission from Si FEAs.

Systematic studies as a function of laser fluence and wavelength revealed that photodecomposition is the main physical mechanism for the removal of the contaminants, suggesting that higher energetic photons

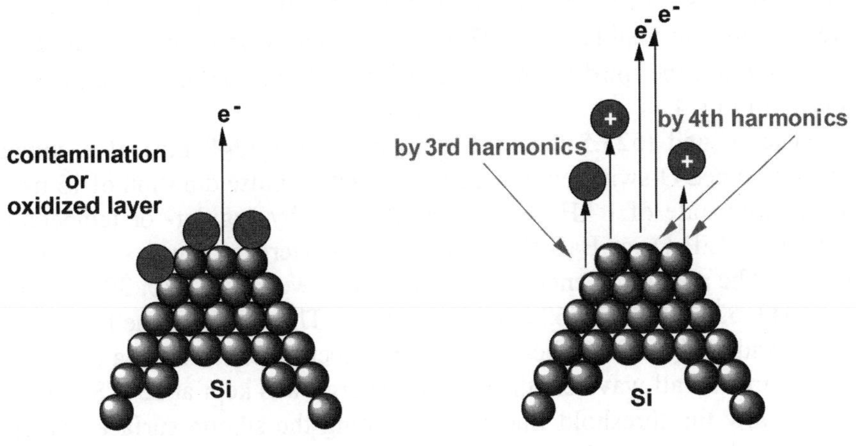

Fig. 1. Laser cleaning of surface contaminants on the top of Si FE.

should be more effective in cleaning the field emitter surfaces and increasing the emission current. In this work, however, the application of shorter laser wavelengths was found to be not desirable due to excessive ionization of the desorption products. Therefore the right choice of the laser wavelength turned out to be an important parameter for an efficient cleaning of the emitter surfaces.

2. Experimental Procedures

Niobium gated silicon FEAs of various tip numbers ranging from single tip to 10 x 10 (100 tips), and 40 x 40 (1600 tips) were fabricated by conventional dry etching and oxidation sharpening techniques on (100)-orientated n-type silicon wafers with a resistivity of 3-5 Ohm-cm as shown in Fig. 2 (Takai, 1995). The sample was placed in a ultra high vacuum (UHV) chamber with a base pressure of 10^{-10} Torr and irradiated by a diode-pumped Q-switched niobium-doped yttrium lithium fluoride (Nd: YLF) laser. Prior to laser irradiation the FEAs were operated for 1 hour using a gate voltage of 100 V and an anode voltage of 200 V. The anode, consisting of an ITO film on a glass substrate, was placed at a distance of 1 cm from the field emitter surface as shown in Fig. 3. The laser wavelength was systematically varied from infrared (λ = 1047 nm; the fundamental wavelength), to visible (λ = 523 nm; the second harmonic), and ultraviolet (λ = 349 nm; the third harmonic, and λ = 262 nm; the fourth harmonic) as shown in Fig. 4.

The second (523.5 nm) and third harmonics (349 nm) of a compact diode-pumped Q-switched Nd: YLF laser with a pulse duration of 15 ns and a repetition rate of 1 kHz have been obtained by angular- or temperature-controlled LiB_3O_5 or BaB_2O_4 crystals and interference mirrors as shown in Fig. 4. The maximum energies of laser pulses were 270, 130, 30, and 10 μJ for 1047, 523, 349 and 262 nm, respectively. The 4 lines of the laser system were irradiated on FEAs before or during electron emission at a pressure of 10^{-10} Torr. At all wavelengths the laser energy was kept at 35 – 85 mJ/ cm^2 well below the threshold fluence for melting the silicon surface, which is known to be at about 300 mJ/cm^2 in the visible wavelength region (Yavas, 1998). The laser-induced temperature rise was numerically calculated to be about 100 K. The effects induced by laser irradiation were conveniently monitored *in-situ* by simultaneous measurement of electron emission under an applied electrical bias as shown in Fig. 3. In case of a single tip emitter, the evolution of the emission pattern on the phosphor screen on the ITO during laser irradiation was monitored using a CCD (charge coupled device) camera. The chamber pressure during laser cleaning was also recorded to confirm any removal of surface contaminants.

3. Laser light irradiation

3.1. Laser irradiation modes

3.1.1. *Laser irradiation without field emission*

Laser irradiation over the FE cathodes without field emission, i.e., without application of gate bias voltage would result in surface cleaning due to the desorption of surface contaminants through direct excitation, when the energy of incident laser photons exceed the binding energy of surface contamitants chemisorbed on the surface. Therefore the effect of cleaning can be confirmed by post field emission measurement after laser irradiation.

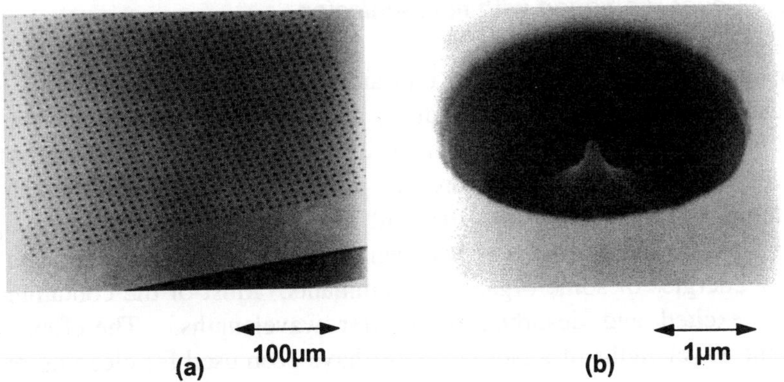

Fig. 2. SEM views of Si FEA with 40 x 40 FEs and a gate opening of 2.5 μm.

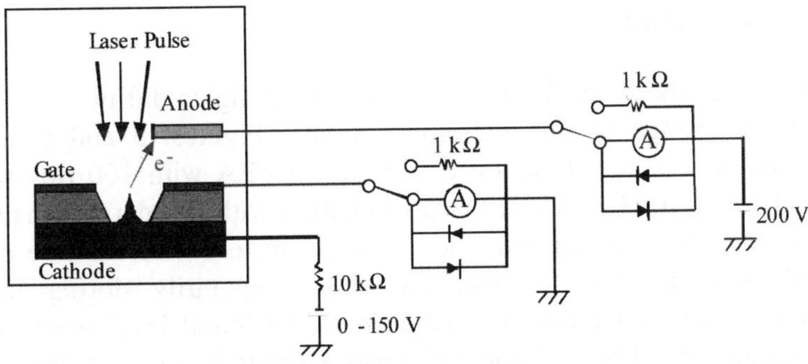

Fig. 3. On-line monitoring of cleaning efficiency. FEAs were operated in a vacuum chamber during laser cleaning.

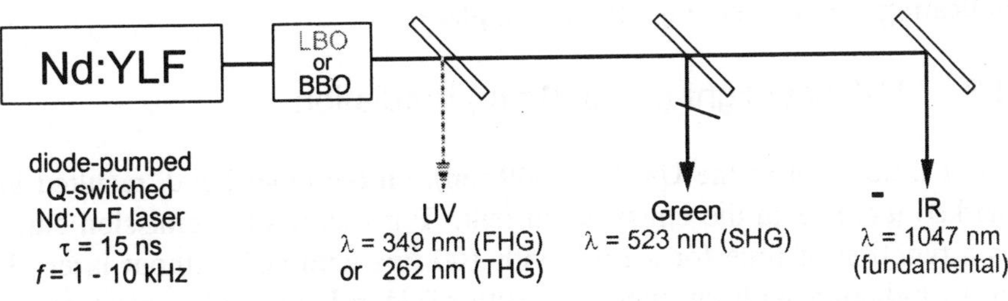

Fig. 4. Optical configuration for a Nd:YAG laser system with the fundamental (1047 nm, the second (523 nm), the third (349 nm), and the fourth harmonics (262 nm).

3.1.2. Laser irradiation with field emission

Laser irradiation during field emission, i.e., during operation with application of gate bias voltage, induces an enhanced electron emission due to the photon induced electron emission and, at the same time, desorption of surface contaminants. Therefore the effect of laser cleaning can be monitored by continuous measurement after turning off the laser light. Table 1 shows the comparison between the laser photon energies and the binding energies of some organic contaminants. Most of the contaminants can be excited and desorbed by shorter wavelengths. Therefore four different wavelengths of a Nd:YLF laser have been used for cleaning of the FEA.

3.2. IR and visible laser light (λ = 1047 and 523.5 nm) irradiation

Irradiation of the FEAs by IR and visible laser lights did not result in any permanent change of the emission current. Figures. 5 and 6 show the emission current as a function of time for a FEA with 100 tips with and without laser (1047 and 523.5 nm) irradiation with an energy density of 35 mJ/cm^2. Within the applied fluence range up to 85 mJ/cm^2 the emission current was observed to increase only temporarily during the laser irradiation and to decrease immediately to the initial level when the laser was turned off. The rise and fall times were less than 1 s, the temporal resolution of our experimental setup, so that this temporary increase could be attributed to photo-excitation of carriers. In fact, contaminated single-tip emitters exhibited a non-symmetrical three-fold emission pattern as discussed in 3.3 both prior to and after the (IR and visible) laser irradiation, indicating that no cleaning effect took place.

3.3. UV laser light (λ = 349 nm) irradiation

Laser irradiation in the UV (λ = 349 nm), on the other hand, resulted in a marked increase in the emission current. Fig.7 shows the emission current as a function of time for a FEA with 100 tips with and without laser (349 nm) irradiation with an energy density of 35 mJ/cm^2. The onset of laser irradiation is marked with a sharp increase in the emission current due to the photoeffect, which, on the contrary to IR or visible laser irradiation, does

Table 1. Chemical bond energies, wavelengths vs laser types.

Covalent Chemical Bond	Energy (eV)	Wavelength (nm)
C-H	3.50	354.3
O-H	4.44	279.3
H-H	4.52	274.3
O-O	5.16	240.0
C-C	6.29	197.0

Laser Type	Energy (eV)	Wavelength (nm)
Nd:YLF(Fundamental)	1.18	1047
Nd:YLF(2nd harmonics)	2.37	523.5
Nd:YLF(3rd harmonics)	3.55	349.0
Nd:YLF(4th harmonics)	4.74	261.8

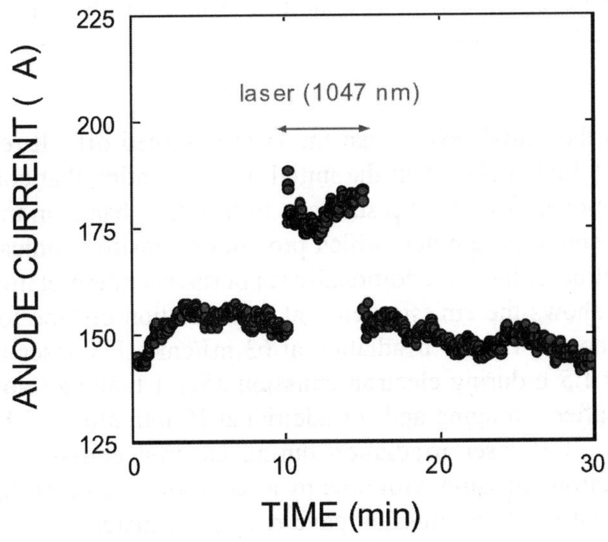

Fig. 5. Electron emission current as a function of time for FEA with and without laser irradiation at 1049 nm.

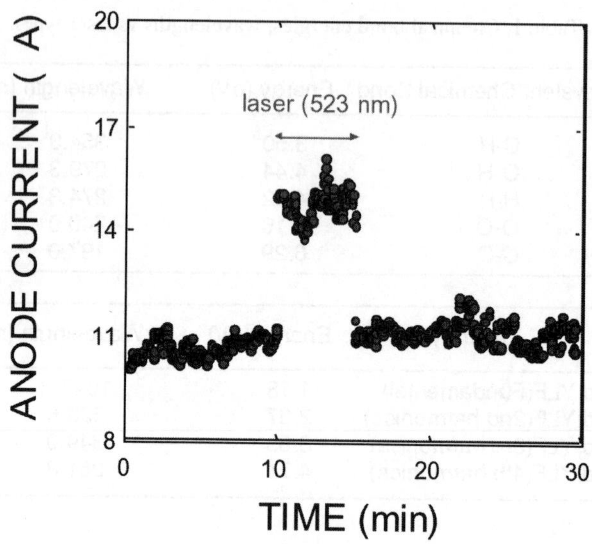

Fig. 6. Electron emission current as a function of time for FEA with and without laser irradiation at 523.5 nm.

not return to the initial level when the laser is turned off. It remains at a level of about 30% higher than the initial one, indicating that contaminants have been removed from the tip surface. In fact, the change in the emission pattern for a single-tip emitter, which provides a sensitive indicator for the surface condition of the tip, additionally supports this interpretation.

Figure 8 shows the emission current as a function of time for Si FEAs with and without UV laser irradiation at 63 mJ/cm^2. UV laser lights were irradiated for 1.5 h during electron emission after 1 h aging (closed circle) or for 5 min after 1 h aging and for additional 10 min after 2.5 h operation (open circle). UV laser irradiation during electron emission resulted in enhanced electron emission with time by a factor of 2 - 2.4, while that for 5 -10 min without electron emission gave rise to the increase in emission by about 25 - 50 %. A marked difference in the emission enhancement was observed between two modes of UV laser irradiation, i.e., with and without electron emission. The emission current gradually decreased with time for

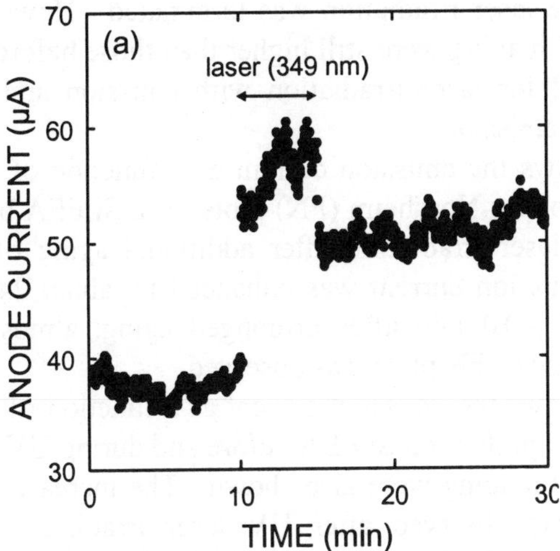

Fig. 7. Emission current as a function of time with laser irradiation (λ = 349 nm).

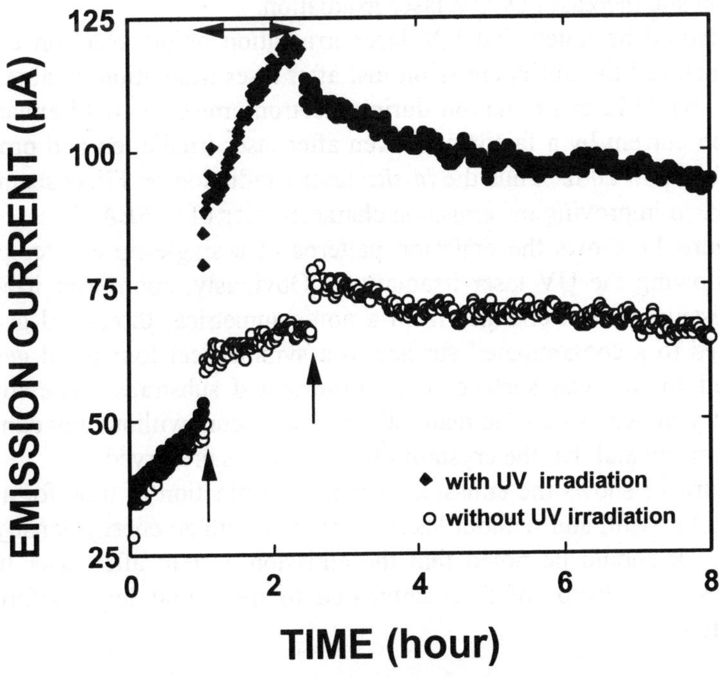

Fig. 8. Emission current as a function of time for 40 x 40 Si FEAs with laser irradiation (λ = 349 nm) with or without electron emission.

both cases as the laser irradiation was terminated. However the emission currents after 5.5 h aging were still higher than those before laser irradiation by a factor of 2 for laser irradiation with emission and by about 30 % without electron emission.

Figure 9 shows the emission current as a function of gate voltage and corresponding Fowler-Nordheim (FN) plots for a Si FEA before and after 5 or 10 min UV laser irradiation after additional aging of 1 h and 20 h. Although the emission current was enhanced by about 24 -32 % by laser irradiation for 5 - 10 min after prolonged aging, almost no remarkable change in the slope of FN plots was observed.

Figure 10 shows the emission current as a function of gate voltage and corresponding FN plots for a Si FEA before and during UV laser irradiation. The data after 20 h aging were also shown. The increase in emission by a factor of 2.2 was observed after UV laser irradiation during electron emission and it slightly decreased after 20 h aging. No marked change in the slope of FN plots was observed. This indicates that the number of emission site increases by UV laser irradiation.

It should be noted that UV laser irradiation before electron emission only enhanced the initial emission just after laser irradiation by about 25 % - 50 %, while laser irradiation during electron emission could enhance the emission current by a factor of 2 even after laser irradiation and prolonged aging for up to 20 h. Thus the *in situ* laser irradiation on FEAs seems quite effective in improving the emission characteristics of Si FEAs.

Figure 11 shows the emission patterns of a single-tip emitter prior to and following the UV laser irradiation. Obviously, upon laser irradiation the emission pattern changes from a non-symmetrical three-fold geometry attributed to a contaminated surface to a symmetrical four-fould geometry expected for a clean surface of a 100-oriented substrate. The four-fold symmetry indicates that the material removal occurs without any damage to the silicon tip and that the crystallinity of the tip is preserved.

Figure 12 shows the emission current as a function of time for a single field emitter with and without laser irradiation with an energy density of 35 mJ/cm^2. It should be noted that the emission current after laser turn-off increases by a factor of 5 as compared to the initial level before laser irradiation.

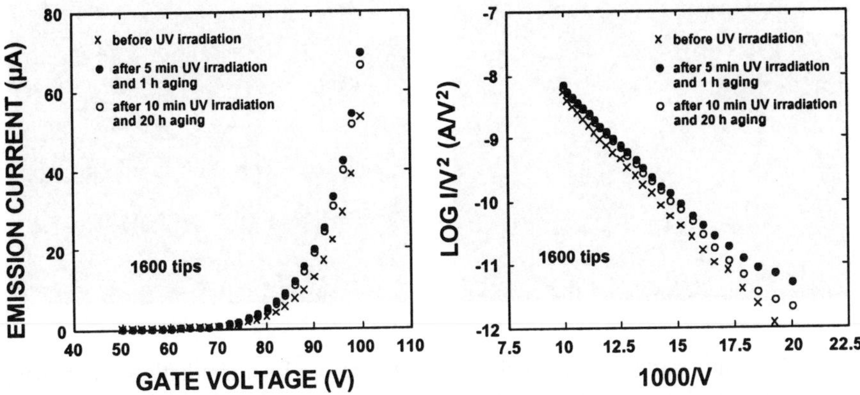

Fig. 9. Emission current as a function of gate voltage and corresponding FN plot for a 40 x 40 Si FEA after UV laser irradiation for 5 min and additional 1 h aging and 10 min and addition 120 h aging.

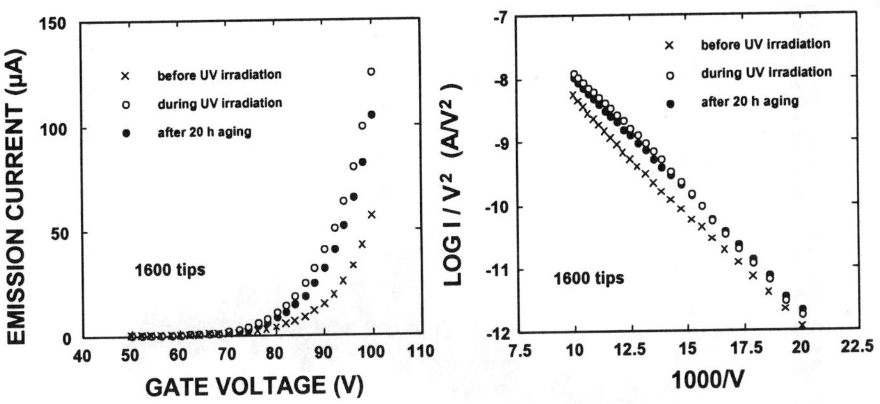

Fig. 10. Emission current as a function of gate voltage and corresponding FN plot for a 40 x 40 Si FEA after UV laser irradiation with electron emission and 20 h aging.

Fig. 11. Emission pattern for a single tip emitter before laser irradiation (a) showing a nonsymmetrical threefold pattern. Emission pattern after laser irradiation (b), showing a symmetrical fourfold pattern.

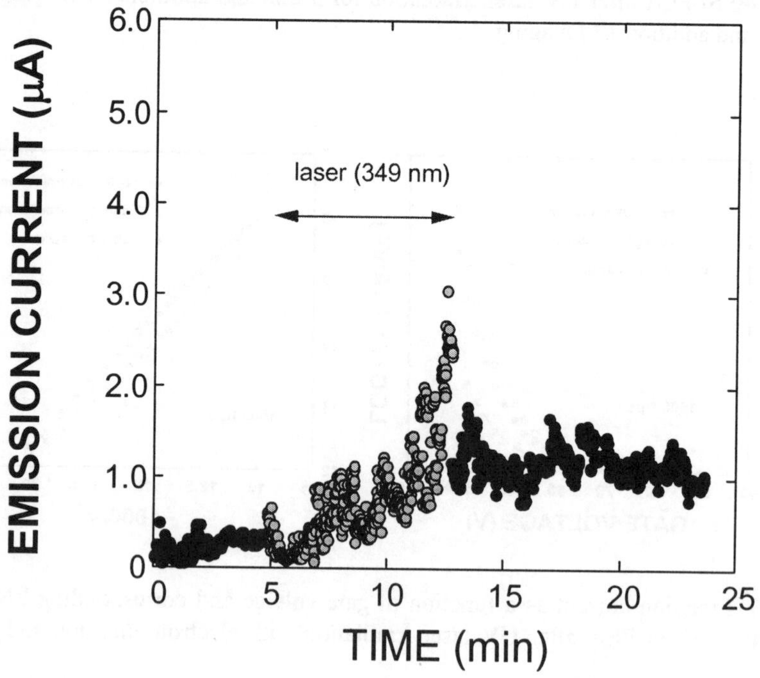

Fig. 12. Emission current as a function of time for a single field emitter with and without laser irradiation.

3.4. UV laser light ($\lambda = 262$ nm) irradiation

For the removal of contaminants the thermal desorption should play only a minor role, since the calculated temperature rise for the silicon surface is only about 100 K (Yavas, 2000). The results could be rather consistently explained by the cleaning of the emitter tip surface via photochemical decomposition of organic contaminants upon UV laser irradiation. The photon energies of the fundamental and the second harmonics (1.18 and 2.37 eV, respectively) are insufficient to break covalent bonds such as C-H (3.5 eV), O-H (4.44 eV), or H-H (4.52 eV) [9], whereas the photon energy of the third harmonic (3.55 eV) is sufficiently high to break C-H bonds.

Since the photochemical decomposition is the dominant mechanism for the removal of contaminants, the cleaning yield is expected to be higher when higher energetic photons are used. However, on the contrary to the expectation, during irradiation with the fourth harmonic at $\lambda = 262$ nm the emission current temporarily decreases and slowly recovers to the initial level when the laser is turned off (Fig. 13(a)). Free-radical formation during laser ablation is a well-known phenomenon, and can account for this unexpected behavior.

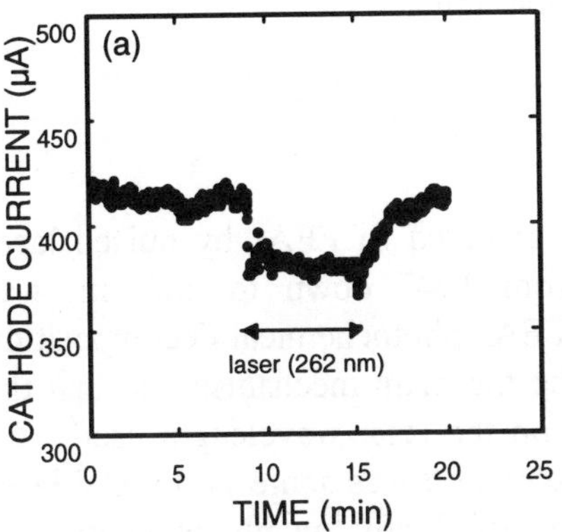

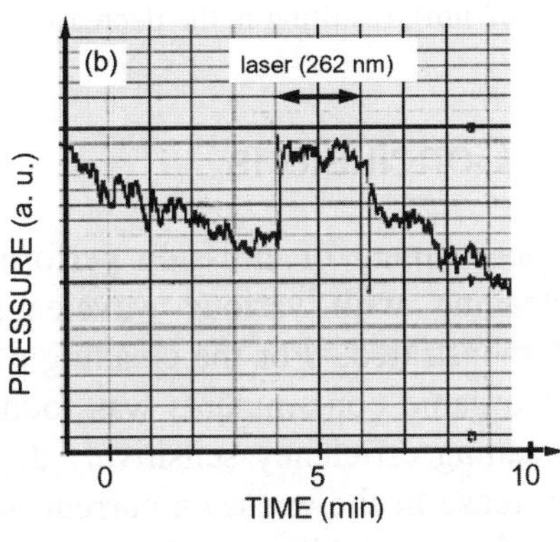

Fig. 13. Cathode current as a function of time for 40 x 40 FEA during UV laser irradiation ($\lambda = 262$ nm) (a) and corresponding chamber pressure as a function of time during laser irradiation.

Excessive ionization of desorption products during laser irradiation is believed to offset the effect of the photo-generated carriers and cause the observed temporary decrease in the emission current (Fig. 14). Material removal has been indeed confirmed by the simultaneous measurement of the chamber pressure, which increases for the duration of laser irradiation (Fig. 13(b)). The recovery of the current level to the initial level despite the material removal could be explained by redeposition of the ionized species on the tip surface due to the presence of high electric field.

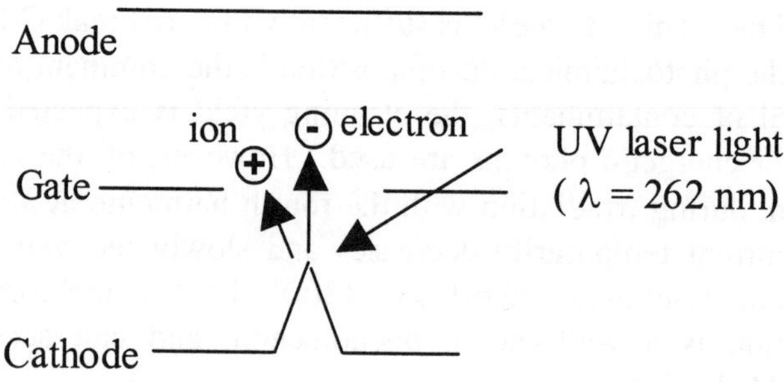

Fig. 14. Ionized particle emission from the cathode induced by laser irradiation (λ = 262 nm), resulting in the decrease of emission current.

4. Conclusions

Improvement of emission performance of gated Si FEAs by pulsed laser cleaning with various wavelengths from 1047 down to 262 nm was demonstrated. For the cleaning of the FEAs photochemical decomposition of organic contaminants was found to be the main mechanism, so that the cleaning efficiency sensitively depends on the laser wavelength used. An increase in the emission current by a factor of 5 was achieved by UV laser irradiation at 349 nm, while, due to the insufficiently low photon energy, no cleaning was induced by IR and visible laser irradiation at 1047 and 523 nm, respectively. Excessive ionization of desorption products by higher energetic photons, e.g. the fourth harmonic in the UV at 262 nm, on the other hand, had a neutralizing effect in emission current and no cleaning was observed.

Acknowledgment

The authors are indebted to Akihiko Hosono and Soichiro Okuda (Mitsubishi Electric Corporation) for collaboration.

References

Lu Y.F., Aoyagi Y., Takai M., Namba S., *Laser surface cleaning in air-Mechanisms and applications*, Jpn. J. Appl. Phys. **33**, pp. 7138 – 7148 (1994a)

Lu Y.F., Takai M., Komuro S., Shiokawa T., Aoyagi Y., *Surface cleaning of metals by pulsed laser irradiations in air*, Appl. Phys. **A59**, pp. 281 - 288(1994b)

Schwoebel P. R., Brodie I., *Surface-science aspects of vacuum microelectronics*, J. Vac. Sci. Technol. **B13**, pp. 1391 – 1410 (1995)

Seidl A., Takai M., Hosono A., Yura S., Okuda S., *Geometry effects arising from anodization of field emitters*, J. Vac. Sci. Technol. **B18**, pp. 929 – 932 (2000)

Takai M., Yamashita M., Wille H., Yura S., Horitaba S., Ototake M., *Enhancement in emission current from dry-processed n-type Si field emitter arrays after tip anodization*, J. Vac. Sci. Technol. **B13**, pp. 441 - 444 (1995)

Takai M., Kishimoto T., Yamashita M., Morimoto H., Yura S., Hosono A., Okuda S., Lipp S., Frey L., Ryssel H., *Modification of field emitter array tip shape by focused ion beam irradiation*, J. Vac. Sci. Technol. **14**, pp. 1973 - 1976 (1996)

Takai M., Suzuki N., Morimoto H., Hosono A., Kawabuchi S., *Effect of laser irradiation on electron emission from Si field emitter arrays*", J. Vac. Sci. Technol. **B16**, pp. 780 - 782 (1998a)

Takai M., Iriguchi T., Morimoto H., Hosono A., Kawabuchi S., *Electron emission from gated silicide field emitter arrays*, J. Vac. Sci. Technol. **B16**, pp. 790 - 792 (1998b)

Takai M., Morimoto H., Hosono A., Kawabuchi S., *Effect of gas ambient emission on improvement of Si field emitter arrays*", J. Vac. Sci. Technol. **B16**, pp. 799 - 802 (1998c)

Tam A.C., Leung W.P., Zapka W., Ziemlich W., J. Appl. Phys. **71**, pp. 3515-3523 (1992)

Temple D., Palmer W.D., Yadon L.N., Mancusi J.E., Vellenga D, McGuire G.E., *Silicon field emitter cathodes: fabrication, performance, and applications*, J. Vac. Sci. Technol. **A16**, pp. 1980 – 1990 (1998)

Temple D., *Recent progress in field emitter array development for high performance applications,* Mat. Sci. Eng. **R24**, pp. 185 – 239 (1999)

Yavas O., Suzuki N., Takai M., Hosono A., Kawabuchi S., *Laser cleaning of field emitter arrays for enhanced electron emission*", Appl. Phys. Lett. **72**, pp. 2797 - 2799 (1998).

Yavas O., Hashimoto T., Suzuki N., Takai M., Higuchi Y., Kobayashi M., Hosono A., Okuda S., *Pulsed laser deposition of diamond-like carbon films on gated Si field emitter arrays for improved electron emission*, Jpn J. Appl. Phys. **38**, 7208 – 7212 (1999)

Yavas O., Suzuki N., Takai M., Hosono A., Okuda S., *Improvement of electron emission of silicon field emitter arrays by pulsed laser cleaning*", J. Vac. Sci. Technol, **B18**, pp. 1081 - 1084 (2000)

Chapter 11

LASER CLEANING OF ORGANIC CONTAMINATION ON MICROELECTRONIC DEVICES AND PROCESS REAL-TIME MONITORING

M.H. Hong, W.D. Song, Y. F. Lu, B. Luk'yanchuk, T.C. Chong

Laser cleaning of organic contamination from microelectronic device surfaces has been studied. It was applied to remove plasma debris accumulated on flexible circuit for inkjet printer cartridge and mold flash on IC packages. Compared to conventional cleaning techniques, pulsed laser cleaning has advantages of being fast speed, flexible, easy control, dry and non-contact processing. An audible acoustic wave, plasma-induced electric field and plasma optical signals were used for real-time monitoring of the cleaning process. This permits to perform complete removal of the organic contamination of the device substrates without any damage.

Keywords: laser ablation, cleaning, organic contamination, microelectronics, signal diagnostics, process control in real-time

PACS: 41.20.-q; 42.62.Cf; 43.35.-c; 52.70.-m; 81.15.Fg; 81.65.Cf; 85.40.-e

1. Introduction

Microelectronic devices are currently developing very fast in the direction of higher densities and smaller circuit dimensions. During the manufacturing processes, the contamination problem of the device surfaces inevitably appears.

The contamination affects the device performance or even causes device failure during its high-speed operation. For a 100 GB/inch^2 hard disk drive, magnetic head slider is flying about 10 nm above the ultra-smooth media surface. This situation is similar to a jumbo jet airplane flying only several meters above the ground. It is not difficult to imagine that a 100-nm particle on the media surface would cause the hard disk crash as the magnetic media disk rotates at a high speed up to 10,000 rpm.

In integrated circuits (IC) packaging, it is a key issue how to dissipate the heat generated during the operation of 1 GHz CPU to avoid the device mal-function. Organic contamination (for example, mold flash) accumulated on the heat sink surface must be removed in the back end of the IC manufacturing lines. It is clear that contamination control becomes one of the most critical problems in modern microelectronics industry (Chang, 1996; Singer 1992).

There are many conventional cleaning techniques in microelectronics manufacturing, such as chemical solution etching, plasma processing, thermal and ultrasonic cleaning (Kern, 1990; Cho, 1991; Yamazaki, 1992). Among them, ultrasonic cleaning is the most frequently used technique. An ultrasonic source with the frequency up to several hundred MHz is applied to generate a strong vibration of the devices submerged inside a liquid solution and detach the contamination from the surfaces.

Since the adhesion force $F \propto R$ (R is the particle radius), then the necessary acceleration a to remove particle from the surface, $a = F/m \propto R^{-2}$ increases greatly with the reduction of particle size. It is a high challenge for the ultrasonic cleaning to remove sub-100 nm particles (Tam, 1992; Zapka, 1991).

Ultrasonic cleaning is a wet processing and it is highly desirable in the industry to find a non-contact and dry cleaning technique (Deal, 1990). In this book, many papers deal with the laser dry cleaning of nanoparticles from the substrate surface. There are new phenomena observed during the dry laser cleaning of the particles, such as near field effects and optical enhancement (Luk`yanchuk, 2000; Lu, 2000; Mosbacher, 2001).

Besides the particle type contamination, thin film contamination on the substrate surface is also frequently encountered in microelectronics industry since there are many organic compounds involved in the manufacturing (Hashimoto, 1991; Kogure, 1993). Ultrasonic cleaning does not show very good cleaning results for this type of contamination. It is because the water solution is very difficult to access underneath the contamination to have enough contact area for the liquid solution acting on the contamination. Pulsed laser ablation becomes one of the most potential techniques to remove the organic contamination (Maravelaki, 1999). It has been extensively demonstrated that with laser-induced chemical bond breaking, organic compounds can be easily ablated (Wu, 1999; Dyer, 1992). During the laser cleaning, laser parameters should be properly controlled so that laser irradiation does not cause the damage of substrate materials during the laser ablation of the organic contamination. Fortunately, the substrates of most microelectronics devices are metals or semiconductors. Ablation threshold fluences for these substrate materials are generally much higher than those for organic contamination. The cleaning objective can be achieved with proper selection of laser fluence inside a threshold fluence window between the contamination and substrate materials. In this paper, we will refer to some practical applications of the pulsed laser dry cleaning of organic contamination on the microelectronic device surfaces. It will provide our industrial partners with some useful hints to consider the laser dry cleaning technique as one of their potential options to cope with the industrial challenge.

Surface contamination is normally distributed randomly over the whole device surface with different sizes and different thickness. It is not a good way to apply the same laser cleaning parameters to the whole surface. It may lead to time-consuming cleaning (too many laser pulses applied to less or no contamination area) or incomplete cleaning (not enough laser pulses applied to more contamination region). Therefore, process monitoring in real time during the laser cleaning is very important to ensure complete removal of contamination and short operation time in the manufacturing lines. It is necessary to find a solution to detect the signals generated during the laser cleaning and correlate the signal information to the contamination removal.

Non-contact signal diagnostics is the best option since it can avoid the introduction of other contamination to the device surfaces and free coupling of sensors to the devices to be cleaned. During the pulsed laser ablation there is shock wave and plasma generation (Gu, 1996; Ventzek, 1991). There is an acoustic wave and optical signals available for the detection. Since the plasma generated is highly dynamic at the early stage of the laser ablation (delay time less than 100 ns), it also emits electric and magnetic waves (Hong, 1999). With a tiny metal probe nearby, plasma-induced electric and magnetic signals can be sensed. With the removal of organic contamination, the acoustic, electric, magnetic and optical signals gradually reduce to zero at the complete cleaning of the organic contamination from the device surfaces. Signal diagnostics can be applied to figure out whether and when the laser cleaning process is finished so that it can be controlled in real time.

2. Experimental setup

Figure 1 shows the system design for the pulsed laser dry cleaning of the organic contamination on the microelectronics devices and process control in real time. A *pulsed laser* (for example, a Nd:YAG laser with a pulse duration (FWHM) of 7 ns and wavelength of 532 nm) is applied to offer a pulse energy output to remove the contamination on the device surfaces (for example, flexible circuit for inkjet printer cartridge). A laser beam emitted from the laser source passes through a *beam sampler*. A small amount of laser energy (for example, 5 %) is directed onto an *energy meter*. The laser pulse energy is monitored in real time. Its value is sent to a *process controller* to be compared with the setting value. If there is some difference, the process controller generates a signal to the laser controller, which modifies the laser output.

The other part of the laser energy goes through an *optical system*, which adjusts the laser beam size, fluence and shape on the device surface. The laser pulse with a proper selection of processing parameters is then irradiated on a *light scanner*. Controlled by the process controller, this scanner can change the laser irradiation position on the sample surface to remove the contamination at different positions. After the laser cleaning of one device, the process controller sends a signal to the *sample convey* and transports the next sample to the irradiated position. It is equipped with a signal diagnostics and real-time monitoring system so that the cleaning process can be controlled.

A suitable *sensor* (for example, a microphone to record audible acoustic waves or a photodiode to detect plasma optical signal) is used to capture the signals generated during the laser irradiation. With a feedback control, the process can be monitored in situ. Surface pattern inspection by a *CCD camera* is the other system to ensure the cleaning quality in the processing. It also provides the information whether the contamination is completely removed and whether there is any damage on the device surface after the laser irradiation.

During the laser cleaning the surface contamination will be changed into a lot of tiny debris ejected out of the irradiation area. However,

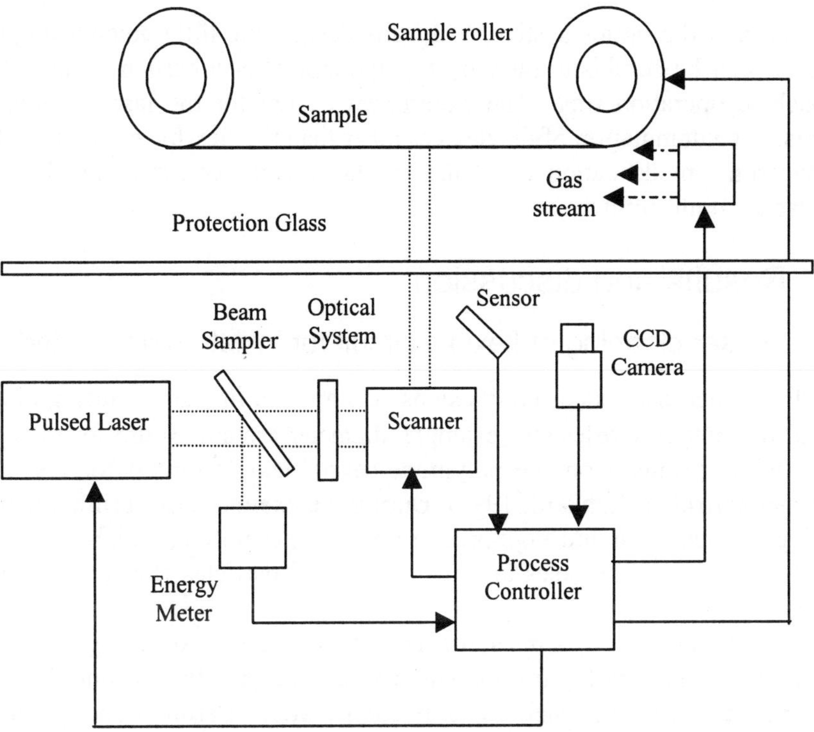

Fig. 1. System design for laser dry cleaning of organic contamination on microelectronics devices and process control in real time.

the contamination debris may re-deposit onto the device surface and re-contaminate the device. Therefore, it is very important to take away the contamination debris before they drop back to the device surface during the laser processing. A *gas stream* or *sucker* system is included in this system. With proper control of gas nozzle position, size and gas stream flowing rate, the contamination debris can be easily taken away. In the system setup, the device (for example, the flexible circuit) is faced down and irradiated by the pulsed laser. With this design the flexible circuit re-contamination by the plasma debris can be greatly lessened due to gravitational attraction on the debris. However, there may be a little amount of the debris breaking through the gas stream and dropping on the optics. It will affect the optics processing performance. A *protection glass* is included between the flexible circuit and the optics system. With this design, the little amount dropping debris can be scrubbed away by the operator after a certain period of the machine operation time. The experiment design for the laser cleaning of organic contamination of the devices other than the flexible circuits needs to have some modification to fit the special device geometry, but the basic ideas are quite similar.

3. Results and discussion

3.1. Laser cleaning of flexible circuit for inkjet printer cartridge

KrF excimer laser is currently extensively applied in manufacturing lines to drill tiny inkjet nozzles (for example, diameter around 40 μm for the 2,400 dpi inkjet printers) on the polyimide-based flexible circuit for the inkjet printer cartridge (Lankard, 1992). During the excimer laser drilling there is a high temperature and high-pressure plasma generation, which consists of CO_2, CO and HCN gases and $C_2 \sim C_{12}$ solid particles (Srinivasan, 1986; Otis, 1989).

In the plasma dynamics the polyimide ablation debris generated is limited between the top transparent layer and polyimide substrate. It results in the deposition of black contamination materials around and inside the tiny inkjet nozzles. The organic contamination affects the printing quality due to improper ink shooting and poor flexible circuit contact with other device components. It is one of the important issues in the manufacturing lines to ensure the printer performance by removing the surface contamination before the cartridge assembly.

As the printer technology develops, the tiny inkjet nozzles become smaller and smaller for the higher resolution printing quality. This makes the contamination removal even more critical. Figure 2 shows images of the flexible circuit with the contamination on the polyimide substrate before the laser irradiation. The dark lines in Fig. 2 (a) are the contamination areas. There are many tiny inkjet nozzles along these lines, as shown in Fig. 2 (b).

There are two available methods to remove the carbon-based polymer debris by chemical solution cleaning and plasma etching respectively. In the case of the chemical solution cleaning, the flexible circuit is put into some chemical solutions for a certain period of time until the contamination is completely etched away. The flexible circuit is then rinsed by de-ionized (DI) water and dried in a subsequence process.

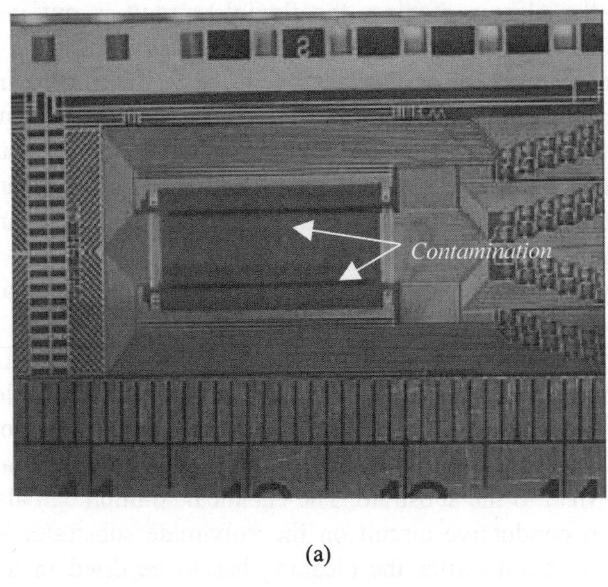

(a)

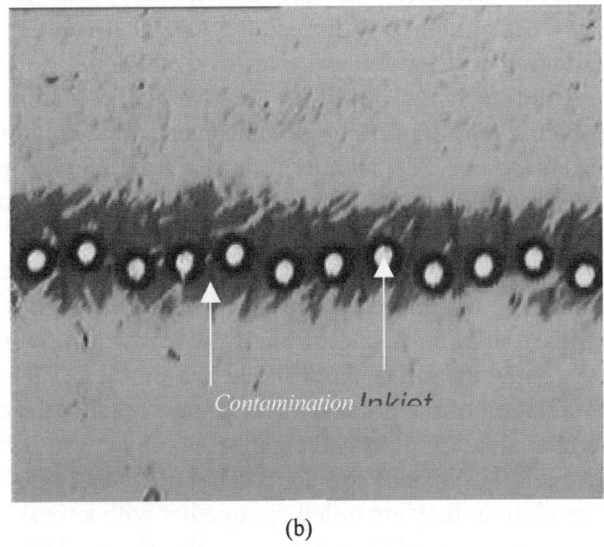

(b)

Fig. 2. Images of the flexible circuit with the contamination on the polyimide flexible circuit before the laser irradiation. The dark lines in the center of the device are the contamination areas needed to be removed.

For the plasma etching, the flexible circuit is put inside a vacuum chamber and then irradiated by the plasma generated from a plasma source. Surface contamination is then removed due to the plasma interaction with the contamination materials. Contamination debris generated during the plasma etching can be taken away by a gas stream or with a sucker system.

Though these two methods mentioned above have cleaning effects on the surface contamination, they have some disadvantages, which significantly affect the productivity of the contamination removal, the reliability of the flexible circuit in manufacturing and bring about the environment protection issue.

For the chemical solution cleaning high cost and subsequence drying of the flexible circuits are the main drawbacks of this technology. When the chemical solutions etch away the contamination, it is also possible to cause some chemical reaction with the polyimide substrate materials and bring in the side effect to the substrate. The chemical solution can also be corrosive to the thin conductive circuit on the polyimide substrate. Meanwhile, the wet flexible circuit after the cleaning has to be dried in the subsequence process, which increases the production cost and processing time. Furthermore, to maintain and dispose the chemical solutions will increase the manufacturing cost for the environment protection.

The second technique used to remove the contamination also has several problems. The processing needs sophisticated equipment to provide a stable plasma source. Meanwhile, it must be carried out inside a vacuum, which increases the cost and manufacturing time. Plasma-induced thermal effect causes shirking to the polyimide substrate. Plasma can also etch away not only the contamination but also the polyimide substrate. Furthermore, the thin conductive circuit may be damaged inside the plasma atmosphere.

Pulsed laser ablation is one of highly potential technologies to remove the organic contamination without the above problems. It is based on the laser photo-ablation of the organic materials by breaking the chemical bonds or the thermal ablation (Luk'yanchuk, 1996; Schmidt, 1998; Pettit GH, 1994; Andrew, 1983). Since the contaminations come from the polyimide debris accumulation, they are pilled up together with a weak adhesion force, which can be easily broken when they absorb the laser photo energy. Appropriate processing parameters (such as light wavelength, pulse duration, laser fluence and pulse number) are selected to remove the contamination.

Though there are many reports on the UV excimer laser cleaning of organic contamination from Si, glass and metal surfaces (Feng, 1999; Lu, 1994; Lu, 1996 a), these UV lasers are not suitable for cleaning the contamination on the polyimide surface. This is because the photo energy of the UV excimer laser (for example, 5 eV for a KrF excimer laser) is higher than the molecular bonding energy (3 ~ 5 eV) for the polyimide materials (Brannon, 1997; Luk'yanchuk, 1997). These UV lasers may cause the surface modification of polyimide substrate (Himmelbauer, 1996). It will cause the mal-function of the device. However, IR lasers are not good light sources either in this application since they may induce much heat effect on the substrate surface. In this cleaning task, a Q-switched 7 ns (FWHM) and 532-nm Nd: YAG laser is the best light source.

Figure 3 shows the microscopic photograph of the flexible circuit surface with and without the laser irradiation. It is clear that the organic contamination can be removed completely after only 5 pulses of the laser irradiation at a laser fluence of 34 mJ/cm^2. Surface roughness before and after the laser cleaning was compared with AFM analysis. It shows that the average surface roughness of the cleaned surface is around 3.02 nm, which is almost equal to the original polyimide substrate roughness of 3.04 nm (Gu, 2001). It demonstrates that there is no substrate damage during the laser cleaning.

Since the cleaning area is located around the tiny inkjet nozzle area (200 μm as shown in Fig. 3 (a)), there is no laser interaction with the thin conductive circuit. Meanwhile, it is a dry process in the air, which does not need the sophisticated equipment and subsequence drying process. There are also no chemical solutions involved. By proper control of the processing parameters, the laser irradiation can remove the organic contamination without damage of the polyimide substrate and thin conductive circuit.

We used the cylindrical lens set to change the laser spot (diameter 8 mm) to a narrow and long laser line (200 μm x 120 mm) for cleaning. Light scanner shifts the laser line at a high speed after the application of 5 laser pulses. This technology offers high productivity and low cost in the manufacturing.

3.2. Laser deflashing of IC packages

In the early days of semiconductor fabrication devices were largely packaged in glass, ceramic or metal. Since 1960s, device manufacturers

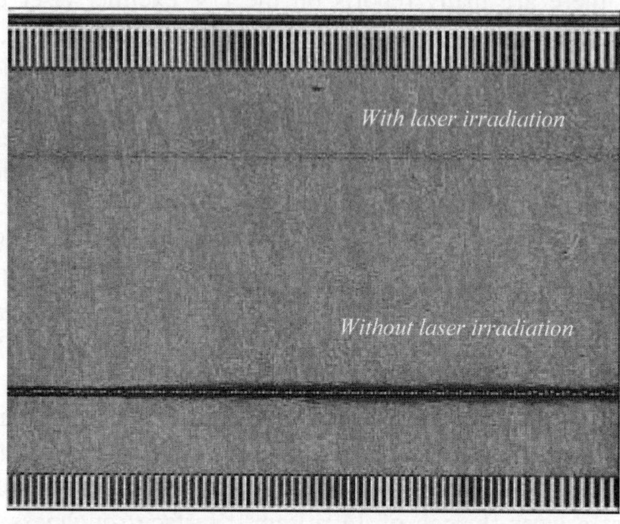

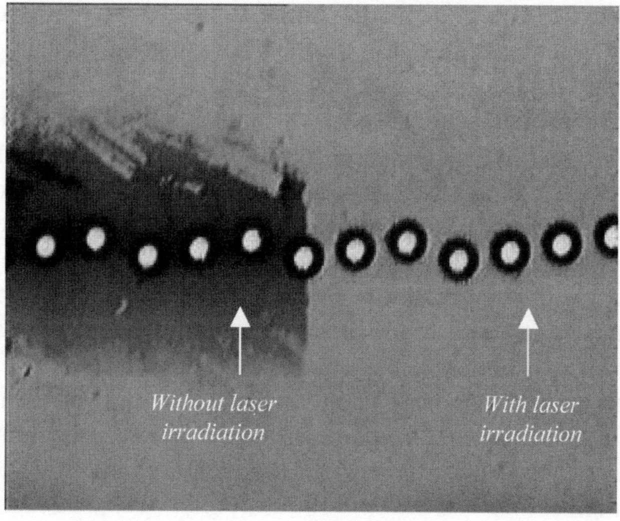

Fig. 3. Microscopic photography of the flexible circuit surface with and without laser irradiation at a laser fluence of 34 mJ/cm^2 and a pulse number of 5. Inkjet hole diameter is about 40 μm.

began to investigate the possibility of encapsulating their products in plastic materials or polymers (Goosey, 1985). The main incentives of the plastic encapsulation were large cost reduction over other material packaging and high compatibility with mass production. It allows several hundred packages to be encapsulated in a single operation, which only takes a few minutes. As IC technology develops, plastic packaging has already received the greatest emphasis and is extensively applied in the IC packaging (Einspruch, 1985). Transfer molding is the most popular technique. In this process the components to be encapsulated are placed into the cavities of a mold. Molding compound liquefied by heat and pressure is forced into the cavities. It solidifies and gives plastic encapsulated devices (Manzione, 1990).

A plastic-encapsulated microcircuit consists of an IC chip physically attached to a lead frame, electrically interconnected to input-output leads and molded in a plastic that is in direct contact with the chip, lead frame and interconnects. With major advantages in cost, size, weight, performance and availability, plastic IC packages have attracted 97% of the market share of worldwide microcircuit sales.

During the molding process, plastic encapsulation is by no means a perfect process. Inevitable wears on encapsulation mold as well as small variances in IC leadframe cause small amounts of molding compounds to leak out onto the lead surfaces or heat sink of the device. In its thin form, this material is known as resin bleed or thin flash. A thicker bleed of material is known as flash. If this material is left on the leads, it will cause problems in the downstream operations of lead trimming, forming, solder dipping and/or plating. In some cases plastic IC packages are designed with an integral heat sink exposed to air to meet high thermal and electrical performance demands. The die is attached directly to the heat sink to minimize the thermal resistance. During the molding process molding compound usually leaks out and forms flash on the heat sink surfaces. This will greatly influence heat dissipation and even cause damage of the plastic IC packages. Therefore, deflashing of the plastic IC packages is one of the critical processes in manufacturing. Figure 4 (a) shows the microscopic image of an IC package with mold flash on the heat sink surfaces.

There are three major deflashing techniques used in manufacturing (Singer, 1985): (1) dry blasting: mold flash is removed by abrasive media propelled at a high speed. Mechanical punching is one of the conventional methods. It removes mold flash by fine SiC needle punching; (2) wet blasting: mold flash is removed by liquid jet propelled at a high speed. Water jet sputtering is often applied to remove mold flash by high-pressure water jet shooting; (3) chemical deflashing: mold flash is removed by soaking IC packages in a chemical bath. After the flash is loosened from the leadframe, IC packages are rinsed with water and then blasted with compressed air or water streams. The typically used chemical solution is N-methyl-2-pyrrolidone. However, these deflashing methods have many disadvantages: (1) mold flash cannot be removed completely, especially for high pin count packages in which the size of interface holes is much smaller; (2) subsequent processing is required, such as drying IC packages after water jet sputtering and frequent changing of tiny broken needles for mechanical punching. It limits the assembling speed of IC packages; (3) the deflashing may damage IC packages and even affect their performance. For high pressure shooting, the IC packages will endure strong vibration. It may make inside wire bonding loosen. The high pressure shooting can even deform the leadframe and affect the solderability of the packages. Since plastic materials are not hermetic against water, water related processing will make moisture filtrate into the packages easily. Chemicals used in deflashing could leach up the leads and find their ways into the packages. It will degrade IC chips inside the packages. Some chemicals used are relatively hazardous. They will create waste disposal problems.

Device manufacturers generally feel that molding equipment should produce flash-free packages in order to avoid the task of mold deflashing. However, this wish has been expressed for a long time without success in the industry applications. As IC technology develops, the interface hole of leadframe becomes much smaller for higher pin count packages. It is more difficult to remove mold flash for conventional processing and to design the flash-free molding equipment. A more practical solution is to find a new deflashing method to replace the conventional methods. Based on its unique effect during the interaction with materials, pulsed laser irradiation can be used to remove mold materials in an order of submicron or micron per pulse. By irradiating the laser light on the mold flash at proper laser parameters, pulsed laser ablation of the molding compound can lead to the removal of the mold flash completely. The laser deflashing has the advantages of being dry and non-contact processing, easy control and high efficiency. It can be a suitable alternative in IC packaging and the related equipment is currently available in the market.

Figure 4 (b) shows the optical macrograph of the plastic IC packages after the pulsed laser deflashing at a laser fluence of 300 mJ/cm^2 and a pulse number of 4. The light source is a Nd: YAG laser with the wavelength of 532 nm and a pulse duration of 7 ns (FWHM). Comparing Fig. 4 (a) with (b), it is found that the mold flash has been removed from the heat sink surfaces after the laser irradiation. Surface profiler scanning shows that the mold flash is in a thickness of around 7 μm. To have a better understanding of the laser deflashing process, pulsed laser ablation of molding compound was investigated. Figure 5 shows the dependence of ablation rate for the molding compound on laser fluence during the Nd: YAG laser irradiation. It can be found that the ablation threshold is 50 mJ/cm^2 and ablation rate at a laser fluence of 300 mJ/cm^2 is about 2.3 μm/pulse. It means that 4 pulses of the laser irradiation can completely remove the mold flash at a depth of 7 μm. The ablation rate increases with laser fluence. At a laser fluence of 550 mJ/cm^2, the deflashing rate reaches to 7.5 μm/pulse. This means only one pulse irradiation can remove the flash with a thickness of 7 μm from the heat sink surfaces. Although higher laser fluence results in higher deflashing rate, the laser fluence selected must be less than the damage threshold of heat sinks in order to avoid their damage.

We are assuming Einstein's evaporation rate (Anisimov, 1995):

$$v = v_0 \exp[-T_a / T_s], \qquad (1)$$

where v_0 and T_a are constants of evaporation law (these constants are given in reference books, e.g. Gray, 1972), and T_s is the surface temperature which varies fast during laser heating.

Integration of (1) near the ablation threshold for the case of volume absorption and smooth laser pulse yields:

$$W = A \exp[-T_a / T_{\max}], \qquad (2)$$

where W is ablation rate [μm/pulse], $T_{\max}$ is maximal temperature of the surface and $A = \sqrt{2\pi}\, v_0 \tau_\ell (T_{\max}/T_a)^{1/2}$, τ_ℓ is duration of laser pulse. When fluence $\Phi = I\,\tau_\ell$ varies due to change of intensity, while $\tau_\ell =$ const, then near the ablation threshold $T_{\max} \propto \Phi$. This yields Arrhenius-like dependence (Luk`yanchuk, 1994):

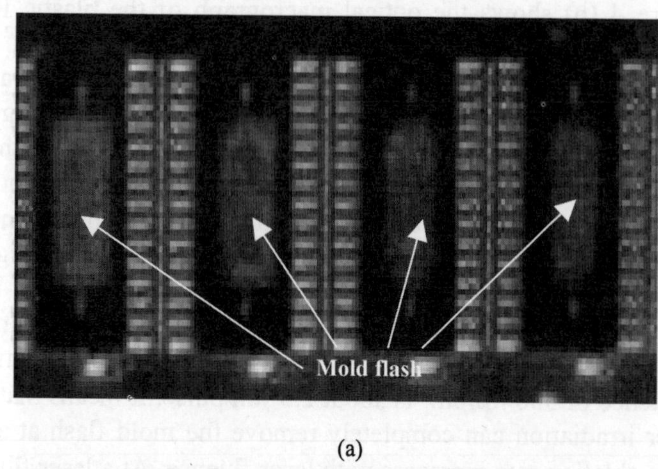

(a)

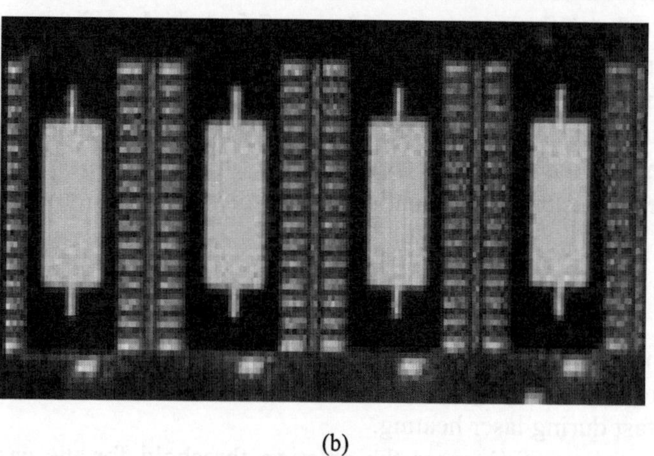

(b)

Fig. 4. Typical optical micrograph of plastic IC packages with mold flash on the heat sink surfaces before (a) and after (b) Nd: YAG laser deflashing at a laser fluence of 300 mJ/cm^2 and a pulse number of 4.

$$W = A\ \exp\left[-B/\Phi\right]. \tag{3}$$

The solid curve in Fig. 5 presents the theoretical interpolation of experimental data with the help of formula (3). Constants $A = 29.5$ μm/pulse and $B = 765.7$ mJ/cm^2 were found by least squares method.

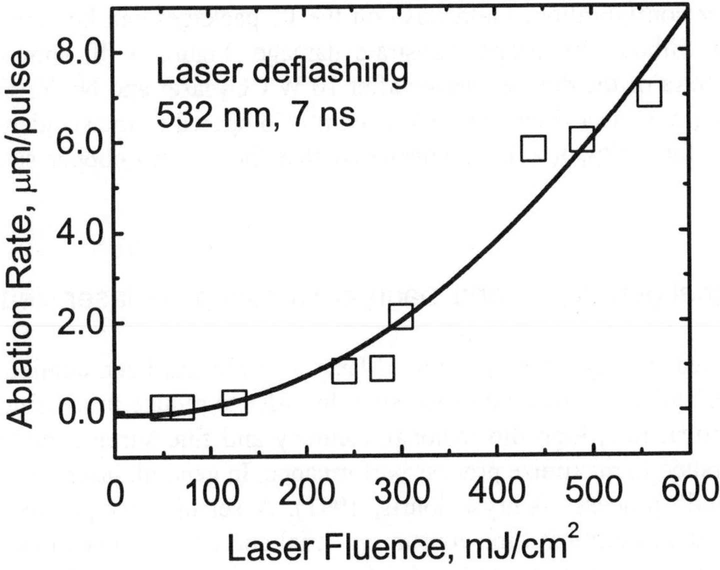

Fig. 5. Dependence of ablation rate for the molding compound on laser fluence during the Nd: YAG laser irradiation.

The other issue needed to be considered during the laser deflashing is the laser-induced surface oxidation on the metal (normally copper) surface. Figure 6 (a) shows X-ray photoelectron spectroscopy (XPS) Cu2p spectra during the Nd: YAG laser deflashing of the IC packages at a laser fluence of 720 mJ/cm^2. Compared to the standard XPS spectrum Cu2p observed in copper, cuprous oxide and cupric oxide, it was found that there are two more peaks in the spectrum. It indicates that there are copper substrate damages during the laser deflashing. Therefore low laser fluence is preferred for laser deflashing. It will take a longer time in the manufacturing, especially for the much thicker mold flash. It is always an industrial requirement to find a low cost, fast speed, flexible and reliable laser technique to fit in the manufacturing lines. We are carrying out a new laser-deflashing scheme with the combination of CO_2 laser with the Nd: YAG laser to achieve the fast speed deflashing. CO_2 laser can easily remove the thick flash but it has no removal effect for the thin flash or bleed. Low fluence Nd: YAG laser is then applied to remove the thin bleed. With this dual laser configuration, mold flash on the IC packages can be completely removed without the copper substrate damage. Figure 6 (b) shows XPS Cu2p spectra of the device surface after 10 W CO_2 laser and Nd:YAG laser deflashing at a laser fluence of 300 mJ/cm^2. Compared to the standard XPS spectrum for Cu2p, it can be concluded that there is no copper substrate damage.

3.3. Signal generation and diagnostics during the laser cleaning

As laser technology develops, pulsed laser ablation has been applied more and more widely in material processing. In order to reduce processing cost, high removal rate, high dimensional accuracy and fine surface quality are often pursued to maximize process performance. In general, laser processing is a “blind” process (Chryssolouris, 1991). A set of laser parameters is normally found through time-consuming “trial and error” procedures. Since the procedures cannot respond to the dynamic variation of the processing, it is difficult to get an optimal operating point. Meanwhile, the amount of the substrate materials removed is the result of time-dependent laser interaction with materials. During the interaction, laser power and thermal properties of materials may be changed. Therefore, it is impossible to predetermine the desired amount of materials ablated accurately. It is highly desired in manufacturing to know at any given time how the device is processed, so that laser parameters can be adapted to the material behavior and achieve the desired processing results (Mao, 1993). It requires the proper in-situ sensing techniques to monitor the laser ablation of the materials.

Figure 7 shows a control scheme for the real-time monitoring of the pulsed laser dry cleaning of microelectronics devices. During the laser processing acoustic waves, optical and electric signals are generated. Proper sensors should be used to detect different signals. The analog signals recorded are digitized through an A/D converter for data processing. To simplify the control loop for high-speed feedback, characteristic parameters of the signal are chosen to replace the whole waveforms. These parameters include the peak amplitude, peak width, peak position, and starting and termination times.

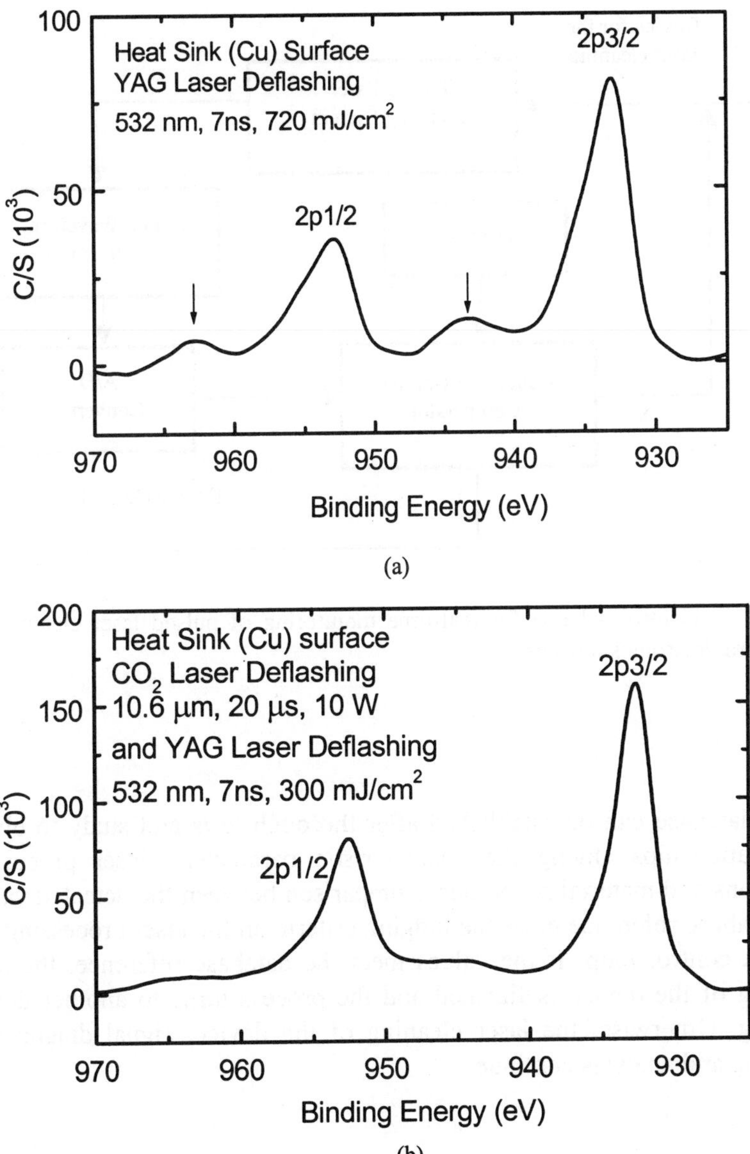

Fig. 6. X-ray photoelectron spectroscopy (XPS) Cu2p spectra during (a) Nd: YAG laser deflashing of the IC packages at a laser fluence of 720 mJ/cm^2 and (b) 10 W CO_2 laser and Nd:YAG laser deflashing at a laser fluence of 300 mJ/cm^2.

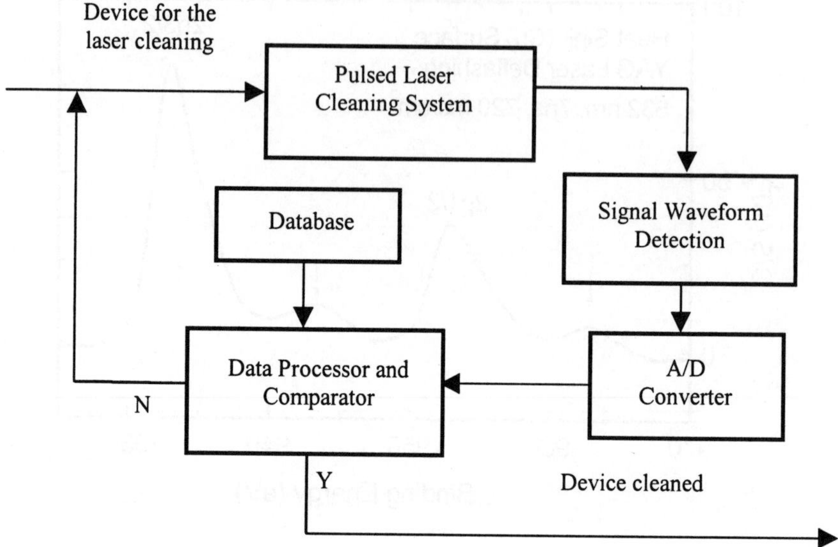

Fig. 7. Control scheme for real-time monitoring of pulsed laser cleaning of microelectronics devices.

A database can be established after thorough tests and study to obtain the relationships among the characteristic parameters, laser processing conditions and material properties. Comparison between the detected values and database reference gives the judging criteria on the laser processing and forms a control loop. If the values meet the database reference, the laser cleaning of the device is finished and the process turns to another device cleaning. Otherwise, the laser cleaning of the device, signal diagnostics, sampling and analysis continue.

3.3.1. Audible acoustic wave generation

Pulsed-laser irradiation is a rich source for acoustic wave emission. It has been extensively applied in scientific research and industrial manufacturing as one of non-contact and non-destructive detection techniques for material characterization and micro-defect testing (Flannery, 1997; Pepper, 1996). The acoustic waves generated have been proven to convey very important information on the laser interaction with substrate materials (Lyamshev, 1991; Lyamshev, 1992). Many studies have been carried out to explain the acoustic wave generation (Bunkin, 1991; Sigrist, 1986). However, mechanisms for the acoustic wave generation by the laser are diverse and are difficult to quantify. In principle there are five important mechanisms responsible for the acoustic wave generation: dielectric breakdown, vaporization or material ablation, thermo-elastic effect, electrostriction and radiation pressure (Sigrist, 1986). Their contributions depend on laser processing parameters as well as on optical and thermal characteristics of the materials. Among these five mechanisms, dielectric breakdown and electrostriction should be considered only for the cases of very high laser intensity and very weakly absorbing materials. For the pulsed laser ablation of the organic contamination for the device cleaning, these two mechanisms can be ignored (Lu, 1996 b).

When a laser beam irradiates on the substrate surface, it is partially reflected and the photon travelling direction is changed. From the momentum conservation law, momentum loss of the photons is transferred to the substrate. Since laser pulse duration is very short, there is very high radiation pressure acting on the surface. It will cause vibration of the substrate surface and induce ultrasonic waves.

On the other hand, the substrate surface absorbs the photon energy at the same time. Energy obtained by the electrons is then transferred to the atomic lattice and causes the rise of the surface temperature. It will induce thermal expansion on the laser-irradiated area due to material elasticity. Since the time interval for the surface temperature variation is very short during the pulsed laser irradiation, thermal-elastic effect will also induce ultrasonic waves, which are transmitted into the substrate.

While for laser fluence above ablation threshold of the substrate materials, the surface temperature will increase to its vaporization point. Since light absorption depths for the solid materials are generally in an order of or less than a micron, laser energy absorption by the substrate is limited in a small region beneath the surface. High power laser irradiation will evaporate the substrate materials within the small confined volume and lead to a rapid increase of the local particle density. It results in a rapid rise of the local pressure in the volume. Such a high local pressure will be released in the form of a shock wave (Zyung, 1989). Fragments generated during the laser ablation will be ejected out at high speeds from the laser-irradiated region. A very strong recoil force will act on the substrate and cause it to vibrate. Meanwhile, the shock wave will transmit inside its surrounding air and decay into acoustic waves by air friction.

The acoustic waves generated during the laser ablation posses both sonic and ultrasonic components (Leung, 1992). The acoustic waves generated during the laser processing are generally detected by piezo-electrical transducer, He-Ne laser deflection probe and ultrasonic microphone (Dyer, 1992; Ventzek, 1991; Grad, 1993). However, these studies were only concentrated on the substrate vibration and ultrasonic waves. The disadvantages of these detection techniques are high cost, complex setup and direct contact with the substrate. In this study, a normal microphone is placed nearby the substrate surface to detect the audible acoustic waves generated during the laser ablation. The signal detection is also applied to correlate the peak amplitudes with laser cleaning efficiency and ablation rate to control the laser cleaning in real time. This very simple audible acoustic wave diagnostics is: 1) suitable for different substrates and 2) has no coupling problem (due to non-contact approach).

The other advantages of this technique are:

1. Applicability for rapid scanning and in process monitoring with its real-time inspection;
2. Applicability for hostile environments and for irregularly shaped objects;
3. Simplicity and low cost compared to ultrasonic system;
4. Acoustic diagnostics is indispensable when ultrasonic waves cannot be easily detected.

Figure 8 shows the first peak-to-peak amplitudes of the audible acoustic waves as a function of pulse number during the laser deflashing of IC packages at a laser fluence of 300 mJ/cm^2. The acoustic waves captured at the first, second and fourth pulses of laser irradiation are plotted as an inset in Fig. 8. The acoustic waves are generated by the laser ablation of the flash

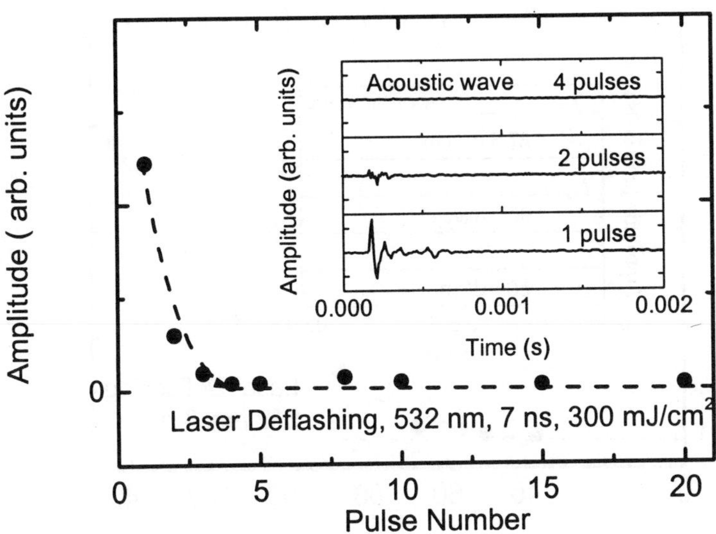

Fig. 8. Acoustic waves captured at first, second and fourth pulse of laser irradiation and first peak-to-peak amplitudes as a function of pulse number during the laser deflashing of IC packages at a fluence of 300 mJ/cm^2.

materials on the heat sink surfaces. It is found that a strong acoustic wave is generated during the first laser pulse of laser irradiation. The acoustic waves and first peak-to-peak amplitudes reduce gradually with pulse number and almost disappear after 4 pulses of laser irradiation. In our previous study, the variation of acoustic waves and the first peak-to-peak amplitude has been proven to be related to the change of contamination level on the device surfaces (Lu, 1995). This implies that the flash on the heat sink surfaces in the IC packages becomes less and less with more pulses of the laser deflashing. As the acoustic wave disappears, a clean surface is obtained by the laser deflashing. Therefore, the acoustic wave detection can be used to monitor the flash removal level and surface cleanness during the laser deflashing process. The first peak-to-peak amplitude of the acoustic waveform can be used as the characteristic parameter for the process monitoring in real time.

Figure 9 shows the first peak-to-peak amplitudes of the audible acoustic waves captured during the laser deflashing of IC packages at different laser fluences. The inset plot shows the audible acoustic waves recorded during

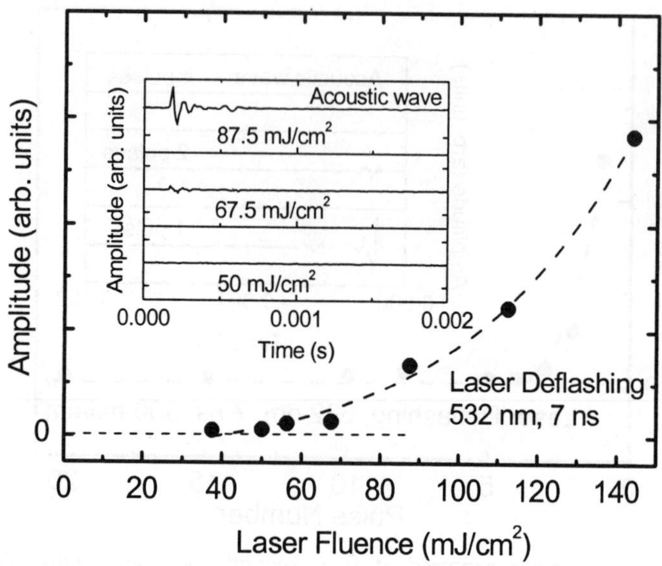

Fig. 9. Audible acoustic waves captured during the laser deflashing of IC packages at laser fluences of 50, 67.5 and 87.5 mJ/cm^2 and first peak-to-peak amplitude as a function of laser fluence

the laser ablation of the flash materials on the heat sink surfaces in the IC packages for the laser fluences of 50, 67.5 and 87.5 mJ/cm^2. It is found that higher fluence results in a stronger acoustic wave and a higher peak-to-peak amplitude, which corresponds to more flash material removal from the heat sink surfaces. It implies that higher fluence results in a higher deflashing rate, which is consistent with the result in Fig. 5.

It is also observed that the first peak-to-peak amplitudes of acoustic waves are zero for the laser fluence below 40 mJ/cm^2. The ablation threshold for the laser deflashing can be estimated from the curve as around 50 mJ/cm^2, same as the threshold fluence value obtained from the curve of the deflashing rate versus laser fluence in Fig. 5. Therefore, the audible acoustic wave detection can also be used to estimate the deflashing threshold and deflashing rate.

Figure 10 shows the main control panel of the real-time monitoring and process control system for laser cleaning with an audible acoustic wave as

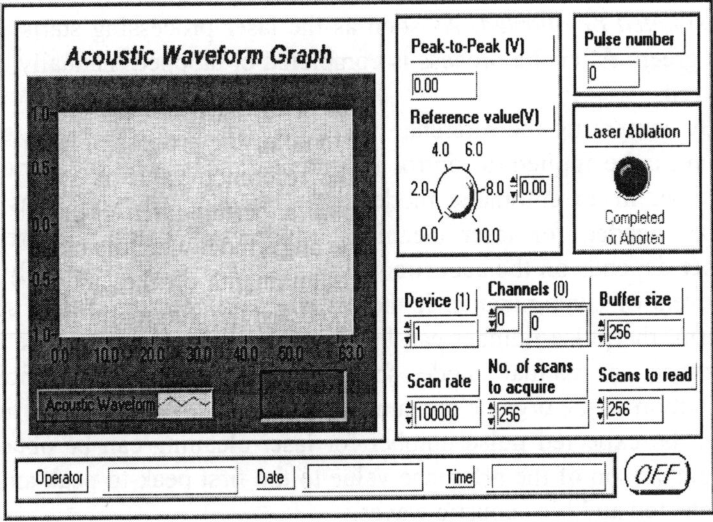

Fig. 10. Main control panel of real-time monitoring and process control system for the laser cleaning with the audible acoustic wave as the feedback signal.

the feedback signal. The system was designed and developed with Software LabVIEW 4.1 under Windows 98 environment together with the necessary hardware and motion control design. Both the acoustic signal and the laser process are visualized with the graphical user interface (GUI). It is the main interface of the program and keeps visual during the whole execution process. Before the execution of the program, key parameters such as *Device*, *Channels*, *Buffer size* and *Scan rate* should be set to make *data acquisition* (DAQ) board work properly. *Reference value* is also required to be set in advance to make the process of laser cleaning smooth. The *Acoustic Waveform* recorded can be displayed on the control panel in real time so that the users can know clearly how the process is going on. Since the acoustic waves involve abundant information of the process during laser material interaction, any abnormal situation can be implied by abnormal waveforms displayed on the panel. The users can stop the process manually and conveniently by pressing the *OFF* button. Key data such as the *Peak-to-Peak* amplitude and *Pulse number* are also displayed on the main panel. As showed on the main panel, there is a button to indicate whether the process is *completed or aborted.* As soon as the laser processing starts, the button turns green. After the process is completed or stopped manually, the button turns red.

By comparing the reference value with the feedback value, the control system can be applied to control and monitor the process of laser cleaning in real time. In an automatic mode, the reference value is set as the pulse number needed for laser cleaning of a certain area. The pre-set pulse number depends on the acoustic wave amplitude, which is closely related to the levels of contaminants. If the contaminants on the sample surface are uniform, the pulse number can be fixed for the automatic mode operation. Otherwise, a manual mode should be selected for the non-uniform distribution of the organic contamination on the device surfaces. In the latter mode, the expected pulse number for laser cleaning can be obtained from the comparison of the reference value to the first peak-to-peak amplitude of the detected audible acoustic waves.

3.3.2. Diagnostics of plasma-induced electric field

The threshold fluence for the organic contamination ablation is generally much lower than those for the solid substrates, mostly semiconductors and metals for the microelectronics devices. With a proper selection of the laser processing parameters, the organic contamination can be easily removed without damage of the substrate surface. During the laser cleaning of the organic contamination, there is a plasma generated above the laser-irradiated area. There is an explosive emission of electrons, ions and neutral atoms during the laser ablation (Ready, 1971; Malvezzi, 1984). These fast-ejected particles interact with each other and also absorb the incident laser energy, which greatly enhances the electron and positive ion generation (Hansen, 1998; Hendron, 1997).

Plasma dynamics during the laser ablation can be studied from the electric signal diagnostics, which reflects the emission of the electrons and positive ions during the laser cleaning process. At an early stage of the laser ablation (delay time < 100 ns), the electrons and ions are highly dynamic. According to classical electrodynamics, these dynamic charges will emit electromagnetic waves (Ohanian, 1988). With a tiny metal probe nearby the device, the pulsed laser cleaning can be sensed in a non-contact mode. As the cleaning process continues, the concentration of the surface contamination reduces gradually. The plasma generated becomes weaker and there are fewer electrons and positive ions emitted. The electric signal is also weaker. When the surface is completely cleaned, there is no laser ablation of the organic materials and plasma generation. The electric signal will disappear. Therefore, the variation of the electric signal reflects the laser cleaning process in real time.

Figure 11 shows the electric signals detected by a tiny metal probe during the excimer laser cleaning of grease contamination on a copper substrate. Laser fluence applied was 0.54 J/cm^2. It is much lower than the ablation threshold for the copper substrate 1.3 J/cm^2. Therefore, there is no laser ablation of the substrate materials during the surface cleaning. The metal probe was placed 2.5 mm away from the laser-irradiated spot. The electric signals for the plasma-induced electric field were recorded at laser pulse number's of 1, 2, 3, 4, 5, 10, 50 and 100, respectively. It is clear that there is a negative peak (peak width: dozens of nanosecond) in the electric signal, which is attributed to the dynamics of the plasma at the early stage of laser ablation. The negative signal peak is resulted from the formation of negative electric dipole constructed by the electrons and positive ions at the early stage of the laser ablation (Lu, 1999). As pulse number increases, both the signal maximum amplitude and peak width reduce gradually. It implies that the electron and positive ion emission becomes weaker and the plasma density decreases as the laser cleaning proceeds. The electric signal vanishes completely after 50 pulses of the laser irradiation. Surface analyses of the copper substrate after the laser cleaning show that the organic contamination on the copper substrate has been completely removed. It is demonstrated that the variation of the electric signal is closely related to the laser cleaning process. The electric signal detection for the plasma-induced electric field can also be applied to control the laser cleaning in real time if there is plasma generation during the contaminant removal.

3.3.3. Detection of plasma optical signal

Besides the audible acoustic wave and plasma-induced electric field signals, the plasma optical signal also carries the important information of the laser ablation of the organic materials. Laser ablation is a dynamic process of the energy transfer from the laser pulse to the substrate materials (Allmen, 1987; Kettani, 1973). The electrons of the ablated materials are excited to higher energy levels or even cause the materials' ionization

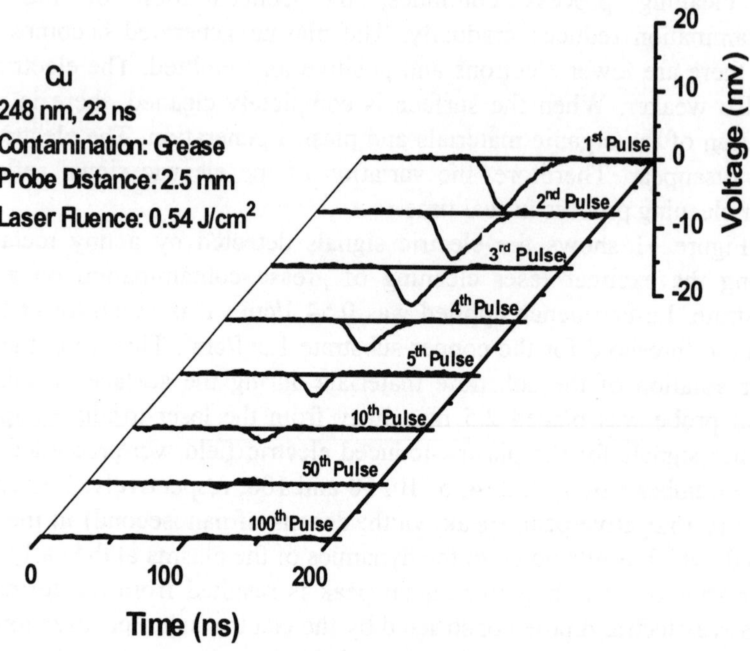

Fig. 11. Electric signals detected during the excimer laser cleaning of grease contamination from the copper substrate for a laser fluence of 0.54 J/cm^2 and pulse numbers of 1, 2, 3, 4, 5, 10, 50 and 100.

through the absorption of laser energy and strong collisions among the plasma species. When the laser pulse is off, these electrons will finally drop back to the lower energy levels, the transition leads to light emission. Its photo energy is equal to the energy difference between the higher and lower energy levels, which corresponds to the emission spectral line position. With an optical sensor, the plasma signals can be detected (Podgornyi, 1971). Laser ablation can be sensed with the optical signal diagnostics and analyses. Optical emission spectrum analysis is one of the most frequently used techniques in the plasma diagnostics (Lochte-Holtgreven, 1995). From the plasma emission spectral line position and intensity, much information on the laser cleaning process can be obtained.

Fig. 12 shows the optical emission spectra captured during the excimer laser ablation of indium–tin-oxide (ITO) thin film (100 nm thick)

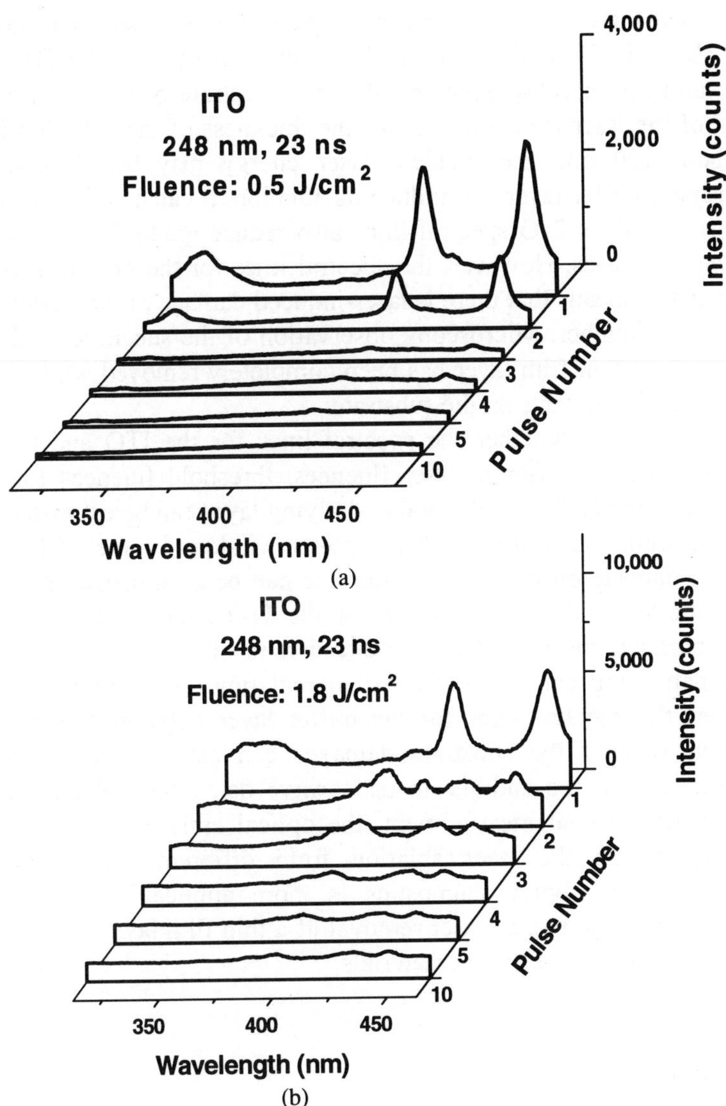

Fig. 12. ITO plasma emission spectra captured during the excimer laser ablation of ITO thin film layer at different pulse numbers. Laser fluences are (a) 0.5 and (b) 1.8 J/cm^2.

found that there exist some emission spectral lines other than those for the ITO ablation. It implies that the laser ablation of both the ITO thin film layer and buffer SiO_2 layer on the glass substrate occurs even at the first pulse of the laser irradiation. Since the thickness of the ITO thin film layer is about 100 nm, the incident laser energy may be absorbed by the underlying buffer layer and induce its ablation. It can also be found that the intensities of the ITO spectral lines also reduce gradually to zero as pulse number increases. However, the spectral lines for the buffer layer ablation still exist. It means that there is laser-induced damage of the buffer layer at a high pulse number. Microscopy observation of the substrate surface shows that the ITO thin film layer has been completely removed while there are a lot of cracking defects on the substrate.

By checking whether the spectral lines for the ITO and buffer layer ablation exist at different laser fluences, threshold fluences for the laser ablation of the ITO thin film and underlying layer can be estimated to be 0.4 and 1.0 J/cm^2, respectively. A parameter window for the LCD patterning without damage on the substrate surface can be established. By evaluating when the spectral line intensities for the ITO ablation drop to zero with pulse number, the pulsed laser-induced removal of ITO thin film for the LCD patterning can be controlled in real time. Meanwhile by checking whether the spectral lines for the buffer layer ablation appear, the laser ablation without the substrate damages can also be monitored in-situ. Compared with the audible acoustic wave detection and plasma-induced electric field signal measurement, the optical emission spectrum analysis can distinguish the laser ablation from different substrate materials. Therefore, this signal diagnostics is more applicable in the real-time monitoring of the pulsed laser removal of a thin film layer from multi-layer structures and multi-element materials.

Conclusions

Pulsed laser ablation for the dry cleaning of the organic contamination from the microelectronics device surfaces is studied. This technology has been successfully applied in the laser cleaning of the plasma debris accumulated on the flexible circuit for the inkjet printer cartridge and the laser deflashing of the mold flash on the IC packages. Compared to the other conventional cleaning techniques, the pulsed laser cleaning has the advantages of being fast speed, flexible, easy control, dry and non-contact processing. It has highly potential applications in the microelectronics industry. Diagnostics of the audible acoustic wave, plasma-induced electric field and plasma optical signals can be used to correlate with the dynamic laser cleaning process and act as a real-time monitoring scheme for the complete removal of the organic contamination without damage on the device substrates.

Acknowledgments

We are thankful to Drs. W. J. Wang and E. Gatskevich for discussions and comments.

References

Allmen M.V., *Laser-Beam Interactions with Materials: Physical Principles and Applications* (Springer-Verlag, New York, 1987)

Andrew J., *Appl. Phys. Lett.*, **43,** 717 (1983)

Anisimov S. I., Khokhlov V. A., *Instabilities in Laser-Matter Interaction*, (CRC Press, Boca Raton, London, Tokyo, 1995)

Brannon J. H., Wassick T. A., *Proc. SPIE,* vol. **2991**, 146 (1997)

Bunkin F. V., Kolomemsky A. A., Mikhalevich V. G., *Lasers in Acoustics* (Harwood Academic Publishers, Paris, 1991)

Chang C. Y., Sze S. M. (Eds.), *ULSI Technology* (McGraw-Hill, New York, 1996).

Cho J., Schneider T. P., Vander Weide J., Jeon H., Nemanich R. J., *Appl. Phys. Lett.*, **59**, 1995 (1991)

Chryssolouris G., Sheng P., Alvensleben F. V., *Trans. ASME*, **113**, 268 (1991)

Deal B. E., McNeilly M. A., Kao D. B., de Larios J. M., *Sol. State Technol.*, 73 (July 1990)

Dyer P. E., *Photochemical Processing of Electronic materials* (Academic, London, 1992)

Dyer P. E., Farrar S. R., Key P. H., *Appl. Surface Sci.*, **54,** 255 (1992)

Einspruch N. G. (Ed.), *VLSI Electronics Microstructure Science,* Vol. **9** (Academic, New York, 1985)

Feng Y., Liu Z., Vilar R., Yi X. S., *Appl. Surface Sci.*, **150**, 131 (1999)

Flannery C. M., Kelly P. V., Beechinor J. T., Crean G. M., *Appl. Phys. Lett.*, **71**, 3767 (1997)

Goosey M. T. (Ed.), *Plastics for Electronics* (Elsevier, London, 1985)

Grad L., Mozina J., *Appl. Surface Sci.*, **69,** 370 (1993)

Gray D. E. (Ed.), *American Institute of Physics Handbook*, 3rd edition (McGraw-Hill, New York, 1972)

Gu H., Duley W. W., *J. Phys. D: Appl. Phys.*, **29**, 556 (1996)

Gu J., Low J., Lim P. K., Lim P., *SPIE*, **4595**, 293 (2001)

Hansen T. N., Schou J., Lunney J. G., *Appl. Phys. Lett.*, **72**, 1829 (1998)

Hashimoto K., Egashira K., Suzuki M., Matsunaga D., *1991 International Conference on Solid-State Devices and materials, Yokohama,* 143 (1991)

Hendron J. M., Mahony C. M. O., Morrow T., Graham W.G., *J. Appl. Phys.,* **81**, 2131 (1997)

Himmelbauer M., Arenholz E., Bäuerle D., Schilcher K., *Appl. Phys. A,* **63,** 337 (1996)

Hong M. H., Lu Y. F., *Appl. Phys. A,* **69**, S605 (1999)

Kern W., *J. Electrochem. Soc.,* **137**, 1887 (1990)

Kettani M. A., Hoyaux M. F., *Plasma Engineering* (Butterworth, London, 1973)

Kogure M., Futatsuki T., Yakano J., Isagawa T., Kimura K., Ogato Y., Tanaka F., Ohmi T., *J. Electrochem. Soc.,* **140**, 3321 (1993)

Lankard J. R., Wolbold G., *Appl. Phys. A,* **54**, 355 (1992)

Leung W. P., Tam A. C., *Appl. Phys. Lett.,* **60**, 23 (1992)

Lochte-Holtgreven W., *Plasma Diagnostics* (AIP Press, New York, 1995)

Lu Y. F., Takai M., Komuro S., Shiokawa T., Aoyagi Y., *Appl. Phys. A,* **59**, 281 (1994)

Lu Y. F., Aoyagi Y., *Jpn. J. Appl. Phys.,* **34**, L1557 (1995)

Lu Y. F., Song W. D., Hong M. H., Chong T. C., Low T. S., *J. Appl. Phys.*, **80,** 499 (1996 a)

Lu Y. F., Hong M. H., Chua S. J., Teo B. S., Low T. S., *J. Appl. Phys.* **79**, 2186 (1996 b)

Lu Y. F., Hong M. H., *J. Appl. Phys.,* **86**, 2812 (1999)

Lu Y. F., Zhang L., Song W. D., Zheng Y. W., Luk'yanchuk B. S., *JETP Letters*, vol. **72**, 457 (2000)

Luk'yanchuk B., Bityurin N., Anisimov S., Bäuerle D., *Photophysical ablation of organic polymers*, In: Excimer Lasers, Ed. by L.D. Laude, (Kluwer Academic Publishers, Dordrecht, 1994)

Luk'yanchuk B., Bityurin N., Anisimov S., Arnold N., Bäuerle D., *Appl. Phys. A,* **62**, 397 (1996)
Luk'yanchuk B., Bityurin N., Himmelbauer M., Arnold N., *Nuclear Instruments and Methods in Physics Research B*, vol.**122**, 347 (1997)
Luk'yanchuk B. S., Zheng Y. W., Lu Y. F., *Proc. SPIE*, vol. **4065**, 576 (2000)
Lyamshev L. M., *Physical Acoustics* (Plenum Press, New York, 1991)
Lyamshev L. M., *Sov. Phys. Acoust.* **38**, 105 (1992)
Malvezzi A. M., Kurz H., Bloembergen N., *Mat. Res. Soc. Symp. Proc.,* **35**, 75 (1984)
Mao Y. L., Kinsman G., Duley W. W., *J. Laser Applications,* **5**, 17 (1993)
Maravelaki P., Kalaitzaki M., Zafiropulos V., Fotakis C., *Appl. Surface Sci.,* **148**, 92 (1999)
Manzione L. T., *AT&T Plastic Packaging of Microelectronic Devices* (Van Nostrand Reinhold, New York, 1990)
Mosbacher M., Münzer H. - J., Zimmermann J., Solis J., Boneberg J., Leiderer P., *Appl. Phys.* **A 72**, 41 (2001)
Ohanian H. C., *Classical Electrodynamics* (Allyn & Bacon, London, 1988)
Otis C. E., *Appl. Phys. B,* **49**, 455 (1989)
Pepper D. M., *Laser Focus World*, 77 (June, 1996)
Pettit G. H., Ediger M. N., Hahn D. W., Brison B. E., Sauerbrey R., *Appl. Phys. A,* **58**, 373 (1994)
Podgornyi I. M., *Topics in Plasma Diagnostics* (Plenum Press, New York, 1971)
Ready J. F., *Effects of High Power Laser Radiation* (Academic, New York, 1971)
Schmidt H., Ihlemann J., Wolff-Rottke B., Luther K., Troe J., *J. Appl. Phys.,* **83**, 5458 (1998)
Sigrist M. W., *J. Appl. Phys.,* **60**, R83 (1986)
Singer P. H., *Semic. Intl.*, 67 (June, 1985)
Singer P. H., *Semic. Intl.* 36 (Dec. 1992)
Srinivasan R., Braren B., Dreyfus R. W., *J. Appl. Phys.,* **61**, 372 (1986)
Tam A. C., Leung W. P., *J. Appl. Phys.,* **71**, 3515 (1992)
Ventzek P. L. G., Gilgenbach R. M., Heffelfinger D. M., Sell J. A., *J. Appl. Phys.,* **70**, 587 (1991)
Wu X., Sacher E., Meunier M., *J. Appl. Phys.,* **86**, 1744 (1999)
Yamazaki T., Miyata N., Aoyama T., Ito T., *J. Electrochem. Soc.,* **139**, 1175 (1992)

Zapka W., Ziemlich W., Tam A. C., *Appl. Phys. Lett.,* **58**, 2217 (1991)
Zyung T., Kim H., Postlewaite J. C., Dlott D. D., *J. Appl. Phys.,* **65**, 4548 (1989)

Subject Index